PARTICLE PHYSICS AND COSMOLOGY

Related Titles from AIP Conference Proceedings

539 Symmetries in Subatomic Physics: 3rd International Symposium
Edited by X.-H. Guo, A. W. Thomas, and A. G. Williams, October 2000, 1-56396-964-5

536 Instrumentation in Elementary Particle Physics: VIII ICFA School
Edited by Sehban Kartal, September 2000, 1-56396-960-2

533 Next Generation Nucleon Decay and Neutrino Detector: NNN99
Edited by Milind V. Diwan and Chang Kee Jung, August 2000, 1-56396-956-4

531 Particles and Fields: Seventh Mexican Workshop
Edited by Alejandro Ayala, Guillermo Contreras, and Gerardo Herrera, July 2000, 1-56396-954-8

490 Particles and Fields: Eighth Mexican School
Edited by Juan Carlos D'Olivo, Gabriel López Castro, and Myriam Mondragón, November 1999, 1-56396-895-9

478 COSMO-98: Second International Workshop on Particle Physics and the Early Universe
Edited by David O. Caldwell, May 1999, 1-56396-853-3

476 3 K Cosmology: EC-TMR Conference
Edited by Luciano Maiani, Francesco Melchiorri, and Nicola Vittorio, May 1999, 1-56396-847-9

444 Particle Physics and Cosmology: First Tropical Workshop/High Energy Physics: Second Latin American Symposium
Edited by José F. Nieves, September 1998, 1-56396-775-8

423 Fundamental Particles and Interactions: Frontiers in Contemporary Physics
Edited by Robert S. Panvini and Thomas J. Weiler, February 1998, 1-56396-725-1

To learn more about these titles, or the AIP Conference Proceedings Series, please visit the webpage **http://www.aip.org/catalog/aboutconf.html**

PARTICLE PHYSICS AND COSMOLOGY

Second Tropical Workshop

San Juan, Puerto Rico   1-5 May 2000

EDITOR
José F. Nieves
University of Puerto Rico

Melville, New York, 2000
AIP CONFERENCE PROCEEDINGS ■ VOLUME 540

Editor:

José F. Nieves
Department of Physics
University of Puerto Rico
P. O. Box 23343
Rio Piedras, PR 00931-3343
USA

E-mail: nieves@ltp.upr.clu.edu

Authorization to photocopy items for internal or personal use, beyond the free copying permitted under the 1978 U.S. Copyright Law (see statement below), is granted by the American Institute of Physics for users registered with the Copyright Clearance Center (CCC) Transactional Reporting Service, provided that the base fee of $17.00 per copy is paid directly to CCC, 222 Rosewood Drive, Danvers, MA 01923. For those organizations that have been granted a photocopy license by CCC, a separate system of payment has been arranged. The fee code for users of the Transactional Reporting Service is: 1-56396-965-3/00/$17.00.

© 2000 American Institute of Physics

Individual readers of this volume and nonprofit libraries, acting for them, are permitted to make fair use of the material in it, such as copying an article for use in teaching or research. Permission is granted to quote from this volume in scientific work with the customary acknowledgment of the source. To reprint a figure, table, or other excerpt requires the consent of one of the original authors and notification to AIP. Republication or systematic or multiple reproduction of any material in this volume is permitted only under license from AIP. Address inquiries to Office of Rights and Permissions, Suite 1NO1, 2 Huntington Quadrangle, Melville, N.Y. 11747-4502; phone: 516-576-2268; fax: 516-576-2450; e-mail: rights@aip.org.

L.C. Catalog Card No. 00-108410
ISBN 1-56396-965-3
ISSN 0094-243X
Printed in the United States of America

This volume is dedicated to Arthur Halprin on the occasion of his retirement from the University of Delaware.

Contents

Preface ix

NEUTRINO THEORIES AND MASSES

Mass Relations for Neutrinos 3
    K. S. Babu
Minimal Model for Neutrino Masses and Mixings 18
    P. H. Frampton
Teeny, Tiny Dirac Neutrino Masses: An Unorthodox Point of View 24
    P. Q. Hung
Neutrino Oscillations in Extended Anti-GUT Model 35
    C. D. Froggatt, H. B. Nielsen, and Y. Takanishi
Neutrino Theories 75
    P. Ramond

NEUTRINO EXPERIMENTS

Final Neutrino Oscillation Results from LSND 93
    W. C. Louis for the LSND Collaboration
The Fermilab Neutrino Oscillation Program 105
    M. H. Shaevitz
Super-Kamiokande's Past, Present, and Future 122
    M. R. Vagins for the Super-Kamiokande Collaboration
SNO Detector Status 193
    R. G. Van de Water for the SNO Collaboration

NEUTRINOS IN COSMOLOGY

Neutrinos in Supernovae and Extra Dimensions 203
    D. O. Caldwell
Relic Neutrino Asymmetries 213
    R. R. Volkas

SUPERNOVA COSMOLOGY AND ASTROPHYSICS

Evidence from Type Ia Supernovae for an Accelerating Universe 227
    A. V. Filippenko and A. G. Riess for the High-z Supernova Search Team
What's New in the Pulsar World? 247
    D. R. Lorimer
**SNAP: Supernova/Acceleration Probe: An Experiment to Measure
the Properties of the Accelerating Universe** 263
    P. Nugent for the SNAP Collaboration

CP, *CPT*, AND *B* PHYSICS

CP Violation—A Brief Review 283
    J. L. Rosner
Experimental Tests of *CPT* Invariance 305
    D. Zavrtanik for the CPLEAR Collaboration
B Physics Prospect at Tevatron Run II 314
    K. Yip for the CDF and DØ Collaborations

SEARCHES FOR NEW PHYSICS IN LEP AND TEVATRON

Searches for New Physics at NuTeV 331
    J. Conrad for the NuTeV Collaboration
Supersymmetry Searches at LEP 341
    L. Duflot
Limits on the Size of Extra Dimensions from LEP Experiments 354
    V. Zhuravlov

ABSTRACT

Neutrino Dispersion in Intense Magnetic Field 372
    E. Elizalde, E. J. Ferrer, and V. de la Incera

Schedule 375
List of Participants 379
Author Index 381

PREFACE

The Second *Tropical Workshop* on Particle Physics and Cosmology took place from 1-5 May 2000, in San Juan, Puerto Rico. This volume is the collection of the written version of the talks that were presented in the meeting.

As I have explained before, and in particular in the first volume of this series of workshops, the idea of holding a workshop on Particle Physics and Cosmology in Puerto Rico was born, to a large extent, as an offspring of the Winter School that had been held here on two previous occasions. The decision to take this idea from conception to production would not have taken place without the encouragement and help from Arthur Halprin, Terry Leung and Qaisar Shafi, whom I thank for serving as the scientific committee.

Several people have asked me what are the aim and motivation of the workshop. Needless to say, I have asked myself the same question, especially since during the period of the organization of these workshops I have had to negotiate with three chancellors of the University of Puerto Rico, two deans of the Faculty of Natural Sciences, and my departure as chairman of the Physics Department.

It is sometimes very difficult to understand our own motivations, and it is much more so to understand the motivation of others. I therefore believe that the answers to such questions must be sought for individually. I can say from my part that, as long as there are people that are willing to participate, and other people willing to support it, then I am willing to organize it.

Thus, this is an appropriate place to acknowledge the help and support from various people. The workshop was sponsored by the Office of the President of the University of Puerto Rico, Norman Maldonado, the Chancellor George Hillyer, and the Dean of the Faculty of Natural Sciences, Brad Weiner. I wish to express my special thanks to Brad for his support, and for courageously announcing publicly, in his welcome message at the workshop, his own willingness to support the next edition of the workshop.

The scientific session of Tuesday, May 2, took place in the splendid site of the Arecibo Observatory, for which we thank the Observatory's Director Daniel Altschuler and the director of its Visitors Center, Jose Luis Alonso, for their personal involvement in the arrangements for that session.

The organization of activities like this requires a team of administrative personnel which can turn the event into a failure or a successful one. With the help from Ileana Desiderio and Carmelo Figueroa from the Department of Physics, and with the backing of Maru's command in the dealings with the workshop site, the rest was smooth sailing.

Finally, the major sponsors of this workshop are the participants themselves. I express my thanks for the effort that all the authors have put in their papers, in spite of many other commitments, to make a speedy and timely publication possible. As I have mentioned many times, I believe that it is important that we document, as best as we can, what we do at conferences. If the quality of the papers that are contained in this volume is a measure of the quality of the workshop, then

The (non-local) Organizing Committee

I believe that we can all feel very satisfied. However, I will let you be the judge of that.

I hope that you will enjoy reading these articles as much as I enjoyed preparing this edition.

José F. Nieves
26 July, 2000
San Juan, Puerto Rico

The 2nd Tropical Workshop on Particle Physics and Cosmology took place two years after the 1st Workshop, again in San Juan, Puerto Rico. The focus of the 2nd Workshop was on "Neutrino and Flavor Physics", an area of active research and rapid experimental developments. We believe the experimental determination of neutrinos' masses and mixing will provide us with the first hint of what may lie beyond the Standard Model. Thanks to the diligence of the speakers and the active participation of all the participants, we feel the Workshop has fulfilled its purpose of updating new results and inspiring new ideas.

This volume records the scientific talks presented at the Workshop. What is not recorded here are the lively physics discussions among the participants and other fond memories the participants enjoyed. What leaps to mind is the surprise birthday celebration for Maru, José's wife, at the Workshop's banquet, as well as the image of Mark Vagin and Richard van der Water being the only persons eating in the rain while the rest of us crowded under a couple of umbrellas during the banquet (somebody made the comment that Mark and Richard were used to water as they both worked with water Čerenkov detectors).

As in the 1st Workshop, this Workshop was run smoothly with the tireless efforts from Maru, Ileana and Carmelo. We extend our heartfelt gratitude for their help. Our thanks also go to members of the Arecibo Observatory, in particular Daniel Altschuler and José Luis Alonso, for their reception and organization of the Tuesday session.

This Workshop would not have been held if not for José Nieves' persistence in securing support from his university. We thank the University of Puerto Rico for sponsoring the Workshop. Special thank goes to Dean Brad Weiner, who voiced his personal support for the next Workshop during his welcoming address. With his encouragement, it was announced at the end of the Workshop that the 3rd Tropical Workshop on Particle Physics and Cosmology would be planned for 20-25 May, 2002.

C. N. Leung and Qaisar Shafi
for the Scientific Committee
July 2000

Janet Conrad's announcement of the Third Tropical Workshop presented by Bill Louis.

NEUTRINO THEORIES AND MASSES

Mass Relations for Neutrinos

K.S. Babu

Department of Physics
Oklahoma State University
Stillwater, OK 74078

Abstract.
A group theoretical generalization of the well–known Georgi–Jarlskog relations $(m_\mu/m_\tau) = 3\,(m_s/m_b)$ and $(m_e/m_\mu) = (\frac{1}{3})^2\,(m_d/m_s)$ to neutrinos is found in the context of $SO(10)$. The new relations are $(m_{\nu_\mu}/m_{\nu_\tau}) = 16\,(m_c/m_t)$, and $(m_{\nu_e}/m_{\nu_\mu}) = (\frac{1}{16})^2\,(m_u/m_c)$, which are consistent with present neutrino data, assuming the MSW solution for solar neutrinos. Inclusion of the LSND result with a constrained set of parameters is analyzed. With Abelian flavor symmetries dictating the form of the neutrino mass matrix we classify all consistent four and five parameter models that accommodate simultaneously the solar, atmospheric and LSND neutrino data.

A Introduction

One of the deficiencies of the Standard Model of strong and electroweak interactions is that all the quark and lepton masses are free parameters, adjusted to their experimentally observed values. When the Standard Model is extended to a higher symmetry, some of these free parameters may get related. A case in point is Grand Unified Theories (GUTs), where quarks and leptons belong to common multiplets. In minimal $SU(5)$ GUT the relation

$$D = L^T \tag{1}$$

arises. Here D is the mass matrix for the down–type quarks, and L^T is the transpose of the charged lepton mass matrix. The equality of mass eigenvalues follows from Eq. (1):

$$m_b^0 = m_\tau^0,\ m_s^0 = m_\mu^0,\ m_d^0 = m_e^0\ . \tag{2}$$

Here the superscript ^{0} indicates GUT scale values, which are not the same as the observed ones. When these relations of Eq. (2) are extrapolated down to low scales, the first one, $m_b^0 = m_\tau^0$, leads to a good prediction for the b–quark mass in terms of the τ lepton mass [1]. The second and the third relations of Eq. (2), on the other hand, are bad, – they predict s and d quark masses that are off quite a bit from their experimental values.

Georgi and Jarlskog found an elegant generalization of the mass relations Eq. (2) [2]. They showed that with a non-minimal Higgs sector, it is possible to preserve the successful relation $m_b^0 = m_\tau^0$, while correcting the bad relations as follows:

$$m_s^0 = \frac{1}{3}m_\mu^0$$
$$m_d^0 = 3m_e^0 \qquad (3)$$

These Georgi–Jarlskog relations work quite well when extrapolated to low energies.

The question we wish to address here is this: Is it possible to find a generalization of the Georgi–Jarlskog relations in the neutrino sector? We will show that it is indeed possible to derive the isospin $+1/2$ analogs of Eq. (3) in the context of $SO(10)$ GUT. Furthermore, these new relations that we derive work quite well to accommodate the solar and atmospheric neutrino data, assuming that the MSW mechanism is relevant for solar neutrinos.

It is not entirely trivial that the $I = +1/2$ analogs of Eq. (3) can actually be derived. We will discuss the potential barriers and ways to overcome them. Now that there appears to be ample information on neutrino masses, any such relation that arises from symmetry considerations can be directly checked against experiment.

B The problem with M_R

The main problem in predicting neutrino masses lies in the fact that they differ in character and presumably in origin from the other fermion masses and are therefore difficult to relate to them. This is evident in the generally favored framework for understanding neutrino mass, the seesaw mechanism [3]. In the seesaw mechanism the neutrino mass matrix M_ν comes from two more basic mass matrices: a Dirac mass matrix N, which couples the left-handed neutrinos to the right-handed ones, and a Majorana mass matrix M_R, which couples the right-handed neutrinos to themselves. The seesaw formula is $M_\nu = -N^T M_R^{-1} N$. There are many models in which the Dirac neutrino mass matrix N is related by symmetry to other mass matrices about which something is known, in particular the Dirac mass matrices of the up quarks (U), down quarks (D), and charged leptons (L). The problem lies with the Majorana mass matrix M_R, about which most theoretical frameworks have very little to say. In the absence of information about M_R it is impossible to predict neutrino masses.

What is usually done is to parametrize the unknown matrix M_R or to make some ansatz for it. What we attempt here is more ambitious, namely to find a framework in which M_R, as well as N, is related by symmetry to the other mass matrices in such a way that a direct and definite prediction of neutrino mass ratios becomes possible.

There is an empirical reason for expecting that M_R might be related to the Dirac mass matrices. The Dirac masses of the quarks and charged leptons exhibit inter-family "hierarchies", and so it seems most likely that the Dirac neutrino mass matrix N has a similar hierarchy. If there were no hierarchy in M_R, then the seesaw formula would imply a hierarchy in M_ν that went as the *square* of the Dirac mass hierarchies. That is why it was long expected that $m_{\nu_\mu}/m_{\nu_\tau} \sim (m_c/m_t)^2$ [4]. However, it is now known that $(m_c/m_t)^2 \sim 6 \times 10^{-6}$, whereas m_{ν_μ}/m_{ν_τ} is in the range 10^{-1} to 3×10^{-3}, depending on which solution (MSW oscillation or vacuum oscillation) of the solar neutrino problem is correct. This suggests that the hierarchy in M_ν goes as the first power of the Dirac mass hierarchies rather than quadratically. From the seesaw formula one sees that this implies that M_R has a hierarchy similar to that of N and the other Dirac matrices. This similarity in structure suggests a direct link among all the mass matrices.

The formulas we obtain for neutrino mass ratios are indeed linear rather than quadratic. In fact, we find that

$$(m_{\nu_\mu}/m_{\nu_\tau}) = 16\ (m_c/m_t)$$
$$(m_{\nu_e}/m_{\nu_\mu}) = \frac{1}{(16)^2}\ (m_u/m_c)\ . \quad (4)$$

These can be thought of as natural generalizations of the well-known Georgi-Jarlskog formulas Eq. (3). The reason for the powers of 16 instead of 3 has to do with the group theory of $SO(10)$ and will be explained later.

C   The Georgi-Jarlskog formulas

Georgi and Jarlskog observed that in non-minimal $SU(5)$ schemes non–trivial group theoretical factors can arise in some of the mass relations. In particular they showed that the relations $m_\mu^0 = 3 m_s^0$ and $m_e^0 = m_d^0/3$, which are quite consistent with present data, can arise in a simple way. Factors of 3 emerge very naturally from the group theory of unification, essentially because of the fact that there are three colors. It has since been shown that the Georgi-Jarlskog formulas can be obtained in a variety of simple and plausible ways in different models [5].

A simple example of $SU(5)$ Yukawa terms that lead to the Georgi-Jarlskog formulas is the following: $h_{33} \mathbf{10}_3 \overline{\mathbf{5}}_3 \mathbf{5}_H + h_{22} \mathbf{10}_2 \overline{\mathbf{5}}_2 \overline{\mathbf{45}}_H + h_{12}(\mathbf{10}_1 \overline{\mathbf{5}}_2 + \mathbf{10}_2 \overline{\mathbf{5}}_1) \overline{\mathbf{5}}_H$. Here the numerical subscripts are family indices, and the subscript H refers to a Higgs field. These terms lead to the matrices

$$D = \begin{pmatrix} 0 & C_d & 0 \\ C_d & B_d & 0 \\ 0 & 0 & A_d \end{pmatrix}, \quad L = \begin{pmatrix} 0 & C_d & 0 \\ C_d & -3B_d & 0 \\ 0 & 0 & A_d \end{pmatrix}. \quad (5)$$

The relative factor of -3 in the (22) elements of L and D comes from a ratio of Clebsch coefficients in the coupling of the Higgs field that is a rank-three tensor of $SU(5)$ (i.e. the $\overline{\mathbf{45}}_H$). This factor of -3 directly gives the prediction $m_\mu^0 = 3m_s^0$. The other Georgi-Jarlskog formula arises from the fact that $m_e^0 = -C_d^2/(-3B_d)$ and $m_d^0 = -C_d^2/B_d$.

It is to be observed, however, that $SU(5)$ relates only L and D to each other. It does not relate the up-quark mass matrix U or the neutrino Dirac mass matrix N to any other mass matrices; they remain quite free. That means that if one is to extend the Georgi-Jarlskog formulas to neutrinos a more powerful symmetry is needed. Thus we turn to $SO(10)$.

D   Extending to $SO(10)$

The group $SO(10)$ relates all four Dirac mass matrices, N, U, D, and L, to each other. In the minimal scheme of Yukawa coupling, in which all masses come from terms of the form $\mathbf{16}_i \mathbf{16}_j \mathbf{10}_H$, $SO(10)$ predicts that $N = U \propto D = L$, with the further requirement that these matrices be symmetric. This would give the predictions $m_c^0/m_t^0 = m_s^0/m_b^0 = m_\mu^0/m_\tau^0$ and $m_u^0/m_c^0 = m_d^0/m_s^0 = m_e^0/m_\mu^0$. There is too much symmetry here and these relations are not realistic. However, as with $SU(5)$, group-theoretical factors different from 1 can be introduced into some of these relations by going to non-minimal Yukawa coupling schemes. For example, the generalization to $SO(10)$ of the simple Yukawa operators that gave rise to Eq. (5) is the following: $h_{33}\mathbf{16}_3\mathbf{16}_3\mathbf{10}_H + h_{22}\mathbf{16}_2\mathbf{16}_2\overline{\mathbf{126}}_H + h_{12}\mathbf{16}_1\mathbf{16}_2\mathbf{10}_H$. This would lead to the same forms for D and L given in Eq. (5) as well as the following forms for the other Dirac mass matrices:

$$U = \begin{pmatrix} 0 & C_u & 0 \\ C_u & B_u & 0 \\ 0 & 0 & A_u \end{pmatrix}, \quad N = \begin{pmatrix} 0 & C_u & 0 \\ C_u & -3B_u & 0 \\ 0 & 0 & A_u \end{pmatrix}, \quad (6)$$

where $C_u/A_u = C_d/A_d$. One sees that the same ratio of -3 between the (22) elements of N and U exists as between the (22) elements of L and D. (This is easily understood in terms of the $SU(4)_c \times SU(2)_L \times SU(2)_R$ decomposition of $SO(10)$, under which the VEV in $\overline{\mathbf{126}}_H$ that breaks the weak interactions is in the $(\mathbf{15}, \mathbf{2}, \mathbf{2})$, and therefore couples proportionally to the $SU(4)_c$ generator $B - L$.) For the same reasons, some of the other ways that the Georgi-Jarlskog factor of 3 can arise in the D-L sector also give rise to parallel factors of 3 in the U-N sector. For example, if the term $\mathbf{16}_2\mathbf{16}_2\overline{\mathbf{126}}_H$ is replaced by $\mathbf{16}_2\mathbf{16}_2\mathbf{45}_{B-L}\mathbf{45}_{I_{3R}}\mathbf{10}_H$, where $\langle \mathbf{45}_{B-L} \rangle \propto B-L$ and $\langle \mathbf{45}_{I_{3R}} \rangle \propto I_{3R}$, both L_{22}/D_{22} and N_{22}/U_{22} are equal to -3 (see the first paper in [5]).

The foregoing only shows that a relation can arise in $SO(10)$ between the up-quark mass matrix U and the Dirac mass matrix of the neutrinos N, with the possibility of non-trivial group-theoretical factors. This by itself is not sufficient to yield predictions for the neutrinos; for that it is also necessary to be able to say something about the Majorana mass matrix M_R. Happily, there does exist the possibility in $SO(10)$ of relating M_R to U and N, as we shall now see.

Consider, for example, the Yukawa operator $\mathbf{16}_i \mathbf{16}_j \overline{\mathbf{126}}_H$. The multiplet $\overline{\mathbf{126}}_H$ contains both weak-doublet components (contained in the $(\mathbf{15}, \mathbf{2}, \mathbf{2})$ of $SU(4)_c \times SU(2)_L \times SU(2)_R$), which can give the Dirac masses for the quarks and leptons, and a weak-singlet component (contained in the $(\mathbf{10}, \mathbf{1}, \mathbf{3})$) that can give superlarge masses to the right-handed neutrinos. Consequently, if both kinds of components acquire non-vanishing vacuum expectation values, the term $\mathbf{16}_i \mathbf{16}_j \overline{\mathbf{126}}_H$ would contribute to all five mass matrices D, L, U, N, and M_R [6]. As noted before, its contributions to N and U would be in the ratio $-3 : 1$.

Other kinds of Yukawa operators also exist that contribute both to M_R and to the Dirac mass matrices. For example, the term $\mathbf{16}_i \mathbf{16}_j \overline{\mathbf{16}}_H \overline{\mathbf{16}}_H$ contributes to U, N, and M_R (but not to D or L). As with the term we previously considered, there is a Clebsch factor between the contribution to N and to U; however, what that factor is depends in this case on how the fields are contracted. There are two independent ways to contract $\mathbf{16}_i \mathbf{16}_j \overline{\mathbf{16}}_H \overline{\mathbf{16}}_H$ to form an invariant. The contraction that we shall consider is $[\mathbf{16}_i \overline{\mathbf{16}}_H]_{45} [\mathbf{16}_j \overline{\mathbf{16}}_H]_{45}$, where we mean that the fields in the brackets are contracted in the adjoint (i.e. $\mathbf{45}$) channel. Such a term can, of course, arise simply by integrating out a fermion field in the adjoint representation of $SO(10)$. With the indices contracted in this way, it is straightforward to show that the contributions to the elements of N and U are in the ratio $3 : 8$.

E Obtaining mass formulas for neutrinos

We now have the ingredients to allow a generalization of the Georgi-Jarlskog formula to neutrinos.

First, we assume that the (22) elements of the matrices N and U arise from a Yukawa operator O_{22} that gives them in the ratio $-3 : 1$. This is motivated by the empirical success of the Georgi-Jarlskog formulas, and by the fact already noted that some simple Yukawa terms that give the Georgi-Jarlskog $-3 : 1$ ratio for the L-D sector give the same ratio for the U-N sector. We also assume that the (22) element of M_R remains zero. (This is automatically the case if $O_{22} = \mathbf{16}_2 \mathbf{16}_2 \mathbf{45}_{B-L} \mathbf{45}_{I_{3R}} \mathbf{10}_H$. It is true for $O_{22} = \mathbf{16}_2 \mathbf{16}_2 \overline{\mathbf{126}}_H$ if the weak-singlet VEV in $\overline{\mathbf{126}}_H$ vanishes.)

Second, there must be enough other terms to make M_R a non-singular matrix. A particularly simple possibility would be that only the (33), (12), and (21) elements of M_R are non-vanishing. We assume that these arise from terms of the form $[\mathbf{16}_i \overline{\mathbf{16}}_H]_{45} [\mathbf{16}_j \overline{\mathbf{16}}_H]_{45}$, already discussed above, with $(ij) = (12)$, and (33).

With just these three terms one obtains the following realistic mass matrices

$$U = \begin{pmatrix} 0 & C_u & 0 \\ C_u & B_u & 0 \\ 0 & 0 & A_u \end{pmatrix}, \quad N = \begin{pmatrix} 0 & \frac{3}{8}C_u & 0 \\ \frac{3}{8}C_u & -3B_u & 0 \\ 0 & 0 & \frac{3}{8}A_u \end{pmatrix}, \quad M_R = \begin{pmatrix} 0 & C_u & 0 \\ C_u & 0 & 0 \\ 0 & 0 & A_u \end{pmatrix} \Lambda, \tag{7}$$

where Λ is a ratio of a GUT-scale VEV to a weak-scale VEV. The ratios of the up quark masses can be directly read off from the form of U: $m_c^0/m_t^0 = B_u/A_u$, and $m_u^0/m_c^0 \cong -C_u^2/B_u^2$. To find the ratios of neutrino masses one must first use the seesaw formula to compute M_ν.

$$M_\nu = -N^T M_R^{-1} N = -\begin{pmatrix} 0 & \frac{9}{64}C_u & 0 \\ \frac{9}{64}C_u & -\frac{9}{4}B_u & 0 \\ 0 & 0 & \frac{9}{64}A_u \end{pmatrix} \Lambda^{-1}. \tag{8}$$

This gives the relations $(m_{\nu_\mu}^0/m_{\nu_\tau}^0) = 16\,(m_c^0/m_t^0)$, and $(m_{\nu_e}^0/m_{\nu_\mu}^0) = \frac{1}{256}\,(m_u^0/m_c^0)$, as given in Eq. (4). The first of these relations is obviously the more practically interesting one, and is quite consistent with what we presently know about neutrino oscillations. If one takes as the central value $m_c^0/m_t^0 = 1/400$ (corresponding to $m_c(m_c) = 1.27$ GeV, $m_t^{\text{physical}} = 174$ GeV, and using beta functions of minimal supersymmetry to extrapolate the masses from m_t to $M_{\text{GUT}} \simeq 2 \times 10^{16}$ GeV with $\tan\beta$ in the range of $3-30$ and $\alpha_s(m_Z) = 0.118$) [7], the prediction is that $m_{\nu_\mu}/m_{\nu_\tau} \cong (25)^{-1}$. Since in this model the neutrino masses are hierarchical, the masses of ν_μ and ν_τ can be gotten directly from the Δm^2 measured in solar neutrino oscillations and atmospheric neutrino oscillations, respectively. For this model the small angle MSW solar solution is the relevant one. If one takes the central value for Δm_{21}^2 to be 5.1×10^{-6}eV^2 [8], and for Δm_{32}^2 to be 3×10^{-3}eV^2 [9], then the central value of the mass ratio is $m_{\nu_\mu}/m_{\nu_\tau} = (24.3)^{-1}$, which is in remarkably good agreement with the prediction. Of course, there are still large uncertainties in the neutrino masses, and some uncertainty in m_t^0 from the running between the unification scale and the low scale, which depends on $\tan\beta$ (this is of order 10% for $\tan\beta = 3-30$; the running of neutrino mass ratio is negligible). But presumably all of these quantities will be well measured in the future, and the mass formula in Eq. (4) well tested.

In our framework, it is possible to predict rather precisely the masses of the heavy ν_R's. We find $M_{\nu_\tau}^R \simeq (9/64)(m_t^2/m_{\nu_\tau}) \simeq 3.7 \times 10^{13}$ GeV and $M_{\nu_e}^R \simeq M_{\nu_\mu}^R \simeq (\sqrt{m_u/m_c})\,(m_c/m_t)\,M_{\nu_\tau}^R \simeq 5.6 \times 10^9$ GeV. This may be relevant for leptogenesis.

How unique is the factor of 16 in Eq. (4)? Clearly, it is no more unique than was the factor of 3 obtained by Georgi and Jarlskog, who could have obtained other factors — such as 1 or 9 — quite easily, but clearly had an eye on the actual quark and lepton masses as well as the group-theoretical possibilities of $SU(5)$. Similarly, by choosing different Yukawa operators we could have gotten other factors than 16. However, we have surveyed the possibilities and find that most other choices lead to factors that are either much too small or much too large to be realistic.

Moreover, the choice that gives 16 also gives mass matrices that look the most similar to the matrices of Georgi and Jarlskog. We should mention, however, one other possibility that is also potentially realistic. If in Eq. (7) one sets U_{22} and N_{22} to zero and takes instead $U_{23} = U_{32} = N_{23} = N_{32}$, as would arise from $\mathbf{16_2 16_3 10}_H$, for example, then the relation $(m_{\nu_\mu}/m_{\nu_\tau}) \simeq (64/3)(m_c/m_t)$ results.

Another point that should be emphasized is that by relating M_R and N to other mass matrices, and thus predicting them precisely, one opens up the possibility of exact predictions for the neutrino mixing angles as well. What these predictions are will depend, of course, on the way that the charged leptons are incorporated in the model, so that different models that give the relation in Eq. (4) can have different predictions for the neutrino angles. We will now show that the charged leptons and down quarks can be accommodated in a simple fashion in the present framework.

F Including the charged leptons and down quarks

There is more than one way to include the charged leptons and down quarks in the present framework. Here we present an example that gives realistic predictions, is fairly simple, and incorporates the Georgi-Jarlskog formulas. The structure is quite similar to the model proposed in [10]. The matrices U, N, and M_R are as given in Eq. (7), the remaining mass matrices are (in a notation where the right–handed singlet fermions f^c multiply on the left and the left–handed doublet fermions f on the right)

$$D = \begin{pmatrix} 0 & C_d & 0 \\ 0 & B_d & A'_d \\ C'_d & 0 & A_d \end{pmatrix}, \quad L = \begin{pmatrix} 0 & 0 & C'_d \\ C_d & -3B_d & 0 \\ 0 & A'_d & A_d \end{pmatrix}. \quad (9)$$

The (22) elements come from the same term as gives the (22) elements of U and N, which can be, for instance, $\mathbf{16_2 16_2 \overline{126}}_H$ or $\mathbf{16_2 16_2 45}_{B-L} \mathbf{45}_{I_{3R}} \mathbf{10}_H$. The "lopsided" (23)[(32)] elements in $D[L]$ come from $[\mathbf{16_2 16}_H]_{10}[\mathbf{16_3 16'}_H]_{10}$, in a notation explained previously. Here the $\mathbf{16}_H$ acquires a GUT scale VEV along the weak singlet direction, while the $\mathbf{16'}_H$ has a VEV along the weak doublet. Such lopsided elements have been shown to explain in a simple way the largeness of the atmospheric neutrino mixing and the smallness of V_{cb} [11]. Similar terms generate the other lopsided off–diagonal entries of D, L.

The hierarchy in masses follows if we assume $C_d, C'_d \ll B_d \ll A_d, A'_d$. Denote the ratios $A'_d/A_d \equiv \sigma$, $B_d/A_d \equiv \epsilon/3$, $C_d/A_d \equiv \delta$, and $C'_d/A_d \equiv \delta'$ with $(\delta \sim \delta') \ll \epsilon \ll (\sigma \sim 1)$. After untangling the large $(s_R - b_R)$ mixing in D and the $\mu_L - \tau_L$ mixing in L (both of which are parameterized by a common angle $\tan\theta \equiv \sigma$), the matrices in Eq. (9) will look very much like those in Eq. (5). All phases in D and L can be removed by field redefinitions; a single phase ϕ will remain in U and an independent phase ϕ' in M_ν. (We take these phases to be in the (12) and (21) elements.) For the mass ratios in the down sector we obtain:

$$b/\tau \simeq 1, s/\mu \simeq 1/3, d/e \simeq 9, \mu/\tau \simeq \epsilon/(1+\sigma^2), e/\mu \simeq \delta\delta'\sigma\sqrt{1+\sigma^2}/\epsilon^2. \qquad (10)$$

(Here and below we denote the mass of a particle at the GUT scale by the particle's name, eg. $s \equiv m_s^0$.) The first of these relations is the well-known successful mass relation of minimal $SU(5)$ [1], while the second and third are the Georgi-Jarlskog relations [2].

For the quark mixing angles we obtain:

$$|V_{us}^0| \simeq \left|\sqrt{d/s}\sqrt{\sigma}(1+\sigma^2)^{-1/4}xe^{i\phi} + \sqrt{u/c}\right|, |V_{cb}^0| \simeq (s/b)\sigma,$$
$$|V_{ub}^0| \simeq (s/b)\left|\sqrt{d/s}\sigma^{-1/2}(1+\sigma^2)^{-1/4}xe^{i\phi} - \sqrt{u/c\sigma}\right|, \qquad (11)$$

where $x \equiv \sqrt{\delta'/\delta}$. From these the values of the model parameters can be extracted, allowing then a prediction of the CP-violating Wolfenstein parameter η, using the following relation:

$$\eta^0 \equiv \text{Im}\left(\frac{V_{ub}^0 V_{cs}^0}{V_{us}^0 V_{cb}^0}\right) \simeq -\frac{1}{|V_{us}^0|^2}\sqrt{u/c}\sqrt{d/s}\sigma^{-3/2}(1+\sigma^2)^{3/4}x\sin\phi. \qquad (12)$$

The parameter σ and ϵ are best determined from $3|V_{cb}^0|/(\mu/\tau)$ and μ/τ respectively. Besides the input value of $|V_{cb}|$ at the weak scale, we need the renormalization factor to extrapolate it to M_{GUT}. For illustration we choose $|V_{cb}| = 0.036$ and the RGE factor to be 0.885, corresponding to $\tan\beta = 3 - 30$ in MSSM. ($|V_{ub}/V_{cb}|$, η, $|V_{us}|$, and m_μ/m_τ are approximately RGE invariant). Then $\sigma \simeq 1.6$ and $\epsilon \simeq 0.21$. This explains naturally why the $\mu-\tau$ mixing angle is large ($\tan\theta = \sigma$) while V_{cb} is small ($V_{cb} \sim s/b$). If we fit $|V_{us}^0| = 0.22$ and $|V_{ub}^0| = 0.0025$ (corresponding to a central value of $|V_{ub}/V_{cb}| = 0.08$), we find $x\cos\phi = -0.64$ and $x = 1.27$. The Wolfenstein parameter η is then predicted to be $\eta = 0.36$, which is right at its central value [7]. This non-trivial relation can be considered as another success of the model.

The fact that M_R has been nailed down, except for the phase ϕ', means that there are predictions for the mixing angles of the neutrinos as well as for their mass ratios. We obtain:

$$|U_{\mu 3}| \simeq \sigma/\sqrt{1+\sigma^2} \simeq 0.85, |U_{\mu 2}| \simeq 1/\sqrt{1+\sigma^2} \simeq 0.53,$$
$$|U_{e2}| \simeq \left|\sqrt{e/\mu}\sigma^{-1/2}(1+\sigma^2)^{-1/4}x^{-1}e^{i\phi'} - \sqrt{\nu_e/\nu_\mu}\right| = 0.031 \pm 0.004,$$
$$|U_{e3}| \simeq |\sqrt{e/\mu}\sigma^{1/2}(1+\sigma^2)^{-1/4}x^{-1}| \simeq 0.05. \qquad (13)$$

For the leptonic phase, we get

$$\eta_\ell \equiv \text{Im}\left(\frac{U_{\tau 1}U_{\mu 2}}{U_{\mu 1}U_{\tau 2}}\right) \simeq \text{Im}(1 - \sqrt{e/\mu}\sqrt{\nu_\mu/\nu_e}\sigma^{-1/2}(1+\sigma^2)^{3/4}x^{-1}e^{i\phi'})^{-1}, \qquad (14)$$

which is bounded to be less than 3×10^{-2} in absolute value.

The solar neutrino oscillation is governed by $4|U_{e2}|^2(1-|U_{e2}|^2) \simeq 3.8 \times 10^{-3}$. This value is nicely consistent with the small angle MSW explanation the solar data [8]. We note that in this model leptonic CP violation is relatively small. As for the atmospheric oscillation angle, we obtain $\sin^2 2\theta_{atm}^{osc} \simeq 0.81$. While this value is in the right range, it is somewhat on the lower side in this illustrative example. If we allow for a 15% reduction in $|V_{cb}|$ due to finite chargino corrections [12] (this is natural if $\tan\beta$ is moderately large; corrections of this order is also suggested by $b-\tau$ unification [13]), then $\sigma \simeq 1.3$, which will change the atmospheric oscillation angle to about 0.93.

While the specific way we adopted for including down quarks and charged leptons is by no means unique, we have illustrated how it becomes possible in our framework to calculate precisely the leptonic mixing parameters simultaneously with the neutrino mass ratios.

G   Incorporating LSND data

So far we have ignored (shamefully) the indication of neutrino masses from the LSND oscillation experiment. Anticipating an independent confirmation of this effect at miniBOONE experiment, we will attempt an explanation of all neutrino data in this section. It is well known that, owing to a lack of overlap of the mass–splittings in the three experiments, a fourth light sterile neutrino will have to be introduced, if all three are to be explained. We will denote this new state as ν_s.

It should be immediately emphasized that a light sterile neutrino (with mass of order 1 eV or less) is a drastic extension of the Standard Model. This is unlike the mild extension that was needed to accommodate small neutrino masses of the three known flavors. Recall that with the three active neutrinos, there was a simple dimensional argument that suggested why their masses are tiny – within the SM no mass term for them can be written down at the renormalizable level. The lowest allowed operator for neutrino mass scales inversely with a heavy mass, which is identified as the scale of the right–handed neutrino in the seesaw mechanism. This beautiful explanation of naturally light neutrinos is lost for the case of a sterile neutrinos. Indeed, its natural mass scale would be a more fundamental scale of the theory, such as the GUT or the Planck scale. In that regard, the confirmation of the LSND result will be quite a major discovery – in conjunction with the solar and atmospheric neutrino data, it will be the discovery of not just neutrino masses, but also of a new type of neutrino state.

The above arguments also suggest that explaining all three oscillation data with a sterile neutrino in a natural way will be a real challenge for theory. Of course, one can do a phenomenological fit to the data, the challenge will arise in deriving the phenomenological model from a more fundamental theory. That task is perhaps for the future. In the interim, we shall analyze predictive ways of accommodating all data with the least number of parameters. We shall make a simplifying assumption, namely that the structure of the light neutrino mass matrix is governed by Abelian flavor symmetries. This is a very natural assumption to make, for much of our success in understanding quark and lepton masses has arisen from such Abelian symmetries.

It turns out that the minimum number of parameters that are needed to accommodate all experiments is four. In this case, we show that the solar neutrino data must be explained via vacuum oscillations. If MSW explanation is relevant, the minimum number of parameters required is five. We analyze both cases. In the case of four parameter vacuum oscillation solutions, there are just two consistent models, in the five parameter MSW scenario, there are three consistent ones. For related discussions see Ref. [15].

H Four parameter fit to neutrino data

We shall impose the theoretical condition that all non-zero elements in the neutrino mass matrix are independent (apart from the symmetric nature dictated by Lorentz invariance). This follows from the assumed Abelian nature of the underlying flavor symmetry. Furthermore, we impose the cosmological constraint on the mixing between an active and a sterile species. Such mixing could bring the sterile state into thermal equilibrium via oscillations during the epoch of nucleosynthesis, making the effective number of neutrino species equal to four, contrary to observations. The limit we impose corresponds to this number not being greater than 3.6, which translates into a condition [14]

$$\Delta m^2 \sin^2 2\theta < 10^{-7} eV^2 \ . \tag{15}$$

There are exceptional situations where this limit need not hold, see e.g. Ref. [16]. We shall however, confine to the limit of Eq. (15), mainly because the exceptions would require generically more parameters in the neutrino mass matrix than we wish to allow.

With these constraints, we find that only two mass matrices can fit all neutrino data. In both cases the solar neutrino puzzle is explained by vacuum oscillation. The two models (I and II) are characterized by the following mass matrices (in the basis $(\nu_s, \nu_e, \nu_\mu, \nu_\tau)$):

$$M_I = \begin{pmatrix} 0 & d & 0 & 0 \\ d & 0 & 0 & b \\ 0 & 0 & 0 & a \\ 0 & b & a & c \end{pmatrix}, \tag{16}$$

$$M_{II} = \begin{pmatrix} 0 & d & 0 & 0 \\ d & 0 & 0 & b \\ 0 & 0 & c & a \\ 0 & b & a & 0 \end{pmatrix}. \tag{17}$$

These mass matrices are written in a basis where the charged lepton mass matrix is diagonal. While this can always be done, in discussing the origin of the neutrino mass matrices, it should be kept in mind that the underlying theory should yield the desired form in a basis where the charged leptons are diagonal. We shall return to this issue in Section J.

Let us analyze Model I first. Note that as $c \to 0$, the eigenvalues become two–fold degenerate. For small c, and for $a \gg b$ (since $b/a \simeq \theta_{\rm LSND} \simeq 0.04$), but allowing for $d \sim a$, we find the eigenvalues to be

$$a + \frac{c}{2}, \quad -a + \frac{c}{2}, \quad d + \frac{3b^2 d^2 c}{4(a^2-d^2)^2}, \quad -d + \frac{3b^2 d^2 c}{4(a^2-d^2)^2}. \tag{18}$$

The resulting Δm^2 are

$$\Delta m^2_{\rm atm} \simeq 2ac,$$
$$\Delta m^2_{\rm LSND} \simeq a^2,$$
$$\Delta m^2_{\rm solar} \simeq \frac{3b^2 d^3 c}{(a^2-d^2)^2}. \tag{19}$$

For illustration, let us choose $a \simeq 1\ eV, b \simeq 0.04\ eV, c \simeq 2.5 \times 10^{-3}\ eV, d \simeq 0.03\ eV$ to get $\Delta m^2_{\rm LSND} \simeq 1\ eV^2$, $\Delta m^2_{\rm atm} \simeq 5 \times 10^{-3}\ eV^2$, $\Delta m^2_{\rm solar} \simeq 3.5 \times 10^{-10}\ eV^2$, all of which fit the data well. In addition, $\sin^2 2\theta_{atm} \simeq 1$, and $\sin^2 2\theta_{solar} \simeq 1$. This implies that solar neutrino anomaly can be solved by vacuum oscillation only in this scenario. Large angle MSW oscillation does not occur in this model, because the ν_s state is predominantly in the lighter eigenstate.

In Model II, as in I, all mass splittings are proportional to c. In the same approximation as I, the eigenvalues are found to be

$$a + \frac{c}{2}; \quad -a + \frac{c}{2}, \quad d(1 - \frac{b^2}{2(a^2-d^2)}) + \frac{ca^2 b^2}{2(a^2-d^2)^2},$$
$$-d(1 - \frac{b^2}{2(a^2-d^2)}) + \frac{ca^2 b^2}{2(a^2-d^2)^2}. \tag{20}$$

Therefore the mass splittings will be

$$\Delta m^2_{\rm LSND} \simeq a^2,$$
$$\Delta m^2_{\rm atm} \simeq 2ac,$$
$$\Delta m^2_{\rm solar} \simeq \frac{2a^2 b^2 cd}{(a^2-d^2)^2}. \tag{21}$$

As an example, choose $a = 1\ eV, b = 0.04\ eV, c = 2.5 \times 10^{-3}\ eV, d = 10^{-5}\ eV$ to get $\delta m^2_{\rm LSND} = 1\ eV^2, \Delta m^2_{\rm atm} = 5 \times 10^{-3}\ eV^2, \Delta m^2_{\rm solar} = 10^{-10}\ eV^2$. The atmospheric oscillation angle is very close to 1, but for the solar angle we need to check. The effective (2,2) entry after diagonalizing the (2-3-4) sector is $b^2 c/a$. Numerically this is $\approx 3 \times 10^{-6}$ eV. So since $d = 10^{-5}$ eV, the mixing angle for solar neutrinos is maximal also [15].

With only four parameters, it is easy to convince oneself that these two are the only allowed matrices within the specified framework. The argument is as follows. We need to mix ν_s with one of the active species. Substantial mixing can occur only with ν_e (from cosmology). This requires the entry d in the mass matrix. The entries a and b are required for maximal mixing in atmospheric oscillation and to explain the required LSND mixing angle. That leaves one free parameter. This parameter, c, must create both the solar and atmospheric mass splittings. It must be in either of the (2,2), (3,3), (4,4) or (2,3) positions. The second and third options are the ones of Models I and II. If c is in (2,2), the atmospheric mass splitting will be $\Delta m^2_{atm} \simeq 2a^3 b^2 c/(a^2 - d^2)^2$, while the solar splitting is $\Delta m^2_{solar} \simeq 2cd$. Clearly, this will not fit the data, since atmospheric splitting is smaller than that of solar. If c is in the (2,3) position, one would have the mass relation

$$[\Delta m^2_{atm}]^2 = 4\Delta m^2_{solar} \Delta m^2_{LSND}. \tag{22}$$

This relation would work well if the solar neutrino mass splitting corresponds to MSW, but by construction, MSW mechanism cannot work here because ν_s is lighter than ν_e.

I Five parameter fit to the neutrino data

There are two motivations to consider a five parameter fit to all neutrino data. As just discussed, a four parameter fit will not admit MSW mechanism. Furthermore, it is not clear how good a fit is obtained with vacuum oscillation of ν_e to ν_s for solar neutrinos.

For the five parameter fits, we demand as before, the cosmological requirement, Eq. (15). We also demand that the solar neutrino anomaly is explained by MSW. (Otherwise, there are trivial extensions of Models I and II obtained simply by adding one more parameter to that system.) There are three models with five parameters. In the basis $(\nu_s, \nu_e, \nu_\mu, \nu_\tau)$, the mass matrices of these models are given as:

$$M_I = \begin{pmatrix} e & d & 0 & 0 \\ d & 0 & 0 & b \\ 0 & 0 & 0 & a \\ 0 & b & a & c \end{pmatrix},$$

$$M_{II} = \begin{pmatrix} e & d & 0 & 0 \\ d & 0 & 0 & b \\ 0 & 0 & c & a \\ 0 & b & a & 0 \end{pmatrix},$$

$$M_{III} = \begin{pmatrix} e & d & 0 & 0 \\ d & c & 0 & b \\ 0 & 0 & 0 & a \\ 0 & b & a & 0 \end{pmatrix}. \qquad (23)$$

In Model I, in the approximation $d \sim c \ll e \ll (a,b)$, the eigenvalues are found to be

$$a + \frac{c}{2}, \quad -a + \frac{c}{2}, \quad e, \quad -\frac{d^2}{e}. \qquad (24)$$

So

$$\Delta m^2_{\text{LSND}} \simeq a^2,$$
$$\Delta m^2_{atm} \simeq 2ac,$$
$$\Delta m^2_{\text{solar}} \simeq e^2. \qquad (25)$$

The mixing angles are $\theta_{\text{LSND}} \simeq b/a$, $\theta_{\text{solar}} \simeq d/e$ and $\theta_{atm} \simeq 45^\circ$. Small angle MSW works well, but if d and e are comparable, large angle MSW will also be possible.

In Model II, the eigenvalues are

$$a + \frac{c}{2}, \quad -a + \frac{c}{2}, \quad e, \quad \frac{(ecb^2 - a^2d^2)}{ea^2}. \qquad (26)$$

So $\Delta m^2_{atm} \simeq 2ac$ and $\Delta m^2_{\text{solar}} \simeq e^2$. Choose $e \sim c \sim 3 \times 10^{-3}$ eV, $d \sim 1 \times 10^{-4}$ eV, $a \simeq 1$ eV, $b \simeq 0.04$ eV. Small angle MSW will work, the solar mixing angle is d/e.

In Model III, the eigenvalues are

$$a + \frac{bc}{a}, \quad -a + \frac{bc}{a}, \quad -\frac{2bc}{a}, \quad e + \frac{ad^2}{2bc}. \qquad (27)$$

Here there is a nontrivial constraint to be checked. $\Delta m^2_{atm} \simeq 4bc$, which fixes $b \simeq 1/40$ eV. The effective (2,2) element after diagonalizing the (2-3-4) sector is $(-2bc/a) \simeq 2 \times 10^{-3}$ eV. This is just right for solar mass splitting. One needs $|e| \geq |2bc/a|$ for MSW resonance to occur. That is just satisfied in this case.

There are no other 5 parameter fits. For example, if c is moved to (2,2) position, from $\Delta m^2_{atm} \simeq 2b^2c/a$, one would need $c \sim eV$. But that is in conflict with MSW solar oscillation. c cannot be put in any other position consistent with cosmology, so Models I, II, and III are the only 5 parameter models.

In all these models, because the number of parameters was small, there is a prediction for the $\nu_e \nu_\tau$ oscillation parameter. For example, in Model I of Section H, the $\nu_e \to \nu_\tau$ oscillation probability is given by

$$P_{\nu_e \to \nu_\tau} \simeq \theta^2_{\text{LSND}} \sin^2\left(\frac{\Delta m^2_{\text{LSND}} L}{4E}\right) + \left[\frac{\Delta m^2_{\text{solar}}}{\Delta m^2_{atm}}\right]^{1/3} \theta^{1/3}_{\text{LSND}} \sin^2\left(\frac{\Delta m^2_{\text{LSND}} L}{4E}\right). \qquad (28)$$

Numerically, the amplitude of the second term is $\sim 5 \times 10^{-5}$, which might be within reach at a neutrino factory.

J Derivation of the mass matrices

We will give some tentative justification of the mass matrices given in Section H and I. We will attempt their derivation in the framework of seesaw mechanism. The heavy ν_R matrix has to be singular, so that one of the ν_R fields survives all the way to eV scale.

Consider as an example, $U(1)_{L_e+L_\mu-L_\tau}$ symmetry, where L_i is the ith lepton number. Since the ν_R fields are singlets of the SM, we have some freedom in choosing their L_i quantum numbers. Let us take the charges of $(\nu_{eR}, \nu_{\mu R}, \nu_{\tau R}) = (\alpha, 1, -1)$ with $\alpha \neq 1$. It is because of this choice that ν_{eR} will not pick up a superheavy mass. Let us also assume that there are singlet fields that break this $U(1)$ symmetry. Specifically, we will need $\eta(-1)$ and $\sigma(-2)$ singlet fields. The most general Dirac and Majorana mass matrices that can be written down with this symmetry are:

$$L = \begin{pmatrix} * & * & 0 \\ * & * & 0 \\ 0 & 0 & * \end{pmatrix}, \quad N = \begin{pmatrix} 0 & * & 0 \\ 0 & * & 0 \\ 0 & 0 & * \end{pmatrix}, \quad M_R = \begin{pmatrix} 0 & A & 0 \\ A & B & C \\ 0 & C & 0 \end{pmatrix}. \qquad (29)$$

Here the $*$ in (L, N) denote nonzero entries. As emphasized earlier, it is important to seek the desired light neutrino mass matrix in a basis where the charged lepton mass matrix is diagonal. With the matrices in Eq. (29), it is evident that when L is diagonalized, the form of D does not change. So one can work in the basis where L is already diagonal.

The entry C in M_R arises from a bare mass term, C from a Yukawa coupling to σ and A from η field. Note that M_R has rank 2. Seesaw diagonalization of N and M_R gives the light neutrino mass matrix of the form given in Model I. If the entry A is of order 100 GeV, while B and C are of order 10^{12} GeV, the ν_s will have a mass of order 1 eV, as required. The entry B being of order 100 GeV may go well within a supersymmetric framework.

Model II is obtained from the above by the interchange of $e \leftrightarrow \mu$.

K Acknowledgments

It was an exciting meeting in San Juan, I wish to thank the organizers and the supporting staff. The work reported here is based on research collaborations with S. Barr (Phys. Rev. Lett. (2000)) as well as with E. Ma and S. Nandi (unpublished). This work is supported in part by the Department of Energy Grant No. DE-FG03-98ER41076 and by a grant from the Research Corporation..

REFERENCES

1. M. Chanowitz, J. Ellis and M. Gaillard, *Nucl. Phys.* **B128**, 506 (1977).
2. H. Georgi and C. Jarlskog, *Phys. Lett.* **B86**, 297 (1979).

3. M. Gell-Mann, P. Ramond, and R. Slansky, in *Supergravity, Proc. Supergravity Workshop at Stony Brook*, ed. P. van Nieuwenhuizen and D.Z. Freedman (North Holland, Amsterdam (1979)); T. Yanagida, *Proc. Workshop on unified theories and the baryon number of the universe*, ed. O. Sawada and A. Sugimoto (KEK, 1979); R.N. Mohapatra and G. Senjanovic, *Phys. Rev. Lett.* **44**, 912 (1980).
4. For review see for e.g: S.A. Bludman, D.C. Kennedy and P.G. Langacker, *Phys. Rev.* **D45**, 1810 (1992).
5. K.S. Babu and R.N. Mohapatra, *Phys. Rev. Lett.* **74**, 2418 (1995); A. Kusenko and R. Shrock, *Phys. Rev.* **D49**, 4962 (1994); L. Hall and S. Raby, *Phys. Rev.* **D51**, 6524 (1995); C.H. Albright, and S.M. Barr, *Phys. Rev.* **D58**, 013002 (1998); K.S. Babu, J.C. Pati and F. Wilczek, *Nucl. Phys.* **B566**, 33 (2000).
6. K.S. Babu and R.N. Mohapatra, *Phys. Rev. Lett.* **70**, 2845 (1993).
7. C. Caso et. al., Review of Particle Physics, *Euro. Phys. J.* **C3**, 1 (1998).
8. J. Bahcall, P. Krastev and A. Yu. Smirnov, *Phys. Rev.* **D58**, 096016 (1998); M.C. Gonzalez-Garcia, P.C. de Holanda, C. Peña-Garay, and J.C.W. Valle, hep-ph/9906469.
9. SuperKamiokande Collaboration, Y. Fukuda et. al., *Phys. Rev. Lett.* **81**, 1562 (1998).
10. C.H. Albright and S.M. Barr, *Phys. Lett.* **B452**, 287 (1999).
11. K.S. Babu and S.M. Barr, *Phys. Lett.* **B381**, 202 (1996); C.H. Albright, K.S. Babu, and S.M. Barr, *Phys. Rev. Lett.* **81**, 1167 (1998); J. Sato and T. Yanagida, *Phys. Lett.* **B430**, 127 (1998); N. Irges, S. Lavignac, and P. Ramond, *Phys. Rev.* **D58**, 035003 (1998).
12. T. Blazek. S. Raby and S. Pokorski, *Phys. Rev.* **D52**, 4151 (1995).
13. L. Hall, R. Rattazzi and U. Sarid, *Phys. Rev.* **D50**, 7048 (1994).
14. R. Barbieri and A. Dolgov, *Phys. Lett.* **B237**, 440 (1990); K. Enquvist, K. Kainulainen and M. Thomson, *Nucl. Phys.* **B373**, 498 (1992).
15. D. Caldwell and R.N. Mohapatra, *Phys. Rev.* **D48**, 3259 (1993); S. Gibbons, R.N. Mohapatra, S. Nandi and A. Raychaudhuri *Phys. Lett.* **B430**, 296 (1998); E. Ma, *Phys. Rev. Lett.* **83**, 2514 (1999); *Phys. Lett.* **B444**, 391 (1998); V. Barger, S. Pakvasa, T. Weiler and K. Whisnant, *Phys. Lett.* **B437**, 107 (1998); J. Gelb and S.P. Rosen, *Phys. Rev.* **D62**, 013003 (2000).
16. R. Foot, M. Thomson and R. Volkas, *Phys. Rev.* **D53**, 5349 (1996).
17. S. Mohanty, D.P. Roy and U. Sarkar, *Phys. Lett.* **B445**, 185 (1998).

Minimal Model for Neutrino Masses and Mixings[1]

Paul H. Frampton

Department of Physics and Astronomy
University of North Carolina
Chapel Hill, NC 27599 USA

Abstract. Working in the framework of three chiral neutrinos with Majorana masses, we investigate a scenario first realized in an explicit model by Zee: that the neutrino mass matrix is strictly off-diagonal in the flavor basis, with all its diagonal entries precisely zero. This CP-conserving ansatz leads to two relations among the three mixing angles $(\theta_1, \theta_2, \theta_3)$ and two squared mass differences. We impose the constraint $|m_3^2 - m_2^2| \gg |m_2^2 - m_1^2|$ to conform with experiment, which requires the θ_i to lie nearby one of four 1-parameter domains in θ-space. We exhibit the implications for solar and atmospheric neutrino oscillations in each of these cases. A unique version of the Zee *ansatz* survives confrontation with experimental data, one which necessarily involves maximal just-so vacuum oscillations of solar neutrinos.

I INTRODUCTION

In this talk I will describe a paper published last year [1] with Shelly Glashow; the work was done at the Korean Institute for Advanced Study, Seoul, in June 1999 and the topic is what is the simplest way to extend the minimal standard model, where neutrinos are by definition massless, to accommodate the compelling evidence especially from the SuperKamiokande experiment for non-zero neutrino mass.

II THE SITUATION OF THE DATA

The minimal standard model involves three chiral neutrino states, but it does not admit renormalizable interactions that can generate neutrino masses. Nevertheless, experimental evidence suggests that both solar and atmospheric neutrinos display flavor oscillations, and hence that neutrinos do have mass. Two very different neutrino squared-mass differences are required to fit the data:

---

[1] This work was supported in part by the U.S. Department of Energy under grant number DE-FG02-97ER-41036

$$10^{-11}\text{eV}^2 \leq \Delta_s \leq 10^{-5}\text{eV}^2 \quad \text{and} \quad \Delta_a \simeq 10^{-3}\text{eV}^2 \quad (1)$$

where the neutrino masses m_i are ordered such that:

$$\Delta_s \equiv |m_2^2 - m_1^2| \text{ and } \Delta_a \equiv |m_3^2 - m_2^2| \simeq |m_3^2 - m_1^2|$$

and the subscripts s and a pertain to solar and atmospheric oscillations respectively. The large uncertainty in Δ_s reflects the several potential explanations of the observed solar neutrino flux: in terms of vacuum oscillations or large-angle or small-angle MSW solutions, but in every case the two independent squared-mass differences must be widely spaced with

$$r \equiv \Delta_s / \Delta_a < 10^{-2}$$

In a three-family scenario, four neutrino mixing parameters suffice to describe neutrino oscillations, akin to the four Kobayashi-Maskawa parameters in the quark sector.[2]

Solar neutrinos may exhibit an energy-independent time-averaged suppression due to Δ_a, as well as energy-dependent oscillations depending on Δ_s/E. Atmospheric neutrinos may exhibit oscillations due to Δ_a, but they are almost entirely unaffected by Δ_s. It is convenient to define neutrino mixing angles as follows:

$$\begin{pmatrix} \nu_e \\ \nu_\mu \\ \nu_\tau \end{pmatrix} = X \begin{pmatrix} \nu_1 \\ \nu_2 \\ \nu_3 \end{pmatrix}$$

where

$$X = \begin{pmatrix} c_2 c_3 & c_2 s_3 & s_2 e^{-i\delta} \\ +c_1 s_3 + s_1 s_2 c_3 e^{i\delta} & -c_1 c_3 - s_1 s_2 s_3 e^{i\delta} & -s_1 c_2 \\ +s_1 s_3 - c_1 s_2 c_3 e^{i\delta} & -s_1 c_3 - c_1 s_2 s_3 e^{i\delta} & +c_1 c_2 \end{pmatrix}$$

with s_i and c_i standing for sines and cosines of θ_i. For neutrino masses satisfying Eq.(1) the vacuum survival probability of solar neutrinos is [4]

$$P(\nu_e \to \nu_e)|_s \simeq 1 - \frac{\sin^2 2\theta_2}{2} - \cos^4 \theta_2 \sin^2 2\theta_3 \sin^2(\Delta_s R_s / 4E) \quad (2)$$

whereas the transition probabilities of atmospheric neutrinos are:

$$\begin{aligned} P(\nu_\mu \to \nu_\tau)|_a &\simeq \sin^2 2\theta_1 \cos^4 \theta_2 \sin^2(\Delta_a R_a/4E) \\ P(\nu_e \leftrightarrow \nu_\mu)|_a &\simeq \sin^2 2\theta_2 \sin^2 \theta_1 \sin^2(\Delta_a R_a/4E) \\ P(\nu_e \to \nu_\tau)|_a &\simeq \sin^2 2\theta_2 \cos^2 \theta_1 \sin^2(\Delta_a R_a/4E) \end{aligned}$$

$$(3)$$

None of these probabilities depend on δ, the measure of CP violation.

---

[2] Two additional convention-independent phases are measurable in principle, but they ordinarily do not affect neutrino oscillations [2,3]

III EXTENDING THE STANDARD MODEL

Let us turn to the origin of neutrino masses. Among the many renormalizable and gauge-invariant extensions of the standard model that can do the trick are:

- The introduction of a complex triplet of mesons (T^{++}, T^+, T^0) coupled bilinearly to pairs of lepton doublets. They must also couple bilinearly to the Higgs doublet(s) so as to avoid spontaneous $B - L$ violation and the appearance of a massless and experimentally excluded majoron. This mechanism can generate an arbitrary complex symmetric Majorana mass matrix for neutrinos.

- The introduction of singlet counterparts to the neutrinos with very large Majorana masses. The interplay between these mass terms and those generated by the Higgs boson—the so-called see-saw mechanism—yields an arbitrary but naturally small Majorana neutrino mass matrix.

- The introduction of a charged singlet meson f^+ coupled antisymmetrically to pairs of lepton doublets, *and* a doubly-charged singlet meson g^{++} coupled bilinearly both to pairs of lepton singlets and to pairs of f-mesons. An arbitrary Majorana neutrino mass matrix is generated in two loops.

- The introduction of a charged singlet meson f^+ coupled antisymmetrically to pairs of lepton doublets *and* (also antisymmetrically) to a pair of Higgs doublets. This simple mechanism was first proposed by Tony Zee [5] and results (at one loop) in a Majorana mass matrix in the flavor basis (e, μ, τ) of a special form:

$$\mathcal{M} = \begin{pmatrix} 0 & m_{e\mu} & m_{e\tau} \\ m_{e\mu} & 0 & m_{\mu\tau} \\ m_{e\tau} & m_{\mu\tau} & 0 \end{pmatrix} \quad (4)$$

We focus on the latter scenario. In particular, we adopt the Zee *ansatz* for $\mathcal{M}$ without committing ourselves to the Zee mechanism for its origin. Related discussions of Eq.(4) appear elsewhere [2,6] The present work is essentially a continuation of [2].

Because the diagonal entries of $\mathcal{M}$ are zero, the amplitude for no-neutrino double beta decay vanishes at lowest order [5] and this process cannot proceed at an observable rate. Furthermore, the parameters $m_{e\mu}$, $m_{e\tau}$ and $m_{\mu\tau}$ may be taken real and non-negative without loss of generality, whence $\mathcal{M}$ becomes real as well as traceless and symmetric. With this convention, the analog to the Kobayashi-Maskawa matrix becomes orthogonal: $\mathcal{M}$ is explicitly CP invariant and $\delta = 0$. But as we have noted, it is well known [4] that the mere existence of a squared-mass hierarchy virtually precludes any detectable manifestation of CP violation in the neutrino sector.

The sum of the neutrino masses (the eigenvalues of $\mathcal{M}$) vanishes:

$$m_1 + m_2 + m_3 = 0 \tag{5}$$

An important result emerges when the squared-mass hierarchy Eqs.(1) is taken into account along with Eq.(5). In the limit $r \to 0$, two of the squared masses must be equal.

IV   THE LOGICAL POSSIBILITIES

There are two possibilities. In case A, we have $m_1 + m_2 = 0$ and $m_3 = 0$. This case arises iff at least one of the three parameters in $\mathcal{M}$ vanishes. In case B, we have $m_1 = m_2$ and $m_3 = -2m_2 > 0$. This case arises iff the three parameters in $\mathcal{M}$ are equal to one another. Of course, r is small but it does not vanish: in neither case can the above relations among neutrino masses be strictly satisfied. But they must be nearly satisfied. Consequently we may deduce certain approximate but useful restrictions on the permissable values of the neutrino mixing angles θ_i. Prior to examining these restrictions, we note that Eqs.(5) and (1) exclude the possibility that the three neutrinos are nearly degenerate in mass. If the Zee *ansatz* is even approximately realized in nature, no neutrino mass can exceed a small fraction of an electron volt in magnitude and neutrinos are unlikely to contribute significantly to the dark matter of the universe.

We first consider case A. The relation $m_1 + m_2 = 0$ may be obtained in three ways depending on which parameter in $\mathcal{M}$ is set to zero. If $m_{e\mu} = 0$, the quantum number $L_\tau - L_e - L_\mu$ is conserved. It follows that $\cos\theta_1 = 0$ and $\theta_3 = \pi/4$. We see from Eq.(3) that atmospheric ν_μ's oscillate exclusively into ν_e's and *vice versa*. This subcase (or any nearby assignment of mixing angles) is strongly disfavored by SuperKamiokande data [7].

Alternatively we may set $m_{e\tau} = 0$ to obtain conservation of $L_\mu - L_\tau - L_e$. For this subcase, we obtain $\sin\theta_1 = 0$ and $\theta_3 = \pi/4$. We see from Eq.(3) that atmospheric ν_μ's do not oscillate at all. This subcase (or any nearby assignment of mixing angles) is also strongly disfavored by SuperKamiokande data [7].

The last and best ([2]) version of Case A has $m_{\mu\tau} = 0$ and leads to conservation of $L_e - L_\mu - L_\tau$. For this subcase, we obtain $\sin\theta_2 = 0$ and $\theta_3 = \pi/4$. We see from Eq.(2) that solar neutrino oscillations are maximal:

$$P(\nu_e \to \nu_e)|_s = 1 - \sin^2(\Delta_s R_s/4E) \tag{6}$$

Moreover, we see from Eq.(3) that atmospheric ν_μ's oscillate exclusively into ν_τ's with the unconstrained mixing angle θ_1:

$$P(\nu_\mu \to \nu_\tau)|_a = \sin^2 2\theta_1 \sin^2(\Delta_a R_a/4E)$$
$$P(\nu_\mu \leftrightarrow \nu_e)|_a = 0 \qquad P(\nu_e \to \nu_\tau)|_a = 0 \tag{7}$$

This implementation of the Zee *ansatz* is compatible with experiment: It predicts maximal solar oscillations (without an energy-independent term) and it is consistent with the just-so vacuum oscillation hypothesis [8]. However, it is evidently not compatible with the small-angle MSW explanation of solar neutrino data. Neither is it compatible with the large-angle MSW solution, because it predicts virtually maximal solar neutrino oscillations. If $m_{\mu\tau}$ is permitted to depart slightly from zero so as to generate a small finite value of $r \equiv \Delta_s/\Delta_a$, the coefficient of the oscillatory term in Eq.(6) will depart from unity by a term of order $r^2 \leq 10^{-4}$. The resulting solar-neutrino oscillations remain nearly maximal: they are energy independent and experimentally disfavored unless Δ_s lies within the just-so domain.

We furthermore note that the $m_{\mu\tau} \simeq 0$ version of the Zee *ansatz* admits atmospheric neutrino oscillations of the type $\nu_\mu \to \nu_\tau$ with any value of the mixing angle θ_1. At the same time, it precludes all oscillations involving atmospheric ν_e. These results are quite in accord with SuperKamiokande data.

Others have noted [9] that the relation $m_1^2 = m_2^2$ is preserved by radiative corrections for case A. This is not necessarily true for case B, where the relations satisfied by the neutrino masses, $m_3 = -2m_2 = -2m_1$, are not a consequence of a symmetry principle. In any event, we argue that case B cannot fit the data. Ref. [2] shows that case B leads to the relation:

$$\tan^2\theta_2 = 1/2 \tag{8}$$

Along with Eqs.(2), this implies:

$$P(\nu_e \to \nu_e)|_a = 1 - (8/9)\sin^2(\Delta_a R_a/4E) \tag{9}$$

That is, we must have large (almost maximal) oscillations of atmospheric ν_e. This result is strongly disfavored by SuperKamiokande data [7], so that case B can be rejected without further ado.

Our conclusion is simple. We find that one (and only one!) of the realizations of the Zee *ansatz* incorporating a squared-mass hierarchy is compatible with both solar and atmospheric neutrino data. It corresponds to the assignments $m_{e\mu} \simeq m\cos\theta_1$, $m_{e\tau} \simeq m\sin\theta_1$ and $m_{\mu\tau} \ll m$, with $\theta_2 \simeq 0$ and $\theta_3 \simeq \pi/4$. Near this domain, atmospheric electron neutrinos oscillate negligibly, while atmospheric muon neutrinos oscillate into tau neutrinos with the arbitrary mixing angle θ_1. Solar neutrino oscillations are very nearly maximal. They can be described by vacuum oscillations, but not by MSW oscillations. It is straightforward to implement the Zee model so as to conserve $L_e - L_\mu - L_\tau$ *exactly* so as to obtain $m_{\mu\tau} = 0$ to all orders [9]. Of course, this must not be done! Some unspecified new physics (beyond the introduction of Zee's f^+ meson) is required to lift the degeneracy between m_1^2 and m_2^2 so as to yield an extreme hierarchy of neutrino squared-mass differences, with $\Delta_s \sim 10^{-8}\,\Delta_a$.

REFERENCES

1. P.H. Frampton and S.L. Glashow, Phys. Lett. **B461**, 95 (1999).
2. C. Jarlskog et al., Phys. Lett. **B449**, 240 (1999);
   M. Matsuda et al., `hep-ph/0005147`.
3. S.L. Glashow, in Atti dei Convegni Lincei, Rome **133**, 69 (1997).
4. H. Georgi and S.L. Glashow, `hep-th/9808293`
5. A. Zee, Phys. Lett. **93B**, 389 (1980).
6. A. Yu. Smirnov and Zhijian Tao, Nucl. Phys. B426 (1994) 415;
   A. Yu. Smirnov and M. Tanimoto, Phys. Rev. D55 (1997) 1665.
7. T. Kajita for the SuperKamiokande collaboration, `hep-ex/9807003`.
8. V. Barger, K. Whisnant and R.J.N. Phillips, Phys. Rev. D24 (1981) 538;
   S.L. Glashow and L.M. Krauss, Phys. Lett. B190 (1987) 199.
9. R. Barbieri, G.G. Ross and A. Strumia, `hep-ph/9906470`.

Teeny, tiny Dirac neutrino masses: an unorthodox point of view[1]

P. Q. Hung

Dept. of Physics, University of Virginia,
382 McCormick Road, P. O. Box 400714, Charlottesville, Virginia 22904-4714
e-mail: pqh@virginia.edu

Abstract. There are now strong hints suggesting that neutrinos do have a mass after all. If they do have a mass, it would have to be tiny. Why is it so? Is it Dirac or Majorana? Can one build a model in which a teeny, tiny Dirac neutrino mass arises in a natural way? Can one learn something else other than just neutrino masses? What are the extra phenomenological consequences of such a model? These are the questions that I will try to focus on in this talk.

WHY SHOULD ONE BOTHER WITH A TEENY, TINY DIRAC NEUTRINO MASS?

A subtitle to this talk should perhaps go like "A see-saw-like mechanism without a Majorana mass". Here, I shall try to present arguments as to why it is interesting and worthwhile to study scenarios in which neutrinos possess a mass which is *pure Dirac* in nature. Along the way, I shall try to argue that one should perhaps try to separate the issue of a see-saw like mechanism from that of a Majorana mass. By 'see-saw-like mechanism", it is meant that a "tiny" mass arises due to the presence of a very large scale.

The suggestions that neutrinos do indeed possess a mass came from three different sources, all of which involve oscillations of one type of neutrino into another type. They are the SuperKamiokande atmospheric neutrino oscillation, the solar neutrino results, and the LSND result [1]. The present status of these three oscillation experiments is well presented in this workshop. The future confirmation of all three will certainly have a profound impact on the understanding of the origin of neutrino masses. In particular, it is now generally agreed that if there were only three light, active (i.e. electroweak non-singlet) neutrinos, one would not be able to

---

[1] This work is supported in parts by the US Department of Energy under grant No. DE-A505-89ER40518.

explain all three oscillation phenomena. The confirmation of *all three results* would most likeky involve the presence of a sterile neutrino.

Whatever the future experiments might indicate, one thing is probably true: If neutrinos do have a mass, it is certainly tiny compared with all known fermion masses. Typically, $m_\nu \leq O(10^{-11})$(Electroweak Scale). Why is it so small? Is it a Dirac or a Majorana mass? This last question presently has no answer from any known experiment. The nature of the mass will no doubt have very important physical implications. The route to a gauge unification will certainly be very different in the two cases. Whether or not the mass is Dirac or Majorana, there is probably some new physics which is responsible for making it so tiny. What is the scale of this new physics? What are the possible mechanisms which could give rise to the tiny mass? In trying to answer these questions, one cannot help but realize that there is something *very special* about neutrinos (specifically the right-handed ones) which make them different from all other known fermions. Do they carry some special symmetry?

One example of new physics which might be responsible for a small neutrino mass is the ever-popular and beautiful see-saw mechanism of Gell-Mann, Ramond and Slansky [2], in which a Majorana mass arises through a lepton number violating process. Generically, one would have $m_\nu \sim m_{D\nu}^2/\mathcal{M}$, with $m_{D\nu} \propto$ Electroweak Scale, and $\mathcal{M} \sim$ some typical GUT scale. Since one expects $\mathcal{M} \gg m_{D\nu}$, one automatically obtains a tiny *Majorana* neutrino mass. The actual detail of the neutrino mass matrix is however quite involved and usually depends on some kind of ansatz. But that is the same old story with any fermion mass problem anyway. The crucial point is the fact that the very smallness of the neutrino mass comes from the presumed existence of a very large scale $\mathcal{M}$ compared with the electroweak scale. This mechanism has practically become a standard one for generating neutrino mass. Why then does one bother to look for an alternative?

First of all, there is so far *no evidence* that, if neutrinos do have a mass, it should be of a Majorana type. If anything, the present absence of neutrinoless double beta decay might indicate the contrary. (Strictly speaking, what it does is to set an upper limit on a Majorana mass of approximately 0.2 eV, although actually it is a bound on $\sum_i U_{ei}^2 m_i$). Therefore, this question is entirely open. In the meantime, it is appropriate and important to consider scenarios in which neutrinos are pure Dirac. The questions are: How can one construct a tiny *Dirac* mass for the neutrinos? How natural can it be? Can one learn something new? Are there consequences that can be tested?

A MODEL OF TEENY, TINY DIRAC NEUTRINO MASS

The construction of the model reported in this talk was based on two papers [3,4]. There exists several other works [5] on Dirac neutrino masses which are very different from [3,4]. The first one [3] laid the foundation of the model. The second one [4] is a vastly improved and much more detailed version, with new results not reported in [3]. In constructing this model, we followed the following self-imposed requirements:

1) The smallness of the Dirac neutrino mass should arise in a more or less natural way.

2) The model should have testable phenomenological consequences, other than just merely reproducing the neutrino mass pattern for the oscillation data.

3) One should ask oneself if one can learn, from the construction of the model, something more than just neutrino masses. This also means that one should go beyond the neutrino sector to include the charged lepton and the quark sectors as well. This last sentence refers to work in progress and will not be reported here.

Description of the model

Before describing our model, let us briefly mention a few facts. First of all, it is rather easy to obtain a Dirac mass for the neutrino by simply adding a right-handed neutrino to the Standard Model. This right-handed neutrino (one for each generation) is an electroweak singlet and, as a result, can have a gauge-invariant Yukawa coupling: $g_\nu \bar{l}_L \phi \nu_R + h.c.$. The Dirac neutrino mass would then be $m_\nu = g_\nu \langle \phi \rangle$. With $\langle \phi \rangle \sim 173 GeV$, a neutrino mass of O(1 eV) would require a Yukawa coupling $g_\nu \sim 10^{-11}$. Although there is nothing wrong with it, a coupling of that magnitude is normally considered to be extremely fine-tuned, if it is put in by hand! Could $g_\nu \sim 10^{-11}$ be *dynamical*? Would the limit $g_\nu \to 0$ lead to some new symmetry? What would it be? This new symmetry would be the one that protects the neutrino mass from being "large".

In choosing such a symmetry, we followed our self-imposed requirement # 3: One should learn something more from it than just merely providing a symmetry to protect the neutrino mass. First, in order to implement the symmetry protection, one should assume that this new symmetry is particular to the neutrinos, in particular the right-handed ones since left-handed neutrinos are weak interaction partners of standard charged leptons. Therefore, it will be assumed that all fermions other ν_R's are *singlets* under this new symmetry.

One of the reasons we adhere to our requirement #3 is the wish to work from the bottom up, instead of from the top down. As a result, we would try to make every increased step in energy as meaningful as possible.

The symmetry chosen in [3,4] is a chiral gauge symmetry. It is $SU(2)_{\nu_R}$, where the subscript ν_R means that *only* ν_R's carry $SU(2)_{\nu_R}$ quantum numbers. Why $SU(2)_{\nu_R}$? Because it is a chiral gauge $SU(2)$ which has a very important property: For Weyl fermions transforming as doublets under such a group, there exists an argument due to Witten [6] that says, in a nutshell, that, because of the presence of a non-perturbative global anomaly the number of such Weyl doublets has to be *even* in order for the theory to be well-defined. (In the language of quantum field theory, this means that the generating functional should be non-vanishing.) Amusingly enough, a long-forgotten fact about the SM is related to the Witten anomaly. The absence of such anomaly for $SU(2)_L$ require an even number of electroweak doublets per family, which is the case there: one lepton and three quark doublets. (From a historical prespective, one might say that, had this constraint be known in the early seventies, the SM, as we now know it, would have had an extra strong argument in its favor, prior to the SLAC experiment.) In our case, let ν_R transform as doublets under $SU(2)_{\nu_R}$, i.e. we now have $\eta_R = (\nu_R, \tilde{\nu}_R)$. (Cosmological issues concerning $\tilde{\nu}_R$'s are discussed in [4].) The absence of the Witten anomaly then requires the number of η_R's to be *even*. If furthermore, η_R's carry some family indices (if a family symmetry exists) then this constraint can have a profound implication on the issue of family replication. Further remarks can be found in [4].

We know that families *do mix*. In consequence, we need some kind of family symmetry. The family symmetry chosen in [3,4] is a gauge symmetry. This choice is pure prejudice: We believe that gauge theories are better choices for a family symmetry because of the fact that they do provide strong constraints on matter representations and because one might want to mimic the vertical symmetry (the electroweak interactions). We choose $SO(N_f)$ as our family gauge group with all fermions transforming as *vector* representations in order to avoid the usual traingle anomaly. Our model is given by the following extension of the SM:

$$SU(3)_c \otimes SU(2)_L \otimes U(1)_Y \otimes SO(N_f) \otimes SU(2)_{\nu_R} \qquad (1)$$

In [3,4], I have discussed the various arguments used to constrain N_f. In this talk, I shall however restrict N_f to be $N_f = 4$. This means that this is a four-family model. Is a 4th generation ruled out by experiment as one often hears? The answer is: Not at all! For instance, the usual question is the following: What about the Z width which tells us that there are only three light neutrinos? This does not apply to the case when the 4th neutrino is more massive than half the Z mass. Why then would it be so heavy when the other three neutrinos are so light? Isn't it unnatural? The answer is NO as we shall see below. Then, what about the 4th generation quarks and charged leptons? There exists a review [7] dealing extensively with this question. A quick summary of that review is the statement that there is plenty of room for the discovery of the 4th generation, either at the next upgraded collider experiments at the Tevatron, or at the LHC. I shall now turn to the basic results of the model.

Basic Reults of the model

I shall describe the results which are based solely on the assumption that only two oscillation results are correct: The solar and atmospheric neutrino oscillation data. I shall mention at the end a possibility in case all three oscillation experiments are confirmed.

In a nutshell, here are the results obtained in [4]. 1) We obtain three light, *near degenerate* neutrinos. 2) The "tiny" masses are obtained *dynamically* at one loop. One will see below the reason for the use of the term "see-saw-like mechanism

TABLE 1. Particle content and quantum numbers of $SU(3)_c \otimes SU(2)_L \otimes U(1)_Y \otimes SO(N_f) \otimes SU(2)_{\nu_R}$

Standard Fermions	$q_L = (3, 2, 1/6, N_f, 1)$
	$l_L = (1, 2, -1/2, N_f, 1)$
	$u_R = (3, 1, 2/3, N_f, 1)$
	$d_R = (3, 1, -1/3, N_f, 1)$
	$e_R = (1, 1, -1, N_f, 1)$
Right-handed ν's	Option 1: $\eta_R = (1, 1, 0, N_f, 2)$
	Option 2: $\eta_R = (1, 1, 0, N_f, 2)$;
	$\eta'_R = (1, 1, 0, 1, 2)$
Vector-like Fermions	$F_{L,R} = (1, 2, -1/2, 1, 1)$
	$\mathcal{M}_{1L,R} = (1, 1, -1, 1, 1)$
	$\mathcal{M}_{2L,R} = (1, 1, 0, 1, 1)$
Scalars	$\Omega^\alpha = (1, 1, 0, N_f, 1)$
	$\rho_i^\alpha = (1, 1, 0, N_f, 2)$
	$\phi = (1, 2, 1/2, 1, 1)$

with Dirac mass". 3) The masses of the light neutrinos, m_{ν_i} ($i = 1, 2, 3$), and Δm^2 are correlated in an interesting way: a) If the MSW solution is chosen for the solar neutrino problem, the masses can be as large as O(few eV's) and can provide enough mass for the Hot Dark Matter (HDM); b) If the vacuum solution is chosen instead, the masses are found to be at most ~ 0.1 eV and, as a result, are too small to be relevant to the HDM. 4) There are a number of phenomenological consequences which can be tested: There is *no* neutrinoless double beta decay since the mass is Dirac; There is a possibility of detection of "light" (a couple of hundreds of GeV's) vector-like fermions; etc... 5) There are a number of possible cosmological consequences: Baryon asymmetry through neutrinogenesis with a pure Dirac neutrino mass; Perhaps some of the very heavy vector-like fermions could be the source of Ultra High Energy Cosmic Rays.

In writing down the Lagrangian for our model, we take into account the fact that our point here is to obtain a pure Dirac mass. Therefore, B-L will be assumed. The particle content is listed in the Table and the Lagrangian is given by

$$\mathcal{L}_{Lepton}^Y = g_E \bar{l}_L^\alpha \phi e_{\alpha R} + G_1 \bar{l}_L^\alpha \Omega_\alpha F_R + G_{M_1} \bar{F}_L \phi \mathcal{M}_{1R} + G_{M_2} \bar{F}_L \tilde{\phi} \mathcal{M}_{2R} + G_2 \bar{\mathcal{M}}_{1L} \Omega_\alpha e_R^\alpha + G_3 \bar{\mathcal{M}}_{2L} \rho_m^\alpha \eta_{\alpha R}^m + M_F \bar{F}_L F_R + M_1 \bar{\mathcal{M}}_{1L} \mathcal{M}_{1R} + M_2 \bar{\mathcal{M}}_{2L} \mathcal{M}_{2R} + h.c. \quad (2)$$

After integrating out the F, $\mathcal{M}_1$, and $\mathcal{M}_2$ fields, the relevant part of the effective Lagrangian below $M_{F,1,2}$ reads

$$\mathcal{L}_{Lepton}^{Y,eff} = g_E \bar{l}_L^\alpha \phi e_{\alpha R} + G_E \bar{l}_L^\alpha (\Omega_\alpha \phi \Omega^\beta) e_{\beta R} + G_N \bar{l}_L^\alpha (\Omega_\alpha \tilde{\phi} \rho_i^\beta) \eta_{\beta R}^i + H.c., \quad (3)$$

where

$$G_E = \frac{G_1 G_{M_1} G_2}{M_F M_1}; \quad G_N = \frac{G_1 G_{M_2} G_3}{M_F M_2}. \quad (4)$$

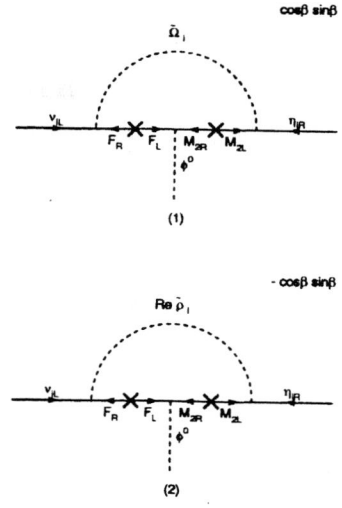

FIGURE 1. Feynman graph showing the computation of $\tilde{G}_\nu$, where $m_\nu = \tilde{G}_\nu \frac{v}{\sqrt{2}}$

As one can see, all neutrinos are *massless* when $SO(4) \otimes SU(2)_{\nu_R}$ is unbroken. Assume $<\Omega> = (0, 0, 0, V)$ and $<\rho> = (0, 0, 0, V' \otimes s_1)$, the 4th neutrino gets a mass $m_N = \tilde{G}_N \frac{v}{\sqrt{2}}$ with $\tilde{G}_N = G_1 G_{M_2} G_3 \frac{V V'}{M_F M_2}$. One can arrange the masses and couplings in such a way that $\tilde{G}_N \sim O(1)$, and $m_N \sim O(100 \text{ GeV})$. There is nothing unnatural about such a choice. It is *natural* in this scenario to have the 4th neutrino having a mass of $O(100 \text{ GeV})$. One point which is worth emphasizing again is the following: The breaking of $SU(2)_{\nu_R}$ (in addition to $SO(4)$) is essential for m_N to be non-vanishing! At this stage (tree-level), there are *three massless* neutrinos.

It turns out, at one loop, that the three formely-massless neutrinos acquire a common mass, i.e. they are *degenerate*:

$$\frac{m_\nu}{m_N} = \frac{\sin(2\beta)}{32 \pi^2} I(\frac{M_F}{M_2}, \frac{M_F}{M_G}, \frac{M_F}{M_P}) \qquad (5)$$

where $M_{F,2,G,P}$ are masses of particles which participate in the loop diagram and where $\tan \beta = V'/V$. This is shown in Fig. 1. Here $I(\frac{M_F}{M_2}, \frac{M_F}{M_G}, \frac{M_F}{M_P})$ is given by

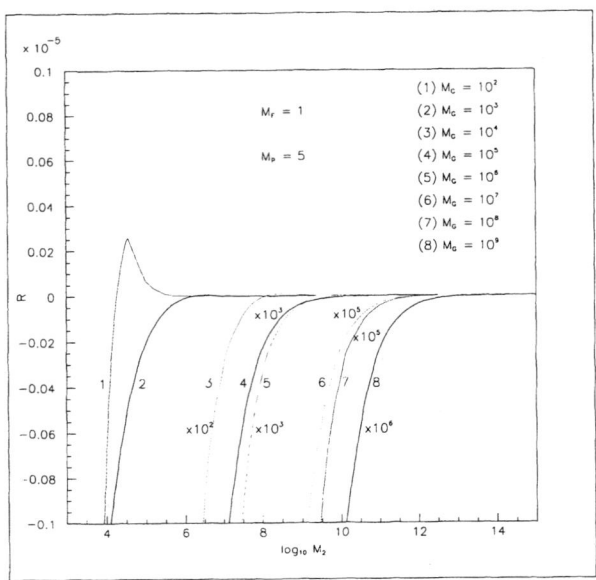

FIGURE 2. The ratio $R \equiv m_\nu/m_N$ (Eq. (23)) as a function of M_2 (in units of M_F, and hence the notation $M_F = 1$), for $M_P = 5$ and for various values of M_G. For visibility purpose, a few curves have been inflated by factors $\times 10^{2,3,5,6}$.

$$I(\frac{M_F}{M_2}, \frac{M_F}{M_G}, \frac{M_F}{M_P}) = \frac{1}{M_F - M_2}\{\frac{M_F[M_F^2(M_G^2\ln(\frac{M_G^2}{M_F^2}) - M_P^2\ln(\frac{M_P^2}{M_F^2})) + M_G^2 M_P^2 \ln(\frac{M_P^2}{M_G^2})]}{(M_G^2 - M_F^2)(M_P^2 - M_F^2)} - (M_F \leftrightarrow M_2)\}. \quad (6)$$

One important remark is in order here. From Eq. (5), one notices that the neutrino mass does not depend explicitely on the value of the masses $M_{F,2,G,P}$ but only on their *ratios*. If one takes M_F as a "base" mass for example, it turns out that one can obtain quite a small mass for the light neutrinos, $m_\nu \leq O(10^{-11})m_N$, as long as one has, e.g., $M_2 \gg M_F$, with M_F being an arbitrary number which can be as low as experimentally allowed i.e. O(200 GeV) [7]. (Remember that F stands for $F = (F^0, F^-)$, where F^0 and F^- are degenerate in mass.) This can be seen from Fig. 2. Just to illustrate this point, a numerical example would be helpful. Take $m_N = 100$ GeV, $M_2/M_F = 10^9$, $M_P/M_F = 5$, $M_G/M_F = 10^4$, we obtain $m_\nu = 1.4$ eV. I wish to emphasize that this is *not* a prediction. It will have to come from some deeper theory which will fix the above mass ratios and hence m_ν/m_N. This is only meant to illustrate the fact that, in our model, it is quite natural to get a teeny, tiny Dirac neutrino mass. To go further, one needs to lift the degeneracy of the light neutrinos. Before showing how one would go about doing it, let us explain what we meant by "see-saw-like mechanism without a Majorana mass".

The function I shown in Eq. (6) has the following limit: $I \to -M_F/M_2$ for $M_2 \gg M_G \gg M_P > M_F$. From Eq. (5), one can see that, in the above limit, $m_\nu \to \frac{m_N M_F}{M_2} \frac{\sin(2\beta)}{32\pi^2}$. If $M_F \sim O(m_N)$ ($M_F \ll M_2$), one would obtain

$$m_\nu \sim \frac{m_N^2}{M_2} \frac{\sin(2\beta)}{32\pi^2}. \tag{7}$$

This is a typical see-saw-like relation! With a "low" mass ($\sim$ Electroweak scale) vector-like fermion, F, one can qualitatively see that m_ν can be *very small* when $M_2 \gg M_F$. This behaviour is very reminescent of the see-saw mechanism, except that, in our case, the mass is *Dirac*.

Now the next step is to introduce mixing among the neutrinos in order to lift the mass degeneracy. This can be acomplished, in our model, by involving mixing in the scalar sector. At this stage, the degeneracy among the three light neutrinos is due to a remaining global $SO(3)$ symmetry. This remaining global symmetry can be explicitely broken by the scalar sector as shown in [4]. The offshoot of all this is, in the end, the fact that this explicit breaking depends on a parameter denoted by b in [4]. It turns out that, in a paper under preparation [8], b itself will be severely constrained when our model is extended to the quark sector. This is because the same scalar sector is also involved in the quarks. It is satisying to see the link between the quark and lepton sectors. This is however not the subject of this talk and I shall now return to the task at hand.

The first case which was investigated in [4] is when the scalar sector is written down in such a way that there is no mixing between the 4th neutrino and the other three in the mass matrix. Something interesting happens here. It turns out that the mass splittings are quasi-degenerate, in the sense that $|m_2^2 - m_1^2| \approx |m_3^2 - m_2^2|$. If this model were to explain both solar and atmospheric oscillation data, this quasi-degeneracy of Δm^2 has to be lifted. Also, the fact that the oscillation data appear to show $\Delta m^2_{Solar} \ll \Delta m^2_{Atmospheric}$ implies, in the context of our model, that indirectly the data suggests the existence of a 4th neutrino whose mixing with the lighter three will lift the quasi-degeneracy of Δm^2. Before showing how this could be done, let us see what these results imply. If the vacuum solution for solar neutrinos is preferred, i.e. $\Delta m^2 \sim 10^{-10} eV^2$, then it is found that the median mass value of three almost degenerate neutrinos is $\bar{m}_\nu \lesssim 0.1 eV$. As stated earlier, this is not enough for the HDM scenario. If the MSW solution is preferred, i.e. $\Delta m^2 \sim 10^{-5} eV^2$, the median mass value could be $\bar{m}_\nu \sim O(\text{few eV's})$, a reasonable value for the HDM scenario.

The fact that $\Delta m^2_{Solar} \ll \Delta m^2_{Atmospheric}$ indicates, in the context of this model, the existence of more than three light neutrinos. In [4] where only the atmospheric and solar data were taken into account, this means that it indicates the existence of a 4th neutrino. Again through the scalar sector, one can construct e.g. a mixing between the 3rd and 4th neutrino (other possibilities exist). The size of the mixing determines the correct mass splittings. It was found that there are strong constraints on the some of the scalar masses when one requires that $|\Delta m_{12}^2| \sim 10^{-5} eV^2$ and $|\Delta m_{23}^2| \sim 10^{-3} eV^2$.

The next question concerns the oscillation angles. To find out what they are, one needs to know the leptonic "CKM" matrix: $V_L = U_l^\dagger U_\nu$. In dealing with the neutrino sector of our model, we have presented a case where U_ν can be computed. It is basically given by:

$$U_\nu^{(3)} = \begin{pmatrix} -\frac{1}{\sqrt{2}} & -\frac{1}{\sqrt{2}} & 0 \\ \frac{1}{\sqrt{2}} & -\frac{1}{\sqrt{2}} & 0 \\ 0 & 0 & -1 \end{pmatrix} \qquad (8)$$

As for U_l which requires a detailed study of the charged lepton sector, a construction is in progress. In the meantime, just for the purpose of illustration, Reference [9] has been used in which a simple ansatz for the charged lepton sector was given. The reason for using this reference is because it contains an ansatz for U_ν which is similar to ours. Therefore the results should be similar: a small angle MSW solution and a large angle atmospheric solution.

EPILOGUE

I have presented in this talk a model which can "naturally" give rise to a teeny, tiny Dirac neutrino mass, without resorting to the concept of a Majorana mass. What was shown was the need to differentiate between the see-saw mechanism and the existence of a Majorana mass. In this model, the smallness of the light neutrino mass arises in a see-saw-like fashion, with the mass being purely Dirac. As we have argued in the Introduction, the reason for constructing such a model is twofold: a) One does not know experimentally whether the mass is Majorana or Dirac; b) The physics is *very* different in the two cases. At this stage of our knowledge, it is perhaps prudent to explore all different possibilities. Since so much has already been worked out with models using a Majorana mass, any new model which takes a different route should have a clear motivation and predictable consequences. For our model, we have presented clearly our motivation: naturally small Dirac neutrino mass, family replication, etc..; and predictions concerning the neutrino sector: Vacuum solution $\not\Leftrightarrow$ HDM, MSW solution $\Leftrightarrow$ HDM, $\Delta m_{Atmospheric}^2 \gg \Delta m_{Solar}^2$ as an indirect indication of a 4th heavy neutrino. Other phenomenological implications include:

1) There is *no* neutrinoless double beta decay because of the fact that we have a Dirac neutrino here.

2) The existence of "long-lived" and "light" (i.e. $\gtrsim 100 GeV$) vector-like leptons (F) whose detection might be possible at the LHC. A study of this kind of search can be found in a comprehensive review [7]. The quark counterparts should also be detectable [8].

3) The existence of several scalars with masses of order TeV's.

There are several other phenomelogical issues to be discussed. For lack of space, a few of those will be briefly mentioned. One is the S parameter for example. It is well-known that, to leading order, vector-like fermions which carry electroweak $SU(2)_L$ quatum numbers *do not* contribute to S if one has a *degenerate SU(2)* doublet. The reason for this being so is because the right-handed contribution cancels exactly the left-handed contribution. Therefore, to leading order, there is no constraint from the S parameter on the mass of the F-fermions. This point and other issues concerning quarks and leptons beyond the third generation are discussed in [7]. Issues such as the decay of the heavy 4th neutrino can be consulted in [4]. Also, another issue such as the magnitude of flavor-changing neutral currents, e.g. $\mu \to e\gamma$, will be discussed in an upcoming paper dealing with the charged lepton sector. However, a preliminary statement can be made. For example, in the case of $\mu \to e\gamma$, there are two kinds of contributions: One coming from the propagation of neutrinos with a non-zero mass inside the loop diagram for the process, and the other one coming from diagrams involving the new vector-like fermions. It turns out that both contributions are negligible: 1) In the first case, it is because $m_\nu \ll M_W$; 2) In the second case, it is because of the cancellations of the type described in [4].

As far as the cosmological implications are concerned, there are:

1) Can the fermion $\mathcal{M}_2$ be the source of Ultra High Energy Cosmics Rays (E $> 10^{11} GeV$)? For example if $M_F \sim 200 GeV$ then $M_2 \sim 2 \times 10^{11} GeV$? Would the decays of $\mathcal{M}_2$ (e.g. $\mathcal{M}_{2R} \to W_L^\pm F^\mp \to$ high energy quarks and leptons) be responsible for UHECR? This deserves a closer look.

2) The possibility of Baryon Asymmetry from neutrinogenesis. This is a scenario of Ref. [10]. The ingredients needed for such a scenario to work are basically: 1) a tiny, pure Dirac neutrino mass; 2) A decay process from some superheavy particles at the GUT (or similar) scale into right-handed neutrinos such that there is an asymmetry between right-handed neutrinos and anti-neutrinos; 3) a B+L violating process from the electroweak sphaleron. Since the Dirac neutrinos have a tiny Yukawa coupling (which is dynamical in our case), the part of B+L which is stored in the right-handed neutrinos, $(B+L)_R$, survived the sphaleron "washout". So if one starts out with B-L=0, this process can generate a *net* baryon number, $n_B = n_L \propto n_{\nu_R}$.

Last but not least, a Dirac neutrino mass would certainly imply a different route to unification, diffrent from the popular scenario such as $SO(10)$.

One last remark is in order here. If all three oscillation experiments were to be confirmed in the future, there seems to be a need for a sterile neutrino. How will it fit in our framework? It turns out that some modifications of the previous analysis will be needed but the basic framework is still the same. This work is in progress.

I would like to thank Jose Nieves, Terry leung, Art Halprin and Qaisar Shafi for a wonderful workshop.

REFERENCES

1. See Peter Fisher, Boris Kayser, and Kevin S. McFarland, hep-ph/9906244, for a comprehensive review and a list of references.
2. M. Gell-Mann, P. Ramond, and R. Slansky in Supergravity, edited by D. Freedman et al. (1979); T. Yanagida, KEK lectures (unpublished) (1979); R. N. Mohapatra and G. Senjanovíc, Phys. Rev. Lett. **44**, 912 (1980).
3. P. Q. Hung, Phys. Rev. **D59**, 113008 (1999).
4. P. Q. Hung, Phys. Rev. **D** (in press) (September, 2000), hep-ph/0003303.
5. B. Balakrishna and R. N. Mohapatra, Phys. Lett. **B216**, 349 (1989); D. Chang and R. N. Mohapatra, Phys. Rev. Lett. **58**, 1600 (1987); G. Barenboim and F. Scheck, Phys. Lett. **B461**, 1235 (1999); J. I. Silva-Marcos, Phys. Rev. **D59**, 091301 (1999).
6. E. Witten, Phys. Lett. **B117**, 324 (1982).
7. P. H. Frampton, P. Q. Hung, and M. Sher, Physics Reports **330**, 263 (2000).
8. P. Q. Hung and A. Soddu, manuscript under preparation (2000).
9. M. Fukugita, M. Tanimoto, T. Yanagida, Phys. Rev. **D57**, 4429 (1998).
10. K. Dick, M. Lindner, M. Ratz, and D. Wright, hep-ph/9907562 v3.

Neutrino Oscillations in Extended Anti-GUT Model[‡]

C.D. Froggatt

*Department of Physics and Astronomy,
Glasgow University, Glasgow, Scotland*

H.B. Nielsen

*Theory Division, CERN
and
Niels Bohr Institute, Copenhagen Ø, Denmark*

Y. Takanishi

Niels Bohr Institute, Copenhagen Ø, Denmark

Abstract. What we call the Anti-GUT model is extended a bit to include also right handed neutrinos and thus make use of the see-saw mechanism for neutrino masses. This model consists in assigning gauge quantum numbers to the known Weyl fermions and the three see-saw right handed neutrinos. Each family (generation) is given its own Standard Model gauge fields and a gauged field coupled to the $B - L$ quantum number for that family alone. Further we assign a rather limited number of Higgs fields, so as to break these gauge groups down to the Standard Model gauge group and to fit w.r.t. order of magnitude the spectra and mixing angles of the quarks and leptons. We find a rather good fit, which for neutrino oscillations favour the small mixing angle MSW solution, although the mixing angle predicted is closest to the upper side of the uncertainty range for the measured solar neutrino mixing angle.

An idea to make a "finetuning"-principle to "explain" the large ratios found empirically in physics such as "why is the weak scale low compared to the Planck scale?" is proposed (solving a hierarchy problem related problem).

Some speculative further extension is supposed to "explain" that we have three families.

I INTRODUCTION

Anti-GUT is the name which we have given to the model based on our favourite gauge group

$$AGUT = SMG \times SMG \times SMG \times U(1)_f. \qquad (1)$$

[‡] talk given by H.B. Nielsen at San Juan.

It began with N. Brene, D. Bennett and I. Picek and others but the inclusion of the $U(1)_f$ and the application to study the masses and mixing angles for quarks and leptons was together with C.D. Froggatt and his students (G. Lowe, D. Smith, M. Gibson). Here the symbol SMG stands for the Standard Model Group - and we may really think about it as a group [1] and not only a Lie algebra. It may then be assigned physical significance via the spectrum of representations it has (but one may just think of it as a Lie algebra, if one wants):

$$SMG = S(U(2) \times U(3)) = SU(2) \times SU(3) \times U(1). \tag{2}$$

In the present talk we shall actually present a slightly extended version of the "original" Anti-GUT model, which could thus be designated the extended Anti-GUT. This model may be described by saying that to each family we assign its own set of gauge fields consisting of a set Standard Model gauge fields, plus a gauged $U(1)$ group coupling to the family in question by coupling to the $B - L$ quantum number (but only to that family, the remaining families being considered having zero value for this $(B - L)$-charge and Standard Model ones related to the considered family gauge boson). Here $B - L$ stands for baryon number minus lepton number. That is to say the gauge group of the extended Anti-GUT - the model we shall consider in the present talk - is taken to be

$$(SMG \times U(1)_{B-L})^3 \approx SU(2)^3 \times SU(3)^3 \times U(1)^6. \tag{3}$$

Here it is meant that the subgroup $(SMG_i \times U(1)_{B-L,i}) = S(U(2) \times U(3)) \times U(1) = SU(2) \times SU(3) \times U(1)^2$ only couples to the ith family and that, in the usual way, the quarks of family i are triplets under the $SU(3)_i$ (but singlets under the other $SU(3)_j$'s, i.e. when $j \neq i$) and so on. The $(B - L)_i$ couples to the $B - L$ = baryon number − lepton number for the i'th family.

This essentially describes the couplings in the model, but there is a little tricky point that was introduced, because it could be shown that otherwise we would have had a no-go theorem for fitting the quark spectra [2]: the right handed up-type ("up-type"= u, c, and t) quarks of the second and third family are permuted. That is to say the right handed components of what is experimentally seen as the third family particle—the top quark—is dominantly equal to the right handed up-type components coupling to the second family gauge fields. Similarly the experimentally seen right handed charm-quark is identified with the formally right handed top-quark of our model, namely the components coupling to the third family gauge fields.

II MOTIVATION FOR THE MODEL

But why should a model of the proposed type be expected to have a chance of being the right one?

Well, the group may be characterised by means of a few relatively simple requirements with some phenomenological support: It is the largest possible gauge group that transforms the known 45 Weyl fermions of the Standard Model and the included three right handed see-saw neutrinos into themselves, without unifying any irreducible representations under the Standard Model and with no anomaly troubles neither gauge anomalies nor mixed anomalies. That is to say it is characterisable as the biggest group for which the gauge symmetry can be upheld without any Green-Schwarz anomaly cancellation, without unifying particles that are not already unified in the Standard Model and without transforming them into "fantasy particles" only existing in the model.

Looking for inspiration for what gauge group to believe in beyond the Standard Model, we are led to consider hints from the following remarkable features of the quark and lepton spectra:

- There are very big mass ratios from family to family and even to some extent within the families. This feature suggests strongly that different family particles should have different quantum numbers, since how else should the quantum number differences between right and left handed components be so different that they can be used to make some particles have more mass protection than others by big factors?

- Pure $SU(5)$ or for that matter all the Grand Unified Theories (GUTs) extending $SU(5)$ predict, unless one helps it by introducing rather big representation Higgs fields, $\underline{45}$ say, GUT-scale mass relations:

$$\frac{m_\tau}{m_b} = \frac{m_\mu}{m_s} = \frac{m_e}{m_d} = 1 \qquad (4)$$

which are, however, not well fulfilled by the renormalisation group evolved experimental values. Only the third family mass ratio m_τ/m_b fits reasonably well with the simple version of $SU(5)$-GUT (or SUSY extension). So we would be much better off w.r.t. fitting with experiment, if we could get the $SU(5)$ mass predictions only as predictions up to factors of order of unity, but not as exact relations. So we really do not want to unify the down-type quarks with the charged leptons for the masses do not match, except in the $\tau - b$ case.

We may classify gauge group proposals, restricted in each case to the factor group which acts on the 45 known Weyl quark and leptons (or 48 if you count the perhaps non-existing right handed neutrinos), according to whether the group is small or big and according to whether it unifies or avoids unifying the various irreducible representations under the Standard Model. Especially we can ask for the four groups obtained by requiring minimal or maximal degree of such group size and of the degree of unification. Under all circumstances we must of course, in order to have a consistent gauge field model, require that the anomalies cancel - in the present article we shall ignore the possibility of Green-Schwarz cancellation, with the excuse that we do not find any Green Schwarz anomaly cancellation needed between the Standard Model gauge fields. Nature has even "carefully" chosen to make equally many quark and lepton families, thereby ensuring simple anomaly cancellation between the Weyl particle contributions without any Green-Schwarz mechanism.

The four corners of the set of gauge groups acting on the Standard Model Weyl fermions can be seen [3] to correspond to the following gauge groups:

- small, separating : SMG.
- small, unifying : $SU(5) (=$ usual GUT$)$.
- large, separating : $SMG^3 \times U(1)_f$ $(=$ Anti-GUT$)$
- large, unifying : $SU(5)^3 \times U(1)_f$ $(=$ Rajpoot Model with $U(1)_f$ [4]$)$.

If we say that we want many approximately conserved gauge quantum numbers in order to produce many different big mass ratios, we are suggested to take "large" groups so as to have many possibilities for separating the particle masses order of magnitudewise. If we decide that really the mass predictions in simple GUT-$SU(5)$ are (except for the third family) not true—because we do not like the big 45 representation—then we favour a separating gauge group.

But then we are just suggested to seek the right model in the large, separating-corner; but that is our Anti-GUT model!

Originally we started from ideas of what we called "confusion" [5]. This means that, if there were different gauge fields attached to isomorphic gauge groups, there is a mechanism (the "confusion") that could cause the gauge group with many identical cross product factors to break down to its diagonal subgroup. Thus only one group would survive from each class of isomorphic ones. The model used at that time was a chaotic lattice gauge theory and the breaking to the diagonal subgroup came about by speculating that, after some going around in the lattice, one would loose track of which group was which. In this way there should in practice only be one–the diagonal subgroup.

Such a model could even "explain" why the Standard Model group is essentially the cross product of the three lowest dimensional (simple) groups from which to build gauge groups, namely $U(1)$, $SU(2)$, and $SU(3)$. The explanation should be that on a very short scale level - Planck scale or fundamental scale say - there are many gauge groups of a lot of types, but the isomorphic ones get "confused" and we only find one representative of each class of isomorphic ones.

We used this picture in connection with the assumption which we today would call the multiple point principle (MPP [6,7]). The MPP says that the coupling constants are—by some presumably yet to be explained speculated effect (perhaps Baby-universe theory)—being finetuned by Nature to make a lot of phases meet each other just for those values of the couplings that are realized in Nature. This idea of a lot of phases meeting is analogous to the feature of a microcanonical ensemble that it often will obtain just the temperature of a phase transition. In English one has a special word for a mixture of ice and water, namely "slush". So such mixtures must be quite common. When in equilibrium as a microcanonical ensemble, such slush has (under one atmosphere pressure) always the temperature that by definition is zero degrees Celsius. The microcanonical ensemble constitutes a model for how it can happen that this phase transition temperature 0° becomes much more common than most other single temperatures; there is a delta function distribution of temperatures with the delta function peaks at the phase transitions. If there were a true analogy between the microcanonical ensemble making specific temperatures much more likely than others and a mechanism making coupling constants in Nature take those values that are just between the different phases of the theory, we would have a mechanism for our MPP.

We used a replacement for this slush analogous MPP-idea to postulate that the gauge couplings should in Nature take just those values that correspond to lattice artifact phase transitions, values that can be read out of lattice calculation literature. Combining this postulate (in an old version) with our favourite Anti-GUT gauge group we fitted the number of families - not yet known at that time - and tried to say that it was so accurately obtainable that we could almost see it was an integer, namely three. While one of us was giving talks about that, it was found at LEP that the neutrino decays of the Z^0-gauge boson showed that there were indeed only three families. So we had made a successful prediction!

III ANTI-GRAND UNIFICATION THEORY AND ITS EXTENSION

In this section we review the Anti-GUT model and its extension describing neutrino masses and mixing angles.

The Anti-GUT model [8-14] has been put forward by myself and my collaborators over many years, with several motivations. It is mainly justified by a very promising series of experimental agreements obtained from fitting many of the SM parameters with rather few Anti-GUT parameters, even though most predictions are only made order of magnitudewise. The Anti-GUT model deserves its name in as far as its gauge group $SMG^3 \times U(1)_f$ which, so to speak, replaces the often-used GUT gauge groups such as $SU(5)$, $SO(10)$ etc. and can be specified by requiring that:

1. It should only contain transformations which change the known 45 (= 3 generations of 15 Weyl particles each) Weyl fermions - counted as left-handed, say, - into each other unitarily, (*i.e.* it must be a subgroup of $U(45)$).

2. It should be anomaly-free even without using the Green-Schwarz [15] anomaly cancellation mechanism.

3. It should NOT unify the irreducible representations under the SM gauge group, called here $SMG = SU(3) \times SU(2) \times U(1)$.

4. It should be as big as possible under the foregoing assumptions.

In the present article we shall, however, allow for see-saw neutrinos - essentially right-handed neutrinos - whereby we want to extend the number of particles to be transformed under the group being specified to also include the right-handed neutrinos, even though they have not been directly "seen".

The extended group, which we shall use as the model gauge group replacing the unifying groups, can be specified by a similar set of assumptions to those used above by two of us [16] to specify the "old" Anti-GUT. We replace assumption 1 by a slightly modified assumption which only excludes unobserved fermions when they have nontrivial quantum numbers under the SM group, so that they are mass-protected. The particles that are mass protected under the SM would namely be rather light and would likely have been seen. But see-saw neutrinos with zero SM quantum numbers could not be mass protected by the SM and could easily be so heavy as not to have been "seen".

The model which we have in mind as the extended Anti-GUT model [17], that should inherit the successes of the "old" Anti-GUT model and in addition have see-saw neutrinos, is proposed to have the gauge group $SMG^3 \times U(1)_f \times U(1)_{B-L,1} \times U(1)_{B-L,23} \approx SU(3)^3 \times SU(2)^3 \times U(1)^6$, and it is assumed to couple in the following way:

The three SM groups $SMG = SU(3) \times SU(2) \times U(1)$ are supposed to be one for each family or generation. That is to say, there is, e.g., a first generation SMG among the three; all the fermions in the second and third generations are in the trivial representations, and with zero charge, while the first generation particles couple to this first generation SMG as if in the same representations (same charges too) as they are under the SM. For example, the left proto-electron and the proto-electron neutrino form a doublet under the $SU(2)_1$ belonging to the first generation (while they are in singlets w.r.t. the other two $SU(2)$'s) and have weak hypercharge w.r.t. the first generation $U(1)_1$, with a value $y_1/2 = -1/2$ analogous to the SM weak hypercharge being $y/2 = -1/2$ for left-handed leptons.

The $U(1)_f$-charge is assigned in a slightly complicated way which is, however, largely the only one allowed modulo various permutations and rewritings from the no-anomaly requirements. It is zero for all first-generation particles and for all particles usually called left-handed. The $U(1)_f$ charge values are the opposite on a "right-handed" particle in the second generation and the corresponding one in the third generation. See Table 1 for the detailed assignment.

The two last $U(1)$-groups, $U(1)_{B-L,1}$ and $U(1)_{B-L,23}$ in our model have charge assignments corresponding to the quantum number $B - L$ (= baryon number minus lepton number) though in such a way that the charges of $U(1)_{B-L,1}$ are zero for the second and third generations, and only non-zero for the first generation, for which they then coincide with the baryon number minus the lepton number. Analogously the $U(1)_{B-L,23}$-charge assignments are zero on the first-generation quarks and leptons, while they coincide with the baryon number minus the lepton number for second and third generations. We will discuss in the next section the anomaly cancellation in the extended Anti-GUT model.

It is then further part of our model that this large gauge group is broken down spontaneously to the SM group, lying as the diagonal subgroup of the SMG^3 part of the group by means of a series of Higgs fields. The quantum numbers of these fields have been selected mainly from the criterion of fitting the masses and mixing angles w.r.t. order of magnitude. The Abelian quantum numbers proposed for the "old" Anti-GUT Higgs fields were:

$$S: \quad (\frac{1}{6}, -\frac{1}{6}, 0, -1) \tag{5}$$

$$W: \quad (0, -\frac{1}{2}, \frac{1}{2}, -\frac{4}{3}) \tag{6}$$

$$T: \quad (0, -\frac{1}{6}, \frac{1}{6}, -\frac{2}{3}) \tag{7}$$

$$\xi: \quad (\frac{1}{6}, -\frac{1}{6}, 0, 0). \tag{8}$$

These four Higgs fields are supposed to have VEVs of the order of between a twentieth and unity compared to the fundamental scale supposed to be the Planck scale. In addition there was then the Higgs field under the Anti-GUT-group which should take the role of finally breaking the SM gauge group down to $U(3) = SU(3) \times U(1)_{em}$, i.e., play the role of the Weinberg-Salam Higgs field

$$\phi_{WS}: \quad (0, \frac{2}{3}, -\frac{1}{6}, 1). \tag{9}$$

Here the quantum numbers were presented in the order of first giving the three different weak hypercharges corresponding to the three generations $y_i/2$ ($i = 1, 2, 3$), and then the $U(1)_f$-charge.

In reference [8] we fitted the parameters, being Higgs fields VEVs, to the masses and mixing angles for charged fermions and the values are as follows:

$$\langle S \rangle = 1 \;, \quad \langle W \rangle = 0.179 \;, \quad \langle \xi \rangle = 0.099 \;, \quad \langle T \rangle = 0.071 \tag{10}$$

In the following we shall often abbreviate by deleting the $\langle \cdots \rangle$ around these Higgs fields, mostly with the understanding that S, W, ... then mean the VEV "measured in fundamental" units.

In the Anti-GUT model, old as well as new, it is assumed at some "fundamental scale" that particles, which can play the role of see-saw with whatever quantum numbers are needed, exist. The fitted "suppression factors" are the VEVs in units of the "fundamental scale" see-saw particles.

It has to be checked that extending the group to have the $U(1)_{B-L,1}$ and $U(1)_{B-L,23}$ does not disturb the successful features of the model. This can be done by only giving the fields ξ and S non-zero charges under these "new" $U(1)$ groups, so as to get:

$$S: \quad (\frac{1}{6}, -\frac{1}{6}, 0, -1, -\frac{2}{3}, \frac{2}{3}) \tag{11}$$

$$\xi: \quad (\frac{1}{6}, -\frac{1}{6}, 0, 0, \frac{1}{3}, -\frac{1}{3}) \tag{12}$$

But now we also want to introduce two new Higgs fields ϕ_{B-L} and χ into the model: the first, ϕ_{B-L}, is a Higgs field to fit the new scale that comes in from neutrino oscillations giving the scale of the see-saw particle masses. When the left-right-transition mass matrix is of the same order as the usual charged fermion mass matrices, this scale is of the order 10^{12} GeV.

We use in our model the gauged $B - L$, in fact the total one, because we break $U(1)_{B-L,1} \times U(1)_{B-L,23} \supseteq U(1)_{B-L,\text{total}}$ at a much higher scale (near Planck scale), to mass-protect the right-handed neutrinos. These are meant to function as see-saw particles, so they can be sufficiently light to give the "observed" left-handed neutrino masses by the see-saw mechanism. The breaking of the $U(1)_{B-L,\text{total}}$ and thereby the setting of the see-saw scale is then caused by our "new" Higgs field called ϕ_{B-L}.

In order to get viable neutrino spectra we shall choose the quantum numbers of ϕ_{B-L} so that it is the effective $\overline{\nu_{\tau R}} C \overline{\nu_{eR}}^t + h.c.$ which gets the direct contribution and thus is not further suppressed.

This is the way to avoid "factorised mass matrices" - $i.e.$ of the form

$$\begin{pmatrix} \phi_1^2 & \phi_1\phi_2 & \phi_1\phi_3 \\ \phi_1\phi_2 & \phi_2^2 & \phi_2\phi_3 \\ \phi_1\phi_3 & \phi_2\phi_3 & \phi_3^2 \end{pmatrix} \tag{13}$$

with different order unity factors, though, on different elements. Such factorised matrices are rather difficult to avoid otherwise. If we get such a "factorised matrix" and, as in our model, have mainly diagonal elements in the ν-Dirac matrix, M_ν^D, the prediction comes out that

$$\frac{\Delta m_\odot^2}{\Delta m_{\text{atm}}^2} \approx (\sin\theta_{\text{atm}})^4 \,, \tag{14}$$

which is not true experimentally. Therefore we choose ϕ_{B-L} to have the quantum numbers of $\overline{\nu_{\tau R}}$ plus those of $\overline{\nu_{eR}}$:

$$\begin{aligned} Q_{\phi_{B-L}} &= Q_{\bar{\nu}_{\tau R}} + Q_{\bar{\nu}_{eR}} \\ &= (0,0,0,0,1,0) + (0,0,0,1,0,1) \\ &= (0,0,0,1,1,1) \,. \end{aligned}$$

The other "new" Anti-GUT Higgs field we call χ and one of its roles is to help the $\langle\phi_{B-L}\rangle$ to give non-zero effective mass terms for the see-saw neutrinos, by providing a transition between $\nu_{\tau R}$ and $\nu_{\mu R}$. It also turns out to play a role in fitting the atmospheric mixing angle (to be of order unity). Its quantum numbers are therefore postulated to be the difference of those of these two see-saw particles

$$\begin{aligned} Q_\chi &= Q_{\bar{\nu}_{\mu R}} - Q_{\bar{\nu}_{\tau R}} \\ &= (0,0,0,1,0,-1) - (0,0,0,-1,0,-1) \\ &= (0,0,0,2,0,0) \,. \end{aligned}$$

IV ANOMALY CANCELLATION

We should introduce here an anomaly-free Abelian extension of the "old" AGUT, which we shall discuss below to obtain the neutrino mass spectra and their mixing angles. The "new" Anti-GUT gauge group is

$$SMG^3 \times U(1)_f \times U(1)_{B-L,1} \times U(1)_{B-L,23} \qquad (15)$$

and is broken by a set of Higgs fields S, W, T, ξ, χ and ϕ_{B-L} down to the SM gauge groups. The SMG will be broken down by the field ϕ_{WS} playing the role of Weinberg-Salam Higgs field into $SU(3) \times U(1)_{em}$.

The requirement that all anomalies involving $U(1)_f$, $U(1)_{B-L,1}$ and $U(1)_{B-L,23}$ then vanish strongly constrains the possible fermion charges (denoting the $U(1)_f$ charges by $Q_f(t_R) \equiv t_R$ etc. and the $U(1)_{B-L}$ charges by $Q_{B-L,1}(u_R) \equiv \bar{u}_F$, $Q_{B-L,23}(t_R) \equiv \tilde{t}_R$ etc. respectively).

The anomaly cancellation conditions then constrain the fermion $U(1)_f$ and $U(1)_{B-L,1}$ and also $U(1)_{B-L,23}$ charges to satisfy the following equations:

$$\text{Tr}\,[SU_1(3)^2 U(1)_f] = 2u_L - u_R - d_R = 0$$
$$\text{Tr}\,[SU_2(3)^2 U(1)_f] = 2c_L - c_R - s_R = 0$$
$$\text{Tr}\,[SU_3(3)^2 U(1)_f] = 2t_L - t_R - b_R = 0$$
$$\text{Tr}\,[SU_1(2)^2 U(1)_f] = 3u_L + e_L = 0$$
$$\text{Tr}\,[SU_2(2)^2 U(1)_f] = 3c_L + \mu_L = 0$$
$$\text{Tr}\,[SU_3(2)^2 U(1)_f] = 3t_L + \tau_L = 0$$
$$\text{Tr}\,[U_1(1)^2 U(1)_f] = u_L - 8u_R - 2d_R + 3e_L - 6e_R = 0$$
$$\text{Tr}\,[U_2(1)^2 U(1)_f] = c_L - 8c_R - 2s_R + 3\mu_L - 6\mu_R = 0$$
$$\text{Tr}\,[U_3(1)^2 U(1)_f] = t_L - 8t_R - 2b_R + 3\tau_L - 6\tau_R = 0$$
$$\text{Tr}\,[U_1(1) U(1)_f^2] = u_L^2 - 2u_R^2 + d_R^2 - e_L^2 + e_R^2 = 0$$
$$\text{Tr}\,[U_2(1) U(1)_f^2] = c_L^2 - 2c_R^2 + s_R^2 - \mu_L^2 + \mu_R^2 = 0$$
$$\text{Tr}\,[U_3(1) U(1)_f^2] = t_L^2 - 2t_R^2 + b_R^2 - \tau_L^2 + \tau_R^2 = 0$$
$$\text{Tr}\,[U(1)_f^3] = 6u_L^3 + 6c_L^3 + 6t_L^3 - 3u_R^3 - 3c_R^3 - 3t_R^3 - 3d_R^3 - 3s_R^3$$
$$- 3b_R^3 + 2e_L^3 + 2\mu_L^3 + 2\tau_L^3 - e_R^3 - \mu_R^3 - \tau_R^3$$
$$- \nu_{eR}^3 - \nu_{\mu R}^3 - \nu_{\tau R}^3 = 0$$
$$\text{Tr}\,[(\text{graviton})^2 U(1)_f] = 6u_L + 6c_L + 6t_L - 3u_R - 3c_R - 3t_R - 3d_R - 3s_R$$
$$- 3b_R + 2e_L + 2\mu_L + 2\tau_L - e_R - \mu_R - \tau_R$$
$$- \nu_{eR} - \nu_{\mu R} - \nu_{\tau R} = 0$$

Similar conditions should be obeyed replacing $U(1)_f$ both by the $U(1)_{B-L,1}$, and $U(1)_{B-L,23}$, i.e. replacing the t_R, b_R, ... by $\tilde{t}_R$, $\tilde{b}_R$, ...:

$$\text{Tr}\,[SU_1(3)^2 U(1)_{B-L,1}] = 2\bar{u}_L - \bar{u}_R - \bar{d}_R = 0$$

$$\text{Tr}\,[SU_2(3)^2 U(1)_{B-L,23}] = 2\tilde{c}_L - \tilde{c}_R - \tilde{s}_R = 0$$
$$\text{Tr}\,[SU_3(3)^2 U(1)_{B-L,23}] = 2\tilde{t}_L - \tilde{t}_R - \tilde{b}_R = 0$$
$$\text{Tr}\,[SU_1(2)^2 U(1)_{B-L,1}] = 3\bar{u}_L + \bar{e}_L = 0$$
$$\text{Tr}\,[SU_2(2)^2 U(1)_{B-L,23}] = 3\tilde{c}_L + \tilde{\mu}_L = 0$$
$$\text{Tr}\,[SU_3(2)^2 U(1)_{B-L,23}] = 3\tilde{t}_L + \tilde{\tau}_L = 0$$

But with several $U(1)$s there will in addition be anomaly conditions for combinations between the different ones. Taking it that $U(1)_{B-L,1}$ charges are zero for all second- and third-generation fermions while $U(1)_{B-L,23}$ charges are zero for the first generation, the further conditions are:

$$\text{Tr}\,[U_1(1)^2 U(1)_{B-L,1}] = \bar{u}_L - 8\bar{u}_R - 2\bar{d}_R + 3\bar{e}_L - 6\bar{e}_R = 0$$
$$\text{Tr}\,[U_2(1)^2 U(1)_{B-L,23}] = \tilde{c}_L - 8\tilde{c}_R - 2\tilde{s}_R + 3\tilde{\mu}_L - 6\tilde{\mu}_R = 0$$
$$\text{Tr}\,[U_3(1)^2 U(1)_{B-L,23}] = \tilde{t}_L - 8\tilde{t}_R - 2\tilde{b}_R + 3\tilde{\tau}_L - 6\tilde{\tau}_R = 0$$
$$\text{Tr}\,[U_1(1) U(1)_{B-L,1}^2] = \bar{u}_L^2 - 2\bar{u}_R^2 + \bar{d}_R^2 - \bar{e}_L^2 + \bar{e}_R^2 = 0$$
$$\text{Tr}\,[U_2(1) U(1)_{B-L,23}^2] = \tilde{c}_L^2 - 2\tilde{c}_R^2 + \tilde{s}_R^2 - \tilde{\mu}_L^2 + \tilde{\mu}_R^2 = 0$$
$$\text{Tr}\,[U_3(1) U(1)_{B-L,23}^2] = \tilde{t}_L^2 - 2\tilde{t}_R^2 + \tilde{b}_R^2 - \tilde{\tau}_L^2 + \tilde{\tau}_R^2 = 0$$
$$\text{Tr}\,[U(1)_{B-L,1}^3] = 6\bar{u}_L^3 - 3\bar{u}_R^3 - 3\bar{d}_R^3 + 2\bar{e}_L^3 - \bar{e}_R^3 - \bar{\nu}_{eR}^3 = 0$$
$$\text{Tr}\,[U(1)_{B-L,23}^3] = 6\tilde{c}_L^3 + 6\tilde{t}_L^3 - 3\tilde{c}_R^3 - 3\tilde{t}_R^3 - 3\tilde{s}_R^3 - 3\tilde{b}_R^3$$
$$+ 2\tilde{\mu}_L^3 + 2\tilde{\tau}_L^3 - \tilde{\mu}_R^3 - \tilde{\tau}_R^3 - \tilde{\nu}_{\mu R}^3 - \tilde{\nu}_{\tau R}^3 = 0$$
$$\text{Tr}\,[(\text{graviton})^2 U(1)_{B-L,1}] = 6\bar{u}_L - 3\bar{u}_R - 3\bar{d}_R + 2\bar{e}_L - \bar{e}_R - \bar{\nu}_{eR} = 0$$
$$\text{Tr}\,[(\text{graviton})^2 U(1)_{B-L,23}] = 6\tilde{c}_L + 6\tilde{t}_L - 3\tilde{c}_R - 3\tilde{t}_R - 3\tilde{s}_R - 3\tilde{b}_R$$
$$+ 2\tilde{\mu}_L + 2\tilde{\tau}_L - \tilde{\mu}_R - \tilde{\tau}_R - \tilde{\nu}_{\mu R} - \tilde{\nu}_{\tau R} = 0$$
$$\text{Tr}\,[U(1)_f^2 U(1)_{B-L,1}] = 6u_L^2 \bar{u}_L - 3u_R^2 \bar{u}_R - 3d_R^2 \bar{d}_R + 2e_L^2 \bar{e}_L - e_R^2 \bar{e}_R - \nu_{eR}^2 \bar{\nu}_{eR} = 0$$
$$\text{Tr}\,[U(1)_f^2 U(1)_{B-L,23}] = 6c_L^2 \tilde{c}_L - 3c_R^2 \tilde{c}_R - 3s_R^2 \tilde{s}_R + 2\mu_L^2 \tilde{\mu}_L - \mu_R^2 \tilde{\mu}_R - \nu_{\mu R}^2 \tilde{\nu}_{\mu R}$$
$$+ 6t_L^2 \tilde{t}_L - 3t_R^2 \tilde{t}_R - 3b_R^2 \tilde{b}_R + 2\tau_L^2 \tilde{\tau}_L - \tau_R^2 \tilde{\tau}_R - \nu_{\tau R}^2 \tilde{\nu}_{\tau R} = 0$$
$$\text{Tr}\,[U(1)_f U(1)_{B-L,1}^2] = 6u_L \bar{u}_L^2 - 3u_R \bar{u}_R^2 - 3d_R \bar{d}_R^2 + 2e_L \bar{e}_L^2 - e_R \bar{e}_R^2 - \nu_{eR} \bar{\nu}_{eR}^2 = 0$$
$$\text{Tr}\,[U(1)_f U(1)_{B-L,23}^2] = 6c_L \tilde{c}_L^2 - 3c_R \tilde{c}_R^2 - 3s_R \tilde{s}_R^2 + 2\mu_L \tilde{\mu}_L^2 - \mu_R \tilde{\mu}_R^2 - \nu_{\mu R} \tilde{\nu}_{\mu R}^2$$
$$+ 6t_L \tilde{t}_L^2 - 3t_R \tilde{t}_R^2 - 3b_R \tilde{b}_R^2 + 2\tau_L \tilde{\tau}_L^2 - \tau_R \tilde{\tau}_R^2 - \nu_{\tau R} \tilde{\nu}_{\tau R}^2 = 0$$
$$\text{Tr}\,[U(1)_1 U(1)_f U(1)_{B-L,1}] = u_L \bar{u}_L - 2u_R \bar{u}_R + d_R \bar{d}_R - e_L \bar{e}_L + e_R \bar{e}_R = 0$$
$$\text{Tr}\,[U(1)_2 U(1)_f U(1)_{B-L,23}] = c_L \tilde{c}_L - 2c_R \tilde{c}_R + s_R \tilde{s}_R - \mu_L \tilde{\mu}_L + \mu_R \tilde{\mu}_R = 0$$
$$\text{Tr}\,[U(1)_3 U(1)_f U(1)_{B-L,23}] = t_L \tilde{t}_L - 2t_R \tilde{t}_R + b_R \tilde{b}_R - \tau_L \tilde{\tau}_L + \tau_R \tilde{\tau}_R = 0$$

From these equations we can get the following solutions:

$$(u_L, u_R, d_R, e_L, e_R, \nu_{eR}) = (0, 0, 0, 0, 0, 0)$$
$$(c_L, c_R, s_R, \mu_L, \mu_R, \nu_{\mu R}) = (0, 1, -1, 0, -1, 1)$$

TABLE 1. All $U(1)$ quantum charges in extended Anti-GUT model

	SMG_1	SMG_2	SMG_3	$U(1)_f$	$U_{B-L,1}$	$U_{B-L,23}$
u_L, d_L	$\frac{1}{6}$	0	0	0	$\frac{1}{3}$	0
u_R	$\frac{2}{3}$	0	0	0	$\frac{1}{3}$	0
d_R	$-\frac{1}{3}$	0	0	0	$\frac{1}{3}$	0
e_L, ν_{e_L}	$-\frac{1}{2}$	0	0	0	-1	0
e_R	-1	0	0	0	-1	0
ν_{e_R}	0	0	0	0	-1	0
c_L, s_L	0	$\frac{1}{6}$	0	0	0	$\frac{1}{3}$
c_R	0	$\frac{2}{3}$	0	1	0	$\frac{1}{3}$
s_R	0	$-\frac{1}{3}$	0	-1	0	$\frac{1}{3}$
μ_L, ν_{μ_L}	0	$-\frac{1}{2}$	0	0	0	-1
μ_R	0	-1	0	-1	0	-1
ν_{μ_R}	0	0	0	1	0	-1
t_L, b_L	0	0	$\frac{1}{6}$	0	0	$\frac{1}{3}$
t_R	0	0	$\frac{2}{3}$	-1	0	$\frac{1}{3}$
b_R	0	0	$-\frac{1}{3}$	1	0	$\frac{1}{3}$
τ_L, ν_{τ_L}	0	0	$-\frac{1}{2}$	0	0	-1
τ_R	0	0	-1	1	0	-1
ν_{τ_R}	0	0	0	-1	0	-1
ϕ_{WS}	0	$\frac{2}{3}$	$-\frac{1}{6}$	1	0	0
S	$\frac{1}{6}$	$-\frac{1}{6}$	0	-1	$-\frac{2}{3}$	$\frac{2}{3}$
W	0	$-\frac{1}{2}$	$\frac{1}{2}$	$-\frac{4}{3}$	0	0
ξ	$\frac{1}{6}$	$-\frac{1}{6}$	0	0	$\frac{1}{3}$	$-\frac{1}{3}$
T	0	$-\frac{1}{6}$	$\frac{1}{6}$	$-\frac{2}{3}$	0	0
χ	0	0	0	2	0	0
ϕ_{B-L}	0	0	0	1	1	1

$$(t_L, t_R, b_R, \tau_L, \tau_R, \nu_{\tau_R}) = (0, -1, 1, 0, 1, -1)$$
$$(\bar{u}_L, \bar{u}_R, \bar{d}_L, \bar{d}_R, \bar{e}_L, \bar{e}_R, \bar{\nu}_{e_L}, \bar{\nu}_{e_R}) = (\frac{1}{3}, \frac{1}{3}, \frac{1}{3}, \frac{1}{3}, -1, -1, -1, -1)$$
$$(\tilde{c}_L, \tilde{c}_R, \tilde{s}_L, \tilde{s}_R, \tilde{b}_L, \tilde{b}_R, \tilde{t}_L, \tilde{t}_R) = (\frac{1}{3}, \frac{1}{3}, \frac{1}{3}, \frac{1}{3}, \frac{1}{3}, \frac{1}{3}, \frac{1}{3}, \frac{1}{3})$$
$$(\tilde{\mu}_L, \tilde{\mu}_R, \tilde{\tau}_L, \tilde{\tau}_R, \tilde{\nu}_{\mu_L}, \tilde{\nu}_{\mu_R}, \tilde{\nu}_{\tau_L}, \tilde{\nu}_{\tau_R}) = (-1, -1, -1, -1, -1, -1, -1, -1)$$

We summarise the Abelian gauge quantum numbers of our model for fermions and scalars in Table 1. However, the following three points should be kept in mind; then the information in Table 1 and these three points describe our whole model:

1. We have only presented here the six $U(1)$-charges in our model. The non-Abelian quantum charge numbers are to be derived from the following rule: find in the table $y_i/2$ ($i = 1, 2, 3$ is the generation number), then find that Weyl particle in the SM for which the SM weak hypercharge divided by two is $y/2 = y_i/2$ and use its $SU(2)$ and $SU(3)$ representation for the particle

considered in the table. But now use it for $SU(2)_i$ and $SU(3)_i$.

2. Remember that we imagine that at the "fundamental" scale (presumed to be $\simeq$ the Planck scale) we have essentially all particles that can be imagined with couplings of order unity. But we do not want to be specific about these very heavy particles in order not to decrease the enormous likelihood of our model being right. We are only specific about the particles in the table and the gauge fields.

3. The 39 gauge bosons correspond to the group (equation (15)) and are also not written in the table.

V IMPROVED CHARGE FORMULATION

The system of charges just presented may seem a little complicated and arbitrary. However the fermion charge combinations are so restricted by the anomaly conditions and the connection to the Standard Model that they can essentially only be permuted in various ways. We can though transform these charges into some linear combinations that come to look nicer and easier to remember; but the physical content of the theory is of course the same in the reformulated version.

Indeed it turns out that the $U(1)_f$-charge, Q_f, contains the information which corresponds to letting even the second and third families have their separate $(B-L)$-charges. We can define generally the second and third family $(B-L)$-charges:

$$(B-L)_2 = \frac{1}{2}(B-L)_{23} + \frac{y_2}{2} - \frac{y_3}{2} - \frac{Q_f}{2} \qquad (16)$$

and

$$(B-L)_3 = \frac{1}{2}(B-L)_{23} + \frac{y_3}{2} - \frac{y_2}{2} + \frac{Q_f}{2} \qquad (17)$$

for the Weyl particles (the fermions in the Standard Model with the charges in our scheme). The $(B-L)_i$ charges then take on their well-known values as given in Table 2. So they should be no problem to remember and one can easily reconstruct the $U(1)_f$-charges from the above two formulae (16,17), in case one should want them.

One can also formally use the same formulae for the quantum numbers of the Higgs fields which we have proposed and obtain their values in the new notation as given in Table 2.

A technical detail that can be useful in checking our tables and the mass matrices is the regularity

$$(B-L)_i = -\frac{y_i}{2} \pmod{\frac{1}{2}} \quad \text{for} \quad i = 1, 2, 3. \qquad (18)$$

TABLE 2. All $U(1)$ quantum charges in re-extended Anti-GUT model

	SMG_1	SMG_2	SMG_3	$U_{B-L,1}$	$U_{B-L,2}$	$U_{B-L,3}$
u_L, d_L	$\frac{1}{6}$	0	0	$\frac{1}{3}$	0	0
u_R	$\frac{2}{3}$	0	0	$\frac{1}{3}$	0	0
d_R	$-\frac{1}{3}$	0	0	$\frac{1}{3}$	0	0
e_L, ν_{e_L}	$-\frac{1}{2}$	0	0	-1	0	0
e_R	-1	0	0	-1	0	0
ν_{e_R}	0	0	0	-1	0	0
c_L, s_L	0	$\frac{1}{6}$	0	0	$\frac{1}{3}$	0
c_R	0	$\frac{2}{3}$	0	0	$\frac{1}{3}$	0
s_R	0	$-\frac{1}{3}$	0	0	$\frac{1}{3}$	0
μ_L, ν_{μ_L}	0	$-\frac{1}{2}$	0	0	-1	0
μ_R	0	-1	0	0	-1	0
ν_{μ_R}	0	0	0	0	-1	0
t_L, b_L	0	0	$\frac{1}{6}$	0	0	$\frac{1}{3}$
t_R	0	0	$\frac{2}{3}$	0	0	$\frac{1}{3}$
b_R	0	0	$-\frac{1}{3}$	0	0	$\frac{1}{3}$
τ_L, ν_{τ_L}	0	0	$-\frac{1}{2}$	0	0	-1
τ_R	0	0	-1	0	0	-1
ν_{τ_R}	0	0	0	0	0	-1
ϕ_{WS}	0	$\frac{2}{3}$	$-\frac{1}{6}$	0	$\frac{1}{3}$	$-\frac{1}{3}$
S	$\frac{1}{6}$	$-\frac{1}{6}$	0	$-\frac{2}{3}$	$\frac{2}{3}$	0
W	0	$-\frac{1}{2}$	$\frac{1}{2}$	0	$-\frac{1}{3}$	$\frac{1}{3}$
ξ	$\frac{1}{6}$	$-\frac{1}{6}$	0	$\frac{1}{3}$	$-\frac{1}{3}$	0
T	0	$-\frac{1}{6}$	$\frac{1}{6}$	0	0	0
χ	0	0	0	0	-1	1
ϕ_{B-L}	0	0	0	1	0	1

which follows from the facts that the quarks have $(B-L)_i = 1/3$ for their family i say (and zero for the other family $(B-L)_j$'s) and that their family weak hypercharge satisfies

$$\frac{y_i}{2} = -\frac{\text{"triality"}}{3} \quad (\text{mod } \frac{1}{2}) \tag{19}$$

As one can see from the table most of the Higgs fields which we have proposed also have quantum number assignments obeying these rules (18), but the two Higgs fields W and T do <u>not</u> obey the rules. In fact they do obey the rule for the first family *i.e.* for $i = 1$, but have deviations of just opposite sign—so that the total $B - L$ and the total weak hypercharge $\frac{y}{2}$ obey the rule (18) even for the W and T fields—for the second and third family. In fact W has $(B-L)_2 + y_2/2 = -1/3 + (-1/2) = 1/6 \,(\text{mod}\, 1/2)$ and $(B-L)_3 + y_3/2 = \frac{1}{3} + 1/2 = 1/3 \,(\text{mod}\, 1/2)$, while T has $(B-L)_2 + y_2/2 = 0 + (-1/6) = 1/3(\text{mod}\, 1/2)$ and $(B-L)_3 + y_3/2 = 0 + 1/6 = 1/6(\text{mod}\, 1/2)$. It then follows with the above assignment of charges that, in order to make transitions between left and right handed quarks or leptons, the fields W and T can only occur in such combinations that the quantities $(B-L)_i + y_i/2$ add up to zero modulo 1/2. That is to say combinations like T^3, $T^2 W^\dagger$, $T(W^\dagger)^2$, $(W^\dagger)^3$, $T^\dagger W^\dagger$, ... are allowed as well as their conjugates, but *e.g.* just T or just W or $W^\dagger$ would not be allowed to occur as mass matrix elements.

VI MASS MATRICES WITHIN THE ANTI-GUT MODEL

In the "old" Anti-GUT model we have only the usual SM fermions at low energies, but in our "new" version we assume that there exist very heavy right-handed neutrinos, all of them having already decayed and not being observable in our world. They shall function as see-saw particles and thus give rise to an effective Majorana-type mass matrix for the left-handed particles. These three "right-handed" neutrinos get masses from the VEV of ϕ_{B-L} (10^{12} GeV), ξ and also χ Higgs fields (having a VEV of order 1/10 in Planck units).

The effective mass matrix elements, left-left, for the left-handed neutrinos - the ones we "see" experimentally - then come about using the ν_R see-saw propagator surrounded by left-right transition neutrino mass matrices. The latter are rather analogous to the charged lepton and quark mass matrices, which are proportional to the VEV of the Weinberg-Salam Higgs field in our model, being components of ϕ_{WS} (with VEV $\sim$ 173 GeV).

Both $B-L$ quantum gauge groups are violated by ϕ_{B-L}, thus the effective Majorana mass terms are added into the Lagrange density using the Higgs field ϕ_{B-L}. The part of the effective Lagrangian we have to consider is:

$$-\mathcal{L}_{\text{lepton-mass}} = \bar{\nu}_L M_\nu^D \nu_R + \frac{1}{2}(\overline{\nu_L})^c M_L \nu_L + \frac{1}{2}(\overline{\nu_R})^c M_R \nu_R + h.c.$$
$$= \frac{1}{2}(\overline{n_L})^c M\, n_L + h.c. \quad (20)$$

where

$$n_L \equiv \begin{pmatrix} \nu_L \\ (\nu_L)^c \end{pmatrix}, \quad M \equiv \begin{pmatrix} M_L & M_\nu^D \\ M_\nu^D & M_R \end{pmatrix} ; \quad (21)$$

M_ν^D is the standard $SU(2) \times U(1)$ breaking Dirac mass term, and M_L and M_R are the isosinglet Majorana mass terms for left-handed and right-handed neutrinos, respectively.

Supposing that the left-handed Majorana mass M_L terms are comparatively negligible, because of SM gauge symmetry protection, a naturally small effective Majorana mass for the light neutrinos (predominantly ν_L) can be generated by mixing with the heavy states (predominantly ν_R) of mass M_{ν_R}. The Dirac mass matrix of neutrinos is similar to the up-type quark mass matrix [18] and therefore has similar magnitude. For no left-left term, the light eigenvalues of the matrix M are

$$M_{\text{eff}} \approx M_\nu^D \, M_R^{-1} \, (M_\nu^D)^t \; . \tag{22}$$

This result is the well-known see-saw mechanism [19]: the light neutrino masses are quadratic in the Dirac masses and inversely proportional to the large ν_R Majorana masses. Notice that if some ν_R are massless or light they would not be integrated away but simply added to the light neutrinos.

We have already given the quantum charges of the Higgs fields, S, W, T, ξ, ϕ_{WS}, ϕ_{B-L} and χ in Table 1. With this quantum number choice of Higgs fields the mass matrices are given for the uct-quarks:

$$M_U \simeq \frac{\langle \phi_{\text{ws}} \rangle}{\sqrt{2}} \begin{pmatrix} S^\dagger W^\dagger T^2 (\xi^\dagger)^2 & W^\dagger T^2 \xi & (W^\dagger)^2 T\xi \\ S^\dagger W^\dagger T^2 (\xi^\dagger)^3 & W^\dagger T^2 & (W^\dagger)^2 T \\ S^\dagger (\xi^\dagger)^3 & 1 & W^\dagger T^\dagger \end{pmatrix} \tag{23}$$

the dsb-quarks:

$$M_D \simeq \frac{\langle \phi_{\text{ws}} \rangle}{\sqrt{2}} \begin{pmatrix} SW(T^\dagger)^2 \xi^2 & W(T^\dagger)^2 \xi & T^3 \xi \\ SW(T^\dagger)^2 \xi & W(T^\dagger)^2 & T^3 \\ SW^2(T^\dagger)^4 \xi & W^2(T^\dagger)^4 & WT \end{pmatrix} \tag{24}$$

the charged leptons:

$$M_E \simeq \frac{\langle \phi_{\text{ws}} \rangle}{\sqrt{2}} \begin{pmatrix} SW(T^\dagger)^2 \xi^2 & W(T^\dagger)^2 (\xi^\dagger)^3 & WT^4 (\xi^\dagger)^3 \chi \\ SW(T^\dagger)^2 \xi^5 & W(T^\dagger)^2 & WT^4 \chi \\ S(W^\dagger)^2 T^4 \xi^5 & (W^\dagger)^2 T^4 & WT \end{pmatrix} \tag{25}$$

the Dirac neutrinos:

$$M_\nu^D \simeq \frac{\langle \phi_{\text{ws}} \rangle}{\sqrt{2}} \begin{pmatrix} S^\dagger W^\dagger T^2 (\xi^\dagger)^2 & W^\dagger T^2 (\xi^\dagger)^3 & (W^\dagger) T^2 (\xi^\dagger)^3 \chi \\ S^\dagger W^\dagger T^2 \xi & W^\dagger T^2 & (W^\dagger) T^2 \chi \\ S^\dagger W^\dagger T^\dagger \xi \chi^\dagger & W^\dagger T^\dagger \chi^\dagger & W^\dagger T^\dagger \end{pmatrix} \tag{26}$$

and the Majorana neutrinos:

$$M_R \simeq \langle \phi_{\text{B-L}} \rangle \begin{pmatrix} S^\dagger \chi^\dagger \xi & \chi^\dagger & 1 \\ \chi^\dagger & S\chi^\dagger \xi^\dagger & S\xi^\dagger \\ 1 & S\xi^\dagger & S\chi\xi^\dagger \end{pmatrix} \tag{27}$$

Note that the random complex order of unity and factorial factors which are supposed to multiply all the mass matrix elements are not represented here. We will discuss these factors in section VIII.

TABLE 3. Typical fit, $\alpha = -1$, $\beta = 1$, $\gamma = 1$, $\delta = 1$

	Fitted	"Experiment"
m_u	3.1 MeV	4 MeV
m_d	6.6 MeV	9 MeV
m_e	0.76 MeV	0.5 MeV
m_c	1.29 GeV	1.4 GeV
m_s	390 MeV	200 MeV
m_μ	85 MeV	105 MeV
M_t	179 GeV	180 GeV
m_b	7.8 GeV	6.3 GeV
m_τ	1.29 GeV	1.78 GeV
V_{us}	0.21	0.22
V_{cb}	0.023	0.041
V_{ub}	0.0050	0.0035
J_{CP}	$1.04 \cdot 10^{-5}$	$(2 - 3.5) \cdot 10^{-5}$
"χ^2"	1.46	

VII CHARGED FERMION SPECTRUM

The matrices for the quarks M_U and M_D happen not to have been changed at all by the introduction of the "new" Higgs fields χ (and ϕ_{B-L}, but that has so little VEV compared to the Planck scale that it could never compete). Even in the charged lepton mass matrix the appearance of χ occurs on off-diagonal matrix elements, which are already small and remain so small as to have no significance for the charge lepton mass predictions, as long as χ is of the order $\langle\chi\rangle \approx 0.07$ as we need for fitting θ_{atm}.

Therefore all the fits of the "old" Anti-GUT model are valid and we can still use the parameter values obtained by these earlier fits to S, W, T, ξ, presented above in equation (10).

The result of fitting is shown in the Table 3.

The best quark and charged lepton fit is for the charge assignments chosen according to the parameters

$$\alpha = 1, \quad \beta = 1, \quad \gamma = -1, \quad \delta = 1. \tag{28}$$

determining details of the discrete quantum numbers, but since the effect of this choice only comes in via a correction which we call "factorial correction" [20] and only make changes of order unity in principle these parameters should not be counted as parmeters - we allowed for them only -1, 0, 1. The parameters for this choice under inclusion of this "factorial correction" are

$$\langle W\rangle = 0.0894, \quad \langle T\rangle = 0.0525, \quad \langle S\rangle = 0.756, \quad \langle\xi\rangle = 0.0247. \tag{29}$$

It is seen that it is rather good fit from the point of view that it is only expected to be up to order of unity factors. The worst case is the strange quark mass since the Jarlskog-triangle-area that is a measure of CP-violation J_{CP} has so many factors that we expect the uncertainty of our prediction to be larger. The trouble with the Strange quark is due to that the model has an order of magnitude family degeneracy built in as long as the masses are dominated by the diagonal elements in the mass matrices - which means except for the charmed and the top quarks which have their masses domnated from off-diagonal matrix elements in their mass matrices-. Then we namely cannot avoid to have *e.g.* the $SU(5)$ relations (4) order-of-magnitudewise. Still having (4) only as an order of magnitude relation is much better than getting it as an exact prediction!

In the article [20] we even predict the accuracy with which we expect our predictions to work - namely that the uncertainty in the logarithm shall be equal to that for the distribution of the logarithm of a Gauss-distribution. Our fits are actually rather doing a bit too well, and even agree with the prediction concerning the skewness of the distribution: that the worst deviation shall be that some mass(es) should be too small experimentally compared to the prediction, the strange quark fit in this way.

VIII CALCULATION OF M_{EFF}

We calculate in this section the effective neutrino mass matrix for left-handed components. Since, strictly speaking, our model only predicts orders of magnitude, a crude calculation is in principle justified. This calculation is presented in the first subsection, and then in the next subsection we make "statistical calculations" with random order-one factors and "factorial factors".

A Crude calculation

From equation (27) we see to the first approximation that there are one massless and two degenerate right-handed neutrinos coming from the VEV of the $B-L$ breaking Higgs field, $\langle \phi_{\text{B-L}} \rangle$.

The splitting between the two almost degenerate see-saw neutrinos would be $M_{31} \langle S \rangle \langle \chi \rangle \langle \xi \rangle$, where $M_{31} \approx \langle \phi_{\text{B-L}} \rangle$ is the approximately common mass of the two heaviest see-saw neutrinos. The third lightest see-saw neutrino is dominantly "proto second generation" and has the mass $\langle \phi_{\text{B-L}} \rangle \langle \chi \rangle \langle \xi \rangle$.

For the left-handed neutrinos to the first approximation we get the effective mass matrix as follows:

$$M_{\text{eff}} \approx \frac{W^2 T^2 \langle \phi_{WS} \rangle^2}{2 \langle \phi_{\text{B-L}} \rangle} \begin{pmatrix} \frac{T^2 \xi^5}{\chi} & \frac{T^2 \xi^2}{\chi} & T\xi^2 \\ \frac{T^2 \xi^2}{\chi} & \frac{T}{\xi} & \frac{T}{\xi} \\ T\xi^2 & \frac{T}{\xi} & \frac{\chi}{\xi} \end{pmatrix}, \quad (30)$$

But we have to emphasise here that this approximation is *good enough to calculate only the heaviest left-handed neutrino*, because all the mass matrix elements to this approximation come from the propagator contribution of the lightest see-saw particle, so that they really form a degenerate matrix of rank one. Using this contribution only would lead to two left-handed massless neutrinos and one massive. But we can still obtain the mixing angles θ_{13} and θ_{23} and the heaviest mass from M_{eff}:

$$\theta_{13} = \theta_{e,\text{heavy}} \approx \frac{T}{\chi}\xi^3 \tag{31}$$

$$\theta_{23} = \theta_{\mu,\text{heavy}} \approx \begin{cases} \frac{T}{\chi} & \text{when } \chi \gtrsim T \\ 1 & \text{when } \chi \lesssim T \end{cases} \tag{32}$$

$$M_{\nu_L \text{heavy}} \approx \begin{cases} \frac{W^2 T^2 \langle \phi_{WS} \rangle^2}{2\langle \phi_{B-L} \rangle} \frac{\chi}{\xi} & \text{when } \chi \gtrsim T \\ \frac{W^2 T^2 \langle \phi_{WS} \rangle^2}{2\langle \phi_{B-L} \rangle} \frac{T}{\xi} & \text{when } \chi \lesssim T \end{cases} \tag{33}$$

From these equations we can restrict the region of χ comparing with Super-Kamiokande experimental data; χ must be almost of the same order as T. Thus we know the mixing angle of the first and third generations must be of the order of ξ^3.

However, to get the much lower masses we cannot use the contribution from the lightest see-saw propagator, but we have to use the propagator terms from the two approximately equally heavy see-saw particles. This contribution to the propagator matrix is

$$M_R^{-1}\bigg|_{\substack{heavy \\ see-saws}} \approx \frac{1}{\langle \phi_{B-L} \rangle} \begin{pmatrix} \chi\xi & \xi & 1 \\ \xi & \chi\xi & \chi \\ 1 & \chi & \chi\xi \end{pmatrix} \tag{34}$$

where the $\chi\xi/\langle \phi_{B-L} \rangle$ comes from the mass difference of the almost degenerate see-saw particles.

Surrounding this propagator contribution with the "Dirac ν"-mass matrix we get

$$M_{\text{eff}}\bigg|_{\substack{heavy \\ see-saws}} \approx M_\nu^D M_R^{-1}\bigg|_{\substack{heavy \\ see-saws}} (M_\nu^D)^t$$

$$\approx \frac{W^2 T^2 \langle \phi_{WS} \rangle^2}{2\langle \phi_{B-L} \rangle} \begin{pmatrix} T^2\xi^6 & T\xi^3 & T\xi^2 \\ T\xi^3 & T^2\chi\xi & T\chi \\ T\xi^2 & T\chi & \chi\xi \end{pmatrix} . \tag{35}$$

It is from this contribution that the two lightest left-handed neutrino masses and their mixing angle, θ_{12}, should be obtained:

$$M_{\nu_L \text{medium}} \approx \frac{W^2 T^2 \chi\xi \langle \phi_{WS} \rangle^2}{2\langle \phi_{B-L} \rangle} \tag{36}$$

$$\theta_{12} = \theta_{e,\text{medium}} \approx \begin{cases} \frac{T}{\chi}\xi & \text{when } \chi \gtrsim T \\ \xi & \text{when } \chi \lesssim T \end{cases} . \tag{37}$$

Note that the lightest mass is quite dominantly the ν_{e_L} neutrino and the small mixing angle goes mainly to the medium mass neutrino ($\theta_{e,\text{medium}}/\theta_{e,\text{heavy}} \approx \xi^{-2} \gg 1$). So we should identify approximately the solar oscillation mixing angle with the mixing to the medium heavy neutrino:

$$\theta_\odot \simeq \theta_{e,\text{medium}} \approx \xi \tag{38}$$

and the solar mass square difference

$$\Delta m_\odot^2 \approx M_{\nu_L\,\text{medium}}^2 \approx \frac{W^4 T^4 \chi^2 \xi^2 \langle \phi_{WS}\rangle^4}{4 \langle \phi_{B-L}\rangle^2} . \tag{39}$$

The atmospheric mixing angle goes between the heaviest and the medium one:

$$\Delta m_{\text{atm}}^2 \approx M_{\nu_L\,\text{heavy}}^2 - M_{\nu_L\,\text{medium}}^2$$
$$\approx \frac{W^4 T^4 \chi^2 \langle \phi_{WS}\rangle^4}{4 \langle \phi_{B-L}\rangle^2 \xi^2} \tag{40}$$

From equations (39) and (40) we find that the ratio of solar and atmospheric mass square differences must be of the order of ξ^4, say, about 10^{-4}.

B Statistical calculation using random order unity factors

In this subsection we will discuss the numerical calculation. The elements of the mass matrices are determined up to factors of order one to be a product of several Higgs VEVs measured in units of the fundamental scale M_{Planck} - the Planck scale.

We imagine that the mass matrix elements for, e.g., the right-handed neutrino masses or the mass matrix M_ν^D, are given by chain diagrams consisting of a backbone of fermion propagators for fermions with fundamental masses, with side ribs (branches) symbolising a Yukawa coupling to one of the Higgs field VEVs.

We know neither the Yukawa couplings nor the precise masses of the fundamental mass fermions, but it is a basic assumption of the naturalness of our model that these couplings are of order unity and that the masses, also deviate from the Planck mass by factors of order unity. In the numerical evaluation of the consequences of the model we explicitly take into account these uncertain factors of order unity by providing each matrix element with an explicit random number λ_{ij} - with a distribution so that its average $\langle \log \lambda_{ij}\rangle \approx 0$ and its spreading is a factor two.

Then the calculation is performed with these numbers time after time with different random number λ_{ij}-values and the results averaged in logarithms. A crude realisation of the distribution of these λ_{ij} could be a flat distribution between -2 and $+2$, then provided also with a random phase (with flat distribution).

Another "detail" is the use of a factor $\sqrt{\#\text{diagrams}}$ multiplying the matrix elements, to take into account that, due to the possibility of permuting the Higgs

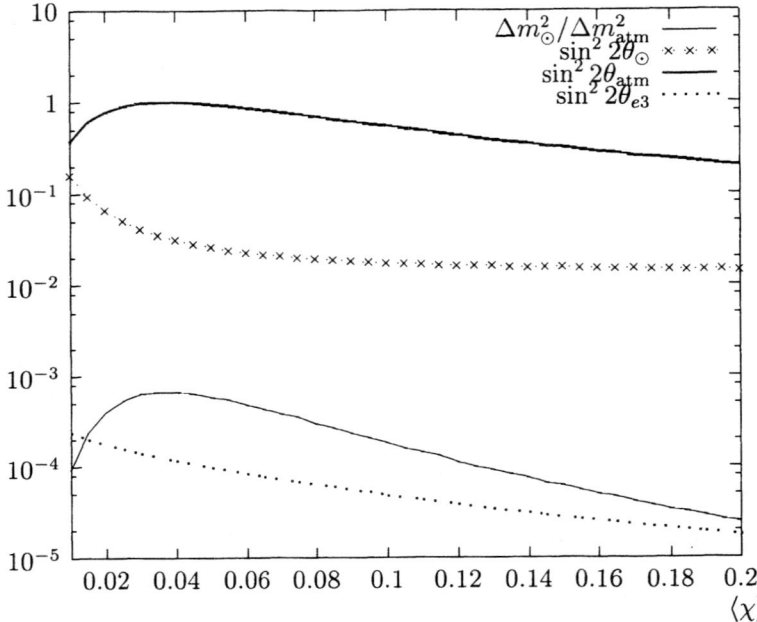

FIGURE 1. The numerical results of the ratio of the solar neutrino mass square difference to that for the atmospheric neutrino oscillation, and the squared sine of the double of the solar neutrino mixing angle, the atmospheric neutrino mixing angle and the mixing angle θ_{e3}.

field attachments in the chain-diagram, the number of different diagrams is roughly proportional to the number of such permutations #diagrams. This is the correction introduced and studied for the charged mass matrices by D. Smith and two of us [20]. In the philosophy of each diagram coming with a random order unity factor, the sum of #diagrams get of the order $\sqrt{\text{\#diagrams}}$ bigger than a single diagram of that sort. But we counted these permutations ignoring the field S. If we allowed both S and $S^\dagger$ in the same diagram, the $\sqrt{\text{\#diagrams}}$ factor could give arbitrarily large numbers. It turns out that these factors are especially important for some elements involving the electron-neutrino in the matrix M_ν^D, which are suppressed by several factors, as then many permutations can be made.

Yet another detail is that the symmetric mass matrices - occurring for the Majorana neutrinos - give rise to the same off-diagonal term twice in the right-handed neutrino matrix in the effective Lagrangian, so we must multiply off-diagonal elements with a factor $1/2$. But in the M_ν^D-matrix, columns and rows are related to completely different Weyl fields and of course a similar factor $1/2$ should not be introduced.

Concerning the $\sqrt{\text{\#diagrams}}$ factor for the diagonal mass matrix terms in the symmetric matrices, *e.g.* M_R, we shall remember that, contrary to what we shall do in non-symmetric matrices such as M_ν^D, and the charged lepton ones, we must count diagrams with the Higgs fields attachment assigned in opposite order as only one diagram. The backbone in the diagram has no orientation and we shall count diagrams obtained from each other by inverting the sequence of the attached Higgs fields as only ONE diagram. Thus the diagonal elements will tend to have only half as many diagrams.

We give in Figure 1 results obtained with 50,000 random combinations averaged as a function of the small VEV $\langle \chi \rangle$ of the new Higgs field χ.

In order to get the atmospheric mixing angle of the order of unity the range for $\langle \chi \rangle$ around the "old" Anti-GUT VEV $\langle T \rangle \approx 0.07$ is suggested, so only this range is presented.

We should present here the order of magnitude of the right-handed neutrino masses and the mixing angle θ_{e3} (see in Figure $VIII\,B$):

$$M_{R_{\nu_1}} \approx 10^{11} \text{ GeV}, \tag{41}$$

$$M_{R_{\nu_2}} \approx 10^{13} \text{ GeV}, \tag{42}$$

$$M_{R_{\nu_3}} \approx 10^{13} \text{ GeV}, \tag{43}$$

$$\sin^2 2\theta_{e3} \approx 10^{-4}. \tag{44}$$

IX THE PROBLEM OF SCALES

When we, as here, just postulate the see-saw scale to have the value needed by introducing an appropriate Higgs field, ϕ_{B-L} in our model, and fit its vacuum expectation value, here to 10^{13} GeV, we have just another scale problem.

This new scale problem is of an exactly analogous character to that of the famous Hierarchy Problem of why the weak energy scale is so low compared to say the Planck scale (or the unified scale in GUT theories). So, one could say, we have got one more hierarchy-problem! It has been popular to identify the see-saw scale with the GUT-scale but, to do so, one has to make use of the big mass ratios occurring between the right handed neutrinos. When it IS fitted, it is always the heaviest of the see-saw neutrinos that gets a mass at the GUT-scale. This heaviest right handed neutrino is that one which does not have any phenomenological consequences for the (left handed) neutrino oscillations, and thus is very model dependent. To stretch the scale of the see-saw neutrinos from the needed scale, around the masses presented here of 10^{11} to 10^{13} GeV, up to the usual GUT-scale at 10^{15} to 10^{17} GeV is possible but not strongly suggested. But phenomenologically it is not possible to push the see-saw scale up to the Planck scale. Thus we must introduce at least one new scale, in addition to the Planck scale, the weak scale and the strong scale. This may be very important to think about in connection with solving the hierarchy problem, or the related scale problem, because we now have to look for a solution that can provide yet another scale—the see-saw scale or, if we drop the see-saw model altogether, the neutrino oscillation scale directly.

Preferably we should have a mechanism that could put these different scales at exponential values relative to, say, the Planck scale, much in the same way mathematically as the strong interaction scale is understood to be exponential due to the renormalization group running of α_S–the QCD-gauge coupling (squared and divided by 4π).

We are indeed working on such a project (in this project is also included L.V. Laperashvli [21]) using the above mentioned multiple point principle (MPP). Once we have such a principle that fixes the couplings and masses in a certain way, we are not really solving the problem of avoiding finetuning, but rather we have postulated a <u>finetuner</u> that finetunes the parameters for us, namely the postulated MPP-mechanism. That should make it much easier to "solve" the scale problem or rather problems–now we include the neutrino oscillations–since what we need is "only" that, when the couplings and masses are being adjusted to make several degenerate minima as the MPP postulates, then as a consequence of that requirement they tend to also organize hierarchies of scales.

The argument on which we work goes roughly like this:

The MPP is taken to postulate that the effective potential $V_{eff}(\phi_{WS})$ shall have (at least) two minima as a function of $|\phi_{WS}|^2$, with the same value of the effective potential, for non-zero values of this Standard Model Higgs field ϕ_{WS}. This is an interpretation of the somewhat vague statement that there shall be "many" minima with the same energy density.

This obviously implies that the second derivative of this effective potential with respect to $|\phi_{WS}|^2$ has two places on the positive $|\phi_{WS}|^2$ axis where it is positive–namely the minima. It then follows that there is one place in between where the second derivative is negative–namely at the maximum that must occur between the two minima, for topological reasons so to speak,. For our argument here the degeneracy of the minima does not seem particularly relevant–it is just that the MPP is a statement about degenerate minima.

Then we go through, loop-order for loop-order, to see if it is possible, under reasonable assumptions, that the second derivative $d^2V_{eff}(|\phi_{WS}|^2)/d(|\phi_{WS}|^2)^2$ could have the alternating sign behaviour, first positive then negative and then positive again, required for the presence of two minima. *In 0-loop approximation*:

$$\frac{d^2V_{eff}(|\phi_{WS}|^2)}{d(|\phi_{WS}|^2)^2} = \frac{\lambda}{2} = constant \qquad (45)$$

i.e. it cannot have the required sign shifts except if equal to zero. *In 1-loop approximation*:

$$\frac{d^2V_{eff}(|\phi_{WS}|^2)}{d(|\phi_{WS}|^2)^2} \approx \frac{\lambda_{run}(|\phi_{WS}|^2)}{2} + constant \qquad (46)$$

In the approximation that we look for $|\phi_{WS}|^2 \gg \mu^2$ (the Higgs mass squared parameter), the running λ is proportional to a single $\log|\phi_{WS}|^2$ with a coefficient that gets positive terms from the bosons in the field theory and negative terms from the fermions. Ignoring strange solutions with both minima very tightly close to each other, we have a single monotonous logarithm that cannot switch back and forth in sign. So the only solution is, like in the 0-loop case, to put the coefficient zero and kill to zero the whole 1-loop contribution to the second derivative. *In 2-loop approximation*: It is possible to achieve an effective potential second derivative $d^2 V_{eff}(|\phi_{WS}|^2)/d(|\phi_{WS}|^2)$ which can switch sign as needed to have two minima.

In fact we have found [22] that it is completely possible to have a scenario with coupling constants and masses leading to two minima. It is even completely within the experimentally allowed range of parameters to have the second minimum 1) at the Planck scale, and 2) degenerate as the MPP requires. We obtain this particular scenario when the top mass is 173 GeV and the Higgs mass is 135 GeV, namely the smallest Higgs mass allowed if the pure Standard Model is valid up to the Planck scale. Since the indirect estimates of the Higgs mass favour a mass lower than experimental limit, they really mean it should be as low as possible and thus just our prediction of 135 GeV, provided no new physics comes in below the Planck scale allowing a lower Higgs mass.

In fact it is a bit surprising that our calculation, even with a rather crude value for the Planck scale only assuming its order of magnitude for the second minimum, yields a top-mass of 173 GeV with an accuracy similar to the experimental accuracy of about ±6 GeV.

So it looks that with two loops we can easily get the two minima required, but with lower loop accuracy it is impossible. However, the various couplings are experimentally so small that the two-loop terms contribution is only a correction and not usually qualitatively important. If now it is needed that the two-loop-terms must be important, in order to have the required two minima (using MPP), then they must be boosted to importance by sufficiently big logarithms: In a leading log expansion, the two loop terms can become comparable or beat out 0-loop and 1-loop approximation terms if the logs are, so to speak, as big as the couplings are weak. But the log here must be, for instance, the logarithm of the ratio of the two minima we required. If that gets as big as the couplings are weak, it means that, say, the lowest minimum field value must go down relative to the other one by the exponential of the inverse couplings.

This would be very similar to what is achieved by means of the renormalisation group for the strong scale, but should now numerically also work for the lower minimum that determines the weak scale. That is potentially a solution of the scale problem!

There is though a little "technical" problem with our "solution" of the scale problem: At the second minimum it turns out, say in our scenario with the Higgs mass being 135 GeV, that the running self-coupling λ_{run} for the Higgs field runs very very small. If it is allowed to have so small a value, the argument of crude order of magnitude character given above does not quite work. Then the logical argumentation for the need of the big ratio of scales fails, even when the requirement

(MPP) of two degenerate vacua is postulated. Interestingly enough there is better hope to come through successfully with several hierarchy problems at once. So the see-saw scale problem could turn out being helpful in bringing our idea for solving the scale problem to work.

X LOOKING FOR TOP-QUARK ANALOGUES

If we accepted the idea that we wanted the postulate of several minima - or even degenerate minima - to explain large scale ratios, then we need to have the zero- and one-loop approximation to cancel approximately [23], so that the two-loop terms can come to be qualitatively significant. Remember it was only the two-loop terms that could produce the required minima. But the one loop contribution was a logarithm multiplied by positive terms from the bosons and negative ones from the fermion loops. This is the reason why we need a strongly coupled top-quark, *i.e.* with a Yukawa coupling of order unity.

With this very speculative requirement, that each time we have a Higgs field with a small vacuum expectation value it is due to our postulated MPP, we can then use it as a guide in seeking the structure of the model that realizes our assumptions.

As already mentioned the top quark does the job of cancelling the one-loop correction so as to allow two minima perfectly, as far as we yet know.

Luckily enough there is, in our above sketched model for the mass of the see-saw neutrinos coming from the field ϕ_{B-L}, just a combination of right handed neutrinos that can "play the role of the top-quark in the Standard Model", because it couples with of order unity Yukawa coupling. In fact the Higgs field ϕ_{B-L} (which has to have a very small VEV compared to the Planck scale, in fact of order 10^{13} GeV) couples in our model to the transition between the third and the first family right handed neutrinos. Really it can convert a right handed electron neutrino to a CP-antiparticle of the right handed τ-neutrino. This coupling is therefore unsuppressed. That is quite non-trivial in our model, since there can be so many suppressions of effective couplings due to some charges needing to be broken to generate the vertex. But nevertheless we have a top-quark-role-player needed for the working of our two minima machinery for the ϕ_{B-L} field.

Now we might say: but even the Higgs fields which in our model give small hierarchies—the fields T, W, ξ, χ having only VEV's of the order of 1/10 in Planck units—strictly speaking also represent hierarchy-like problems. Why are their expectation values so small compared to the fundamental (presumably Planck) scale. Even a small number of the order of 1/10 needs an explanation. So they should also have their "top-quark-role-players". Also they should have unsuppressed couplings to some fermion, which is mass protected but gets a mass from just the Higgs field in question. Then that fermion could help cancel the one-loop contribution, so as to make the two-loop one become important.

Such a rule of requiring a "top-quark-role-player", for each Higgs field with very small expectation value relative to the presumed fundamental scale, will require the existence of some chiral fermions that then can get mass from the Higgs field in question. But chiral fermion fields, in turn, will give rise to anomalies that must be cancelled. So we obtain a series of consistency requirements that can be useful in building a model from the bottom up.

We start the construction by seeking Higgs fields to explain the quark-lepton mass and mixing angle suppressions. Then, in turn, we look for "top-quark-role-players" for these Higgs fields and, next, for further Weyl-fermions to cancel the anomalies for the "top-quark-role-players". The idea then is that the many consistency requirements shall be used to find a scheme that works self-consistently. Then all the small-VEV-Higgs-fields will have their degenerate minima and fermions, which they give mass to and which play the top quark role. Furthermore the gauge and mixed anomalies of these fermions will cancel.

One might feel that the story of the "top-quark-role-players" has a little too many speculative assumptions to be trustable. It could then be that there is a slightly different reason for the same sort of fermion fields to exist—a reason that could actually be a reformulation of the same model—each Higgs boson could be imagined and required to be a bound state of a pair of chiral fermions. That would lead to about the same requirements as the 'top-quark-player" rule: For each low mass (or low VEV) Higgs/scalar field, there must be some constituents able to form it w.r.t. quantum numbers. They should typically be fermions and need to be mass protected to the scale of the mass of the boson to be constructed. The coupling has to be strong for the constituents to the bound state.

The requirements for a boson being a bound state of a couple of chiral fermions become very similar indeed to the requirements that the boson have them as its "top-quark-role players".

XI AN EXTRA FAMILY OF WEYL FERMIONS

But what fermions could play the "top-quark-role" for the scalar boson fields W, T, ξ, χ? It would have to be fermions that actually obtain a mass from the fields for which they do the job of the top quark. But that means they end up having masses of the order of the Higgs field scale in question (the VEV). So we need, in our model, Weyl particles combining themselves to each other, so as to obtain masses at the scale of our Higgs fields W, T, ξ, ... which is about 1/10th of the Planck scale. The mass protection for these Weyl particles must be due to some of the gauge quantum numbers in our model. However they should not be mass protected by the Standard Model quantum numbers, because then they would have much much lower masses and perhaps should have been seen. They must not even have the total $B - L$ as a mass protecting quantum number, because then they would get masses of the see-saw scale and not just a factor of 10 under the Planck scale, as we need them. They can though have these quantum numbers in a vectorial way, *i.e.* there could be equally many right and left Weyl particles with a given combination of the Standard Model quantum numbers and the total $B - L$.

For instance we could look at the Higgs field χ, which was introduced for the sake of fitting the neutrino oscillations. Since it gives rise to a suppression of the order of a factor 10 to 15, it must have mass and VEV at about a factor 10 to 15 under the Planck scale (taken as the fundamental one). Its quantum numbers in our model are quite simple: It has no weak hypercharges and then, according to our rule for the non-abelian quantum numbers, also no non-abelian couplings. It only has the family-$(B-L)$-quantum numbers: $(B-L)_2 = -1$ and $(B-L)_3 = 1$, the rest being zero.

In order for a pair of Weyl fermions to couple to the χ field in an unsuppressed way, it is needed that their quantum numbers add up to those of the χ field modulo certain quantum number combinations. These quantum number combinations correspond to the ones sitting on the Higgs fields, like the S field in our model, having expectation values of the order of the fundamental scale. Otherwise there would be a need for some Higgs field VEV to provide the lacking quantum number/charge in the coupling and it would be suppressed. In that case the field would couple too weakly and could not play its top-quark-role.

If we seek to keep the quantum numbers from being too large, the simplest would be to have one fermion with all quantum numbers zero except for say $(B-L)_2 = -1$, and another fermion with all zero except for $(B-L)_3 = 1$. But we can also have that these two Weyl fermions in addition carry other quantum numbers, but such that one of these two particles carry just the opposite further charges to those of the other one.

In fact we are forced to assume that the two particles that should play the top-quark-role for χ have other charges. Otherwise they would be mass protected by the total $(B-L)$-charge and get masses at the see-saw scale, rather than the required one-tenth of the Planck scale. Indeed they must each have a $(B-L)$-charge of yet another family so as to get the total $(B-L)$ be zero. The highly suggested quantum numbers for these Weyl particles, counted as left handed, are thus

$$(0,0,0;-1,0,1) \text{ and } (0,0,0;1,-1,0). \tag{47}$$

Now is it possible to build such a pair Weyl particles into a scheme with mass protected particles cancelling the anomalies and not carrying the charges that would bring them to the low mass level, where they would be directly or indirectly seen? That is to say, we should not give them mass protection with Standard Model charges, so that they most easily all have the total weak hypercharge $y/2 = y_1/2 + y_2/2 + y_3/2 = 0$. If we also want them not to show up at the see-saw scale, we should not let them be mass protected by the total $(B-L) = (B-L)_1 + (B-L)_2 + (B-L)_3$. Again the most easy way of ensuring this would be to let them not have the total $B-L$ at all, $i.e.$ to have $B-L = 0$ for them.

It is in fact possible to find a set of Weyl-particles, cancelling anomalies and containing the two particles suggested. This relatively elegant scheme of particles contains 15 Weyl particles—just the same number as the particles in a Standard Model family without the right handed neutrino.

TABLE 4. Quantum numbers of added set of particles

$y_1/2$	$y_1/2$	$y_3/2$	$(B-L)_1$	$(B-L)_2$	$(B-L)_3$	
0	1	-1	0	0	0	(-1;0)
-1	0	1	0	0	0	(-2;0)
1	-1	0	0	0	0	(-3;0)
0	0	0	0	1	-1	(0;-1)
0	1	-1	0	1	-1	(-1;-1)
-1	0	1	0	-1	1	(-2;1)
1	-1	0	0	-1	1	(-3;1)
0	0	0	-1	0	1	(0;-2)
0	-1	1	-1	0	1	(1;-2)
1	0	-1	1	0	-1	(2;2)
-1	1	0	1	0	-1	(3;2)
0	0	0	1	-1	0	(0;-3)
0	-1	1	1	-1	0	(1;-3)
1	0	-1	-1	1	0	(2;3)
-1	1	0	-1	1	0	(3;3)

It can easily be checked that the here proposed particles have anomalies that cancel among themselves. The last column in the table contains a shorthand notation for the quantum numbers, in the way that the first signed number from 0 to 3 enumerates some simple combinations of family weak hypercharges, so that:

$$\mathbf{0} = (0, 0, 0) \tag{48}$$
$$\mathbf{1} = (0, -1, 1) \tag{49}$$
$$\mathbf{-1} = (0, 1, -1) \tag{50}$$
$$\mathbf{2} = (1, 0 - 1) \tag{51}$$
$$\mathbf{-2} = (-1, 0, 1) \tag{52}$$
$$\mathbf{3} = (-1, 1, 0) \tag{53}$$
$$\mathbf{-3} = (1, -1, 0) \tag{54}$$

The same shorthand notation is used for the ordered sets of the three family $(B-L)_i$'s. So that e.g. $(\mathbf{-2}; \mathbf{1})$ means $(y_1/2, y_2/2, y_3/2; (B-L)_1, (B-L)_2, (B-L)_3 = (-1, 0, 1; 0, -1, 1)$.

This shorthand notation can be used to see a certain amount of regularity in the scheme found and can be used to see quickly the cancellation of many of the anomalies. You should have in mind that we have so many anomaly constraints that it is a bit of an art to find any chiral scheme with cancellations of the very many choices of three charges to look for anomalies for. In fact the scheme is written - moderately - nicely in a 4 by 4 array, like Table 5:

The regularity in this table is already a little bit complicated: There are fifteen Weyl particle quantum numbers listed in as far as, except for the totally sterile thinkable particle $(\mathbf{0}, \mathbf{0})$ with no charges at all, we have sixteen combinations. These

TABLE 5. The fifteen Weyl particle quantum numbers as array

-	$(-1;0)$	$(-2;0)$	$(-3;0)$
$(0;-1)$	$(-1;-1)$	$(-2;+1)$	$(-3;+1)$
$(0;-2)$	$(+1;-2)$	$(+2;+2)$	$(+3;+2)$
$(0;-3)$	$(+1;-3)$	$(+2;+3)$	$(+3;+3)$

are obtained by combining one of the following four combinations of a number for the hypercharges and a sign related to the $(B-L)_i$'s:

$$[0;-], \ [1;-], \ [2;+], \ [3;+] \qquad (55)$$

with one from the following series of a sign related to the family weak hypercharges and a number for the family $(B-L)_i$'s:

$$[-;0], \ [-;1], \ [+;2], \ [+;3] \qquad (56)$$

The idea here is that, for example, when we combine $[2;+]$ from the first series with $[-;1]$ from the second series, we construct the shorthand symbol $(-2;+1)$. This then, in turn, is translated into the combination of charges supposed to be sitting on that left handed Weyl particle with quantum numbers $(-2;+1) = (y_1/2, y_2/2, y_3/2; (B-L)_1, (B-L)_2, (B-L)_3) = (-1,0,1;0,-1,1)$.

If we now, for example, will check that there are no anomalies corresponding to a triangle diagram with the three external gauge boson couplings being the ones coupling to the weak hypercharges for family 1 and family 2 and the $(B-L)_2$, it means that we want to check the no anomaly condition

$$\sum_{\text{the 15 Weylp.}} \frac{y_1}{2} \cdot \frac{y_2}{2} \cdot (B-L)_2 \qquad (57)$$

$$= \sum_{\{(-3;+1),(+3;+3)\}} \frac{y_1}{2} \cdot \frac{y_2}{2} \cdot (B-L)_2 \qquad (58)$$

$$= -1 \sum_{\{+1,+3\}} (B-L)_2 = 0 \qquad (59)$$

It is, namely, easily seen that we can only get contributions to the product $\frac{y_1}{2}\frac{y_2}{2}$ from the weak hypercharge shorthand symbols that are neither having the number ± 2 nor ± 3. Further to have a $(B-L)_2$ non-zero factor, we need to avoid the $(B-L)$-related shorthand symbol being ± 2. In our formulation this cancellation takes place because the sign is the same for all the $\pm 3 = \pm(-1,1,0)$ which are the combinations to which the $y_1y_2/4$ can give a contribution.

In the fifteen Weyl-particle system just presented, one also find candidates for a "top-quark-role-player" for a field with the quantum numbers of the field combination $\xi S^\dagger$, which has the simple quantum numbers $(0,0,0;1,-1,0)$.

In our fit the Higgs field S has a vacuum expectation value of order unity. So, in first approximation, there would be essentially no phenomenological consequences to replace the model by a related one, in which the field ξ had got the quantum numbers of this combination $\xi S^\dagger$. Then, when there is use for a Higgs field with the ξ quantum numbers, one would make use of an extra S-field. Since that does not cost any extra suppression, that would make no great difference and it could be an equally good model. But it could have the advantage that we could now find, among the particles which we are tempted to postulate to exist, some "top-quark-role-players" for the slightly modified ξ-field.

But now we must admit that further manipulation of the model into a related one seems to be needed, if there should be any chance of getting the suggested requirements of "top-quark-role-players", no anomaly fermions and all that to work. However, that some rudiments of the requirements come about, by means of almost phenomenologically invented fields, is a good sign. It suggests that further investigation in that direction could yield a model that hangs together and satisfies the requirements.

Let us also remark that this system of fifteen Weyl particles has some similarity with a usual family of Weyl particles in the Standard Model. At least it has the similarity that it consists of just 15 particles, just the number in a usual family. In this way there would be some sense in talking about such a set of particles as a "crossing fourth family". However it is, of course, really quite misleading, in as far as this "crossing family" couples to completely different gauge fields compared to what a proper fourth family should do.

XII ON WHY THREE FAMILIES

It should be noted that the above speculated, and very speculatively phenomenologically called for, system of 15 particles with masses of the order of a factor ten or so under the Planck mass scale has some connection to the question of why we have just three families:

Indeed the construction of a system of just a number of particles corresponding to the single family, in the way we did it above, is specific to the case of just three families! We might think of the abelian gauge groups that were used in the suggested system of 15 particles as analogous to the Cartan algebra of the Standard Model group. If we do that we see that it is natural that there be made use of just four linearly independent abelian charges. Well, at first it looks that we used 6 charge-species, namely 3 weak hypercharges and 3 family $(B - L)_i$'s. However, in order to avoid to let the system of the 15 particles risk to get mass protection by the Standard Model charges or the total $(B - L)$-charge, we had to let the charge assignments for these 15 particles obey the rules that the Standard Model quantum numbers as well as the total $(B - L)$ be zero. That imposes two constraints - of zero diagonal coupling - upon the charges and there are only $6 - 2 = 4$ independent charges.

Now note that we here got the condition, for the working of the "crossing fourth family" as constructed above, that the number of families multiplied by two - one for the weak hypercharge and one for the $(B - L)_i$ - shall be just two more than the rank of the Standard Model group. This constraint would only be satisfied for there being just 3 families!

Once we take it as a good idea that there be such a "crossing fourth family", we may again argue for the three families by a slightly different method: In the following section we shall shortly put forward an idea of thinking of some especially "nice" linear combinations of the Cartan algebra generators, "Han-Nambu like charges". We thereby get an argument that it is not accidental that the number of Weyl-particles in a family in the Standard Model is just a power of two minus unity - or simply a power of two if we include a right handed neutrino.

Once we observe that there are just $2^4 - 1 = 15$ particles in a family, we may ask ourselves whether, by analogy, it would not be expected that also the number of families and/or the total number of under Planck scale Weyl particles should be of the form $2^n - 1$.

In this way we propose a picture of the Weyl fermion charge assignments for all the particles that are mass protected by some charge in the system:

In addition to the three known families there is a bunch of Weyl-particles with the scheme of charge assignments given in the tables above, they are equal in number to - a family without its - for the other families for the sake of see-saw mechanism suggested right handed neutrino, namely 15.

Together this makes $3(15 + 1) + 15 = 63 = 2^6 - 1$. In other words: The three known families each associated to one speculated see-saw neutrino (= right handed neutrino) makes up $3(15+1) = 48$ Weyl particles. Adding to them the phantasy extra family - that is not at all a real family, so we should not call it so - suggested by the "top-quark-role-player" arguments consisting of 15 particles we reach 48+15 = 63 which can be considered remarkable by being a power of 2 minus one, just in analogy to the number of Weyl-particles in Standard Model generation.

But why should these $2^n - 1$ systems of particles be so important and likely to be the system occuring in nature? We have - of course - no really convincing argument for the moment, but hope that there is a weak suggestion in what is really an anomaly counting modulo 2 (see below).

XIII HAN-NAMBU-LIKE-CHARGES

In this section we should like to mention an idea of how one might make more specific a statement saying that the Standard Model representations of the gauge group realized in nature are *very small representations.*

If you look at the non-abelian representations in the Standard model you find except for the Gauge fields themselves *nothing but the trivial and the lowest dimensional representations after the trivial.* This can be said to mean that the representations chosen by Nature are really remarkably *small* as far as the non-abelian groups $SU(2)$ and $SU(3)$ are concerned!

Can we say the same thing in some sense when we think of abelian group(s) ? For abelian groups you cannot so easily use the dimension of the representation, since abelian groups always have one-dimensional representations and thus dimension makes no distinction; in stead the natural suggestion would be to use the ratio of the charges of the representation realized compared to the quantum of charge - the Millikan-like charge value of which all charges must be an integer multiplum. One might now think of looking at the unique abelian *invariant* subgroup in the Standard Model Gauge group, the weak hypercharge; but taken literary that idea does *not* work well if you hope to show that the Standard Model charges are remarkably small. The problem is that the quantum for the half hypercharge $y/2$ is $1/6$ while the right handed charged leptons have values $y/2 = -1$. So it is 6 times as large, but 6 is not the smallest number after zero.

But there may be no good reason for looking only at the invariant abelian subgroups, so we are suggesting to take into account also the non-invariant abelian subgroups. But then there are a lot of abelian subgroups of the Standard Model group, really the Standard Model group has infinitely many abelian subgroups of course. Now most of these have no quantum of for the charge at all and thus would make the question of the number of quanta on the actually realized particles ill defined. Think for example that we considered two non-invariant abelian subgroups of the Standard Model group as for example the one generated by the electric charge and the one generated by the Gell-Mann λ_8 or say $\sqrt{3}\lambda_8$. The electric charge has really in the Standard Model a quantum that is $1/3$ in units of the Millikan quantum, because of the quarks having non-integer charges, while $\sqrt{3}\lambda_8 = \text{diag}(1, 1, -2)$ has a quantum of charge equal to 1. If we, however, now make a linear combination with complicated rational coefficients we get easily a combined charge or generator in the Standard Model Lie algebra which can have a very small quantum of charge. So it is easy by forming linear combinations of Lie algebra generators (essentially the same as charges) to get such generators that have very small quanta of charge and then the actual particles turn out to have a huge number of quanta. Thus in the definition we discuss the representation w.r.t. such abelian charges will be "very big " *i.e.* have lots of quanta for the realized representations ($\approx$ particles). In formulating a statement about whether the Standard Model have small or big abelian representations it is therefore necessary to keep in mind that the different non-invariant charges we can think of have in general very different charge quanta sizes and thus are not at all represented on equally "small" representations.

The complicated linear combinations or let alone linear combinations with irrational coefficients are not so attractive to study or use for the definition of the representations being small or large; so it is rather suggested to define the question of whether a model in question - say the Standard Model - has small or large representations depending on whether it has many or few (for that purpose chosen) charges which have only very low charges realized compared to the quantum. The smallest non-trivial charge compared to the quantum you can hope for is that the charge is only 0 or 1 or -1 measured in quanta. So we should say that the abelian representations being small should by definition mean that it has relatively many charges (*i.e.* generators in the Lie algebra) with the property of having only realized values 0, 1 or −1.

So the question then is: will a reasonable "counting" of the charges (Lie algebra generators) that have this property of only having the three possibilities of the number of quanta on a particle (for a representation) 0, 1 and −1 lead to that the Standard Model has especially many of such charges? The answer we want to suggest is that it is indeed so that the Standard Model has a relatively big family of such charges, and thus one should be able to say that also w.r.t. the abelian charges and not only w.r.t. the non-abelian ones is the Standard Model a model with "small" representations!

Let us for simplicity consider the Caran algebra of the Standard Model gauge group (or rather a Cartan algebra of course) and look for how many Lie algebra generators (or charges) we can find inside this Cartan algebra which have the property of only eigenvalues $1, 0, -1$.

Actually we can present a family of ten such generators, and several of them are already known in the literature as some variations of the Han-Nambu charges. Remember [24] that the Han-Nambu charges were proposals for what the electric charge could be in a model for quarks etc. which because of the deep inelastic experiments has lost support. It has the property that even the quarks have integer Millikan quanta of charges of this Han-Nambu type (which were suggested as a candidate for electric charge, but here is just considered a certain formal charge we can construct if we like). In fact the idea is that one adds to the by now generally accepted form of the electric charge in the Standard Model $Q = y/2 + t_3/2$ a colour dependent term $\lambda_8/\sqrt{3}$ which has eigenvalues $1/3, 1/3$ and $-2/3$. If the last quark that has the $-2/3$ eigenvalue is red one, we can say we get the "red Han-Nambu charge".

It is clear that we even inside the Cartan algebra chosen by using diagonal $t_3/2$ and Gell-Mann λ_8's and λ_3 we can find Han-Nambu charges of the two other colours the blue and the yellow. This already makes up three Han-Nambu charges inside the Cartan algebra. In addition we can take "anti Han-Nambu Charges" which are obtained by replacing the now a day believed electric charge by the the charge $y/2 - t_3/2$ *i.e.* with a minus sign on the $t_3/2$-term. This gives us three more charges which again are easily seen to have no other eigenvalues than $0, 1, -1$. Now it is actually easily understood that if we have two charges say Q_A and Q_B which have only the three eigenvalues $-1, 0$ and 1, then there is an enhanced chance that the sum or the difference would again be a charge with this property, but it is by no means guaranteed, since we could easily risk that say the sum $Q_A + Q_B$ has also double charges. However it turns out that we can indeed construct sums and differences of some of the Han-Nambu and anti-Han-Nambu charges which again have this $-1, 0, 1$ property. For instance subtraction of the Han-Nambu charges from each other leads to λ_3 type colour generators, of which we have in Cartan algebra three combinations (counting as one a charge and its opposite). By subtracting analogous anti Han-Nambu and Han-Nambu charges we can also get t_3 which is 1, or −1 on left handed particles and 0 on the right handed (in the conventional thinking of particles but ignoring anti particles).

This makes up the following 10 charges in the Cartan algebra with our $-1, 0, 1$ property:

$$Q_{\text{HN red}} = y/2 + t_3/2 + \lambda_{8(\text{red})}/\sqrt{3} \tag{60}$$

$$Q_{\text{HN blue}} = y/2 + t_3/2 + \lambda_{8(\text{blue})}/\sqrt{3} \tag{61}$$

$$Q_{\text{HN yellow}} = y/2 + t_3/2 + \lambda_{8(\text{yellow})}/\sqrt{3} \tag{62}$$

$$Q_{\text{HN red}} = y/2 - t_3/2 + \lambda_{8(\text{red})}/\sqrt{3} \tag{63}$$

$$Q_{\text{HN blue}} = y/2 - t_3/2 + \lambda_{8(\text{blue})}/\sqrt{3} \tag{64}$$

$$Q_{\text{HN yellow}} = y/2 - t_3/2 + \lambda_{8(\text{yellow})}/\sqrt{3} \tag{65}$$

$$\lambda_{3\,\text{red, blue}} \tag{66}$$

$$\lambda_{3\,\text{blue, yellow}} \tag{67}$$

$$\lambda_{3\,\text{yellow, red}} \tag{68}$$

If the sum of two of these 10 charges happens to be one of the other ones the difference will not be one but have double charge eigenvalues. They form together with 5 generators that though have eigenvalues ± 2 also an algebra that is to be considered a modulo 2 algebra since you have to avoid to distinguish between sum and difference in order to really have the set of these charges closed under addition (because sometimes it is the sum sometimes the difference that belongs again to the set).

A Modulo 2 considerations

In the light of that the algebra of the Han-Nambu-like charges tend to be a modulo 2 algebra it is naturally considered also a Z_2-vector space and we may study first the anomaly restrictions for the Z_2-vector space informations hoping that this study is simpler but nevertheless can give interesting informations, even if some information has to be obtained afterwards.

This means that we suggest to study first the anomaly restrictions on a set of generally imagined set of Han-Nambu-like charges having to as large extend as possible only the number of charges on the particles being $-1, 0, 1$. By such a study we can think of seeking to derive the Standard Model pattern of particles partly if we can arrive at getting the Standard Model pattern partly out by such a study.

So let us thus here imagine a series "Han-Nambu charges" $Q_{HN\,i}$ ($i = 1, 2, 3, ..., n$) having the property that for all particles in the Model hoped to be shown to suggestively be the Standard Model and for most of the charges the values are always $Q_{HN\,i} = -1, 0, 1$.

In Z_2 algebra language we throw away the information of whether the charge is 1 or -1 but keep the information of whether it is even or odd *i.e.* of whether it is 0. We can think of the no anomaly conditions

$$\sum_{\text{the Weyl particles}} Q_{HN\,i} Q_{HN\,j} Q_{HN\,k} = 0 \qquad (69)$$

and the mixed anomaly condition

$$\sum_{\text{the Weyl particles}} Q_{HN\,i} = 0 \qquad (70)$$

either as concerning the true charges or as being written only modulo 2, so that the symbols $Q_{HN\,i}$ only take the values "even" or "odd". Then of course one shall use the algebra of the field Z_2. Considered this Z_2 way the mixed anomaly condition is really superfluous.

Postulating that the representations being small shall really mean that we for a lot -although may be not all - combinations by + or − (which in Z_2 algebra is just +) of a couple of Han-Nambu-like charges, say $Q_{HN\,i} + Q_{HN\,j}$ again get one of this type (closing of the algebra with a minimal number of exceptions), it is suggested that there be an under the Z_2-algebra closed set of charges, most of which have only the three possible values $-1, 0, 1$.

We shall take this to mean that if we look for gauge field theory model with "smallest possible representations" we shall search among models having a system of abelian charges (that may though be built into non-abelian groups very likely) forming in the way described a Z_2 field vector space.

From the point of view of the Z_2-algebra the Weyl-Fermions of the model - we are looking for - have their charges in the dual vector space of the charges: given a particle p and a charge we get an inner product $\langle Q_{HN\,i}|p\rangle$ = "the (eigen)value of the charge $Q_{HN\,i}$ for the particle p" = "the charge of the particle p of kind $Q_{HN\,i}$". Again we may think of this algebra as being counted modulo 2 if we like to do so, and that we do just now.

It would be most elegant and we could also claim that it would mean the biggest number of charges of this "Han-Nambu-like"type if we had so many that we used up all modulo 2 independent charge assignments. Let us take this as an excuse for assuming that the mass protection of the particles we consider - we are after low energy physics particles so they should be mass protected - is revealed in the modulo 2 counting by there being with some combination of the charges $Q_{HN\,i}$ specified (*i.e.* in a certain vector of the dual vector space of the space of these charges) an odd number of particle kinds (flavours).

Translating the no-anomaly conditions modulo 2 into the language of Z_2 field vector-spaces one can easily see that the condition (69) means that there is an even number of particles with charge assignments in the subspace in the particle-vector-space (the dual of that of the set of charges) lying in the intersection of the "displaced hyperplanes" characterised as the subsets of this particle-vector-space on which the three charges $Q_{HN\,i}$ $Q_{HN\,j}$, and $Q_{HN\,k}$ respectively provide an "odd" inner product. If the Z_2-vector spaces are of dimension less than or equal to three, this intersection can be made just one point and if all the charges are gauged we can make it any point and thus with our assumptions it is impossible to have charge and particle space of dimension less than or equal to 3.

The smallest dimension after the excluded up to 3 is that the space of particles as well as that of charges $Q_{HN\,i}$ should be four. In this case one quickly sees that we have to have all charge combinations - we now only work mod 2 - for an odd number of particles if just one combination as assumed has an odd number of particles, except though of course that a particle with all charges 0 makes no help in solving the anomaly conditions and can be left out. This means that we have to fill $2^4 - 1 = 15$ possible charge assignments modulo 2 with an odd number of particles.

The number 15 is interesting: there are just 15 Weyl particles in a Standard Model family, and the 10 (good with only $-1, 0, 1$) + 5 (bad, with double charges) = 15 Han-Nambu-like charges in the Standard Model can be seen to be assigned to the 15 particles in a family just so that counting only even and odd the 15 Weyl particles are distributed on all the 15 non-zero (non-even) vectors in the particle-vector-space dual to that of the charges!

In other words the Standard Model structure modulo 2 for the Han-Nambu like charges is obtained with a few "minimally of representation assumptions" as the smallest dimensional Z_2-space that can cancel the anomalies but still be mass protected!

Crudely speaking this means that we can claim - may be letting the definition of the concept of "small representations" being replaced by a little series of assumptions - that indeed the Standard Model is having smallest representation as far as the charges in the Cartan algebra goes.

If one take the above argumentation to suggest more generally a Z_2-vector space structure for both a set of Han-Nambu-like charges and the set of Weyl particles it is suggested that we should find in the true model a number of Weyl particles being a power of 2 perhaps minus 1. The latter subtraction because we can in no way mass protect a super-sterile particle with all quantum numbers zero. So the particles that are totally "even" in the Z_2-formulation may be totally sterile and impossible to mass protect and thus we would count them only as belonging to to the "garbage at the Planck scale" where everything is to be found in our philosophy.

The next power of 2 minus one to the number of Weyl fermions already observed namely 45, or 48 (or 47) if you believe in the see-saw mechanism so much as to claim some observed right handed neutrinos, is 63. So if you would uphold such an idea there would be a need for speculating an extra set of 15 Weyl particles that are only so weakly mass protected that they have become so heavy that we do not see them. With a scale in our model a single order of magnitude under the Planck scale it is of course not unexpected that there could be Weyl particles mass protected only down to this scale - counting Planck scale as the *a priori* mass scale. Having in analogy with usual family a system of 15 particles just using some other gauge fields with more strongly broken/Higgsed gauge symmetry is of course not unlikely. If they are with masses of the order of 1/10 of the Planck mass they are safely to stay in the phantasy sector for long yet.

Once you work with Z_2 vector spaces and somehow imagined the known families to go in in some not so clear way in a Z_2 vector space with four elements *i.e.* of dimension 2, it should be remarked to support for fitting 3 families that there is a permutation symmetry of the Z_2 vector field structure among the three non-zero elements while the zero-vector is - of course - special w.r.t. the Z_2-algebra.

XIV CONCLUSION

In this article we have made an extension of the Anti-GUT model to neutrinos by including see-saw ν_R at a scale of mass around 10^{12} GeV. By this extension we introduced two more parameters, namely the vacuum expectation values of two additional Higgs fields, ϕ_{B-L} and χ. But from the neutrino oscillation data one extracts two mixing angles $\theta_\odot$ and θ_{atm} and two mass square differences $\Delta m_\odot^2$ and Δm_{atm}^2, so in this sense we have two predictions:

$$\sin^2 2\theta_\odot \approx 3 \times 10^{-2} \tag{71}$$

$$\frac{\Delta m_\odot^2}{\Delta m_{\text{atm}}^2} \approx 6 \times 10^{-4} \tag{72}$$

These results are *only order of magnitude* estimates, and we shall count something like an uncertainty of 50% for mixing angles and masses and thus for the square, $\sin^2 2\theta$, 100% and i.e. a factor 2 up or down and for the $\Delta m_\odot^2/\Delta m_{\text{atm}}^2$, $\sqrt{2} \cdot 100\%$ meaning roughly a factor 3 up or down,

$$\sin^2 2\theta_\odot = (3^{+3}_{-2}) \times 10^{-2} \tag{73}$$

$$\frac{\Delta m_\odot^2}{\Delta m_{\text{atm}}^2} = (6^{+11}_{-4}) \times 10^{-4} \ . \tag{74}$$

These two small numbers both come from the parameter ξ - the VEV in "fundamental units" of one of the 7 Higgs fields in our model - which has already been fitted to the charged fermions in earlier works and which is essentially the Cabibbo angle measuring strange to up-quark weak transitions ($\xi \simeq 0.1$ essentially $\sin\theta_c \simeq 0.22$). But it is also important for the success of our model that there has been room to put in the χ field, with which we could fix the atmospheric mixing angle to be of order unity (by taking $\chi \sim T$), as well as a parameter ϕ_{B-L}, the Higgs field VEV for breaking the gauged $B-L$ charge to fit the overall scale of observed neutrino masses. These factors in front of equations (75), (76) and (77) are results of our rather arbitrary averaging over random order unity factors and inclusion of diagram counting square root factors as put forward in reference [20]. But in principle the factors in front are just of order unity:

$$\sin^2 2\theta_\odot = 3\,\xi^2 \tag{75}$$

$$\sin\theta_c = 1.8\,\xi \tag{76}$$

$$\frac{\Delta m_\odot^2}{\Delta m_{\text{atm}}^2} = 6\,\xi^4 \tag{77}$$

We want to emphasise here that our model - extended Anti-GUT as well as "old" Anti-GUT - is itself a good model in the sense that all coupling constants are order of unity except for Higgs fields VEVs, and thereby also Higgs masses giving rise to these VEVs. In the SM the most remarkable non-natural feature is the tremendously small Weinberg-Salam Higgs VEV compared to Planck or realistic GUT scales. If somehow we have to accept that there must be a mechanism in nature for making the Weinberg-Salam Higgs VEV very small, we also should admit that the other Higgs VEVs could be very small. In our model we manage to interpret the second non-natural feature of most Yukawa couplings, namely, that they are very small to be also due to small Higgs field VEVs. In this way all small numbers come from VEVs in our model; the rest is put to unity in Planck units. In this sense it is "natural" by the fact that it has only one source of small numbers, VEVs. Even the gauge couplings are interpretable as being of order of unity, if we follow our assumption, MPP, which goes extremely well together with the present model.

Concerning this question of the Higgs fields having numerically often small values compared to the scale - in our philosophy the Planck scale - that is the a priori scale in which the fundamental physics is written with numbers of order unity we presented an idea for how these different scales order of magnitudewise could come in: The proposal was that there was a principle working in nature putting the coupling constants to be finetuned to arrange for several degenerate vacuum states. We argued that requiring that, there were a good chance that exponentially small VEV-scales could come out from this just postulated finetuning principle (called MPP = multiple point principle). Such a picture presupposes some dynamics so to speak in the coupling constants much like in baby universe theory.

In this sense we may say that we did not really solve the hierarchy problem in the sense of getting the scale out *without* finetuning, but rather proposed a model for finetuning, namely the MPP. We have earlier claimed some success with such a principle w.r.t. getting finestructure constants and predicting the top-mass to 173 ± 6 GeV, and a true prediction the Higgs mass being 135 ± 9 GeV (so watch out in the future).

For the working of our scheme for getting scales of highly different orders of

TABLE 6. Number of parameters

	"Yukawa"	"Neutrino"	# of parameters	# of predictions
Standard Model	13	4	17	—
"Old" Anti-GUT	4	—*	4†	9
"New" Anti-GUT	6		6	11

* The "old" Anti-GUT cannot predict the neutrino oscillation.
† Here we have not counted the neutrino oscillation parameters.

magnitude it is important to have a fermion like in the case of the Standard model weak to Planck scale the top-quark which couples by rather large - unsuppressed - coupling. Using the speculation based on that we want a similar unsuppressed fermion for the other cases of Higgs fields it is very speculatively called for some heavy fermions at *e.g.* 1/10 under Planck scale mass.

For this purpose a set of 15 suggestive Weyl particles with anomaly cancellation and carrying family $(B - L)$ and weak hypercharges only were proposed, since they were meant to be at this mass scale of course with zero diagonal quantum numbers, *i.e.* with zero total $(B - L)$ and zero total $y/2$.

In this connection we also had some extremely weak pointing out that the number 3 for the number of families had some special significance in these constructions as well as some suggestion that the Standard Model could be considered having not only as far the non-abelian groups are concerned but also w.r.t. to (non-invariant) abelian (sub)groups very small, one could say minimal, representations.

At the end it would be worthwhile to call attention to that our model mass matrices and mixing angles has managed to fit order of magnitudewise about 17 quantities (11 observed fermion masses or mass square differences, 5 mixing angles and CP-violating phase of quarks) with 6 parameters - the Higgs field VEVs. We can find from Table 6 that we have used by the inclusion of the neutrinos two parameters to fit four more quantities, thus gaining two predictions (the solar mixing angle and the ratio of the neutrino oscillation masses).

Also it should be mentioned that this model we really developed from seeking to look at the fermion spectrum in a model that D.L. Bennett and I. Picek with one of us used to predict the fine structure constants using postulates later replaced by the the above mentioned MPP, and that these finestructure constants actually agree with an uncertainty of plus minus 6 in the inverse fine structure constants. When you add to this that this same MPP is promising for some finetuning problems, we can claim that the combined MPP and the (extended) Anti-GUT model provides a fitting of a very large number of the Standard Model parameters!

Note that our model is very successful in describing neutrino oscillations and their mixing angles, but this model does not have any good candidate for dark matter; the monopoles could be such a candidate. We will study this problem in a forthcoming article.

ACKNOWLEDGEMENTS

We would like to thank M. Gibson and S. Lola for useful discussions. Two of us (H.B.N. and Y.T.) wishes to thank W. Buchmüller and T. Yanagida for an important discussion of the see-saw mechanism. Concerning Han-Nambu charges S.E. Rugh and D.L. Bennett are thanked, and L.V. Laperashvili concerning the problem of scales, especially. One of us (Y.T.) thanks the Theory Division of CERN for the hospitality extended to him during visits. H.B.N. wishes to thank the EU commission for grants SCI-0430-C (TSTS), CHRX-CT-94-0621, INTAS-RFBR-

95-0567 and INTAS 93-3316(ext). Y.T. thanks the Scandinavia-Japan Sasakawa foundation for grants No.00-22.

REFERENCES

1. L. O'Raifeartaigh, Group structure of gauge theories, Univ. Pr., 1986. (Cambridge monographs on mathematical physics)
2. C. D. Froggatt, G. Lowe and H. B. Nielsen, Phys. Lett. **B311** (1993) 163; C. D. Froggatt, G. Lowe and H. B. Nielsen, GUTPA-93-1-2.
3. C. D. Froggatt, M. Gibson, H. B. Nielsen and D. J. Smith, Int. J. Mod. Phys. **A13** (1998) 5037; C. D. Froggatt, M. Gibson, H. B. Nielsen and D. J. Smith, proceeding paper of the 29 th International Conference on High-Energy Physics (ICHEP 98), Vancouver, Canada, 23-29 July 1998.
4. S. Rajpoot, Phys. Rev. **D24** (1981) 1890.
5. H. B. Nielsen and N. Brene, NBI-HE-84-47, *Elaborated version of lecture given at Symp. on High Energy Physics, Ahrenshoop, GDR, October 21-26, 1984.*
6. D.L. Bennett, C.D. Froggatt and H.B. Nielsen, in the Proceedings of the APCTP-ICTP Joint International Conference (AIJIC 97) on Recent Developments in Nonperturbative Quantum Field Theory, Seoul, Korea, 26-30 May 1997; D.L. Bennett, Ph.D. thesis, Niels Bohr Institute, 1996; hep–ph/9607341.
7. D.L. Bennett and H.B. Nielsen, Int. J. Mod. Phys. **A14** (1999) 3313.
8. C.D. Froggatt, M. Gibson, H.B. Nielsen and D.J. Smith, Int. J. Mod. Phys. **A13** (1998) 5037.
9. C.D. Froggatt, M. Gibson and H.B. Nielsen, Phys. Lett. **B446** (1999) 256.
10. C.D. Froggatt and H.B. Nielsen, Nucl. Phys. **B147** (1979) 277.
11. D.L. Bennett, C.D. Froggatt and H.B. Nielsen in Proceedings of the 27th International Conference on High Energy Physics (Glasgow, 1994), eds. P. Bussey and I. Knowles, (IOP Publishing Ltd, 1995) p. 557; *Perspectives in Particle Physics '94*, eds. D. Klabučar, I. Picek and D. Tadić, (World Scientific, 1995) p. 255; hep–ph/9504294.
12. C.D. Froggatt and H.B. Nielsen, Phys. Lett. **B368** (1996) 96.
13. C.D. Froggatt, H.B. Nielsen and D.J. Smith, Phys. Lett. **B385** (1996) 150.
14. C.D. Froggatt and H.B. Nielsen, Nucl. Phys. **B164** (1979) 114.
15. M.B. Green and J. Schwarz, Phys. Lett. **B149** (1984) 117.
16. C.D. Froggatt and H.B. Nielsen, in Proceedings of the International Symposium on Lepton and Baryon Number Violation, Trento, Italy, 20-25 April 1998; hep–ph/9810388.
17. H. B. Nielsen and Y. Takanishi, to be published in Nucl. Phys. B.; hep–ph/0004137.
18. B. Stech, Talk at 23rd Johns Hopkins Workshop on Current Problems in Particle Theory: Neutrinos in the Next Millennium, Baltimore, MD, 10-12 June 1999; hep–ph/9909268.
19. T. Yanagida, in Proceedings of the Workshop on Unified Theories and Baryon Number in the Universe, Tsukuba, Japan (1979), eds. O. Sawada and A. Sugamoto, KEK Report No. 79-18; M. Gell-Mann, P. Ramond and R. Slansky in Supergravity, Pro-

ceedings of the Workshop at Stony Brook, NY (1979), eds. P. van Nieuwenhuizen and D. Freedman (North-Holland, Amsterdam, 1979).
20. C.D. Froggatt, H.B. Nielsen and D.J. Smith, in progress.
21. L.V. Laperashvili and H.B. Nielsen, "Phase Transition in Higgs Model of Scalar Fields with Electric and Magnetic Charges " in progress.
22. C. D. Froggatt and H. B. Nielsen, Phys. Lett. **B368** (1996) 96.
23. M. Veltman, Acta Phys. Polon. **B12** (1981) 437; B. Stech, in Proceedings of Workshop on What Comes Beyond the Standard Model, Bled, Slovenia, 29 June - 9 July 1998.
24. M. Y. Han and Y. Nambu, Phys. Rev. **139** (1965) B1006.

Neutrino Theories[1]

P. Ramond

Institute for Fundamental Theory
Physics Department, University of Florida
Gainesville, Fl 32611

Abstract. After a short historical review, the present experimental situation is reviewed and several theoretical schemes are discussed, including a general description of Kaluza-Klein sterile neutrino towers

I HISTORICAL INTRODUCTION

In neutrino physics, it is very difficult, even for the great ones, to get it completely right. On the experimental side, it took more than fifteen years, from the initial measurements of a discrete β-electron spectrum [1] to the definite experiment [2,3] which established a continuous spectrum, creating a difficult and profound conceptual problem (Debye said this was something he did not want to think about, like the new taxes!).

In December 1930, W. Pauli tries to solve two problems with one new particle: the continuous β spectrum and the wrong symmetry of some nuclear wave functions. In a letter that starts with nerdy panache, "*Dear Radioactive Ladies and Gentlemen...*", W. Pauli puts forward a "*desperate*" way out: there exists a neutral particle which lives in the nucleus, and is ejected in the β process. It is bound to the proton through its magnetic moment, and therefore has a mass. He calls it the neutron, and goes on to say: "*I don't feel secure enough to publish anything about this idea,..., but... only those who wager can win...*". This is a telling remark of the science culture of the time: introducing a new particle was almost a capitulation, like saying one cannot explain things with the tools at hand; it also shows the hesitancy to publish, a charming and quaint feeling which has long since disappeared in our field! It is to be noted that this was also a very difficult time in Pauli's life, having lost his mother a year or so before, and having just divorced his first wife, actress Kate Deppner, five days before he wrote the letter! A year later, he was in analysis with Carl Jung!

---

[1] This research was supported in part by the department of energy under grant DE-FG02-97ER41029.

Two years later, Chadwick discovers a heavy nuclear particle, exactly Pauli's neutron, but as heavy as a proton, and not the particle produced in β decay. It was Fermi, who in his theory of the four-fields (not yet named, I assume, four-fermion) interaction, posits that Pauli's particle exists on its own, and is produced only in the dacay. He calls it the neutrino, a mercifully short name which remains today.

Pauli's instinct was right, but he tried to explain two riddles with one particle! This is also refreshing when comparing with today's physics where one riddle is usually explained by many particles, or even a string of them!

After such a spectacular beginning, the neutrinos were to play crucial roles in our understanding of the fundamental interactions. In 1945, B. Pontecorvo [4] proposes the unthinkable, a way to detect neutrinos: an electron neutrino transforms a ^{37}Cl atom to the inert radioactive gas ^{37}Ar, which can be stored and then detected through radioactive decay. Pontecorvo did not publish the report, perhaps because of the times, or because Fermi thought the idea ingenious but not immediately relevant.

It is only in 1956 that Cowan and Reines [5] discover electron antineutrinos through the reaction $\bar{\nu}_e + p \to e^+ + n$. The positron is annihilated almost immediately and yields two photons, while the neutron is absorbed a bit later (using the large Cd cross-section), releasing three photons, producing a unique signature. Cowan passed away before 1995, the year Fred Reines was awarded the Nobel Prize for their discovery. In neutrino physics, both patience and longevity are required: it took 26 years from birth to detection and then another 39 for the Nobel Committee to recognize the achievement! Like owners of fine wineries, neutrino physicists should train their children, perhaps as early as in the womb, to follow their footsteps.

Pontecorvo [6] introduced the idea of antineutrino-neutrino oscillations in 1956, motivated by the rumor that Davis [7] had found evidence for neutrinos coming from a pile. The rumor went away, but the idea of neutrino oscillations was born; it has remained with us ever since.

Progress in neutrino physics is slow: the first helicity measurement was in 1958 by M. Goldhaber [8], but it took 40 more years for experimentalists to produce convincing evidence for the existence of neutrino masses [9]. The second neutrino, the muon neutrino is detected [10] in 1962, and that same year, Maki, Nakagawa and Sakata [11] introduce two crucial ideas: flavor mixing and flavor oscillations, which are possible only between neutrino flavors of different masses.

In 1968, Davis et al, using a 100,000 gallon tank of cleaning fluid deep underground at the Homestake mine, reported [12] a deficit in the solar ^8B neutrino flux, a result that stands to this day as a truly remarkable experimental *tour de force*. Shortly after, Gribov and Pontecorvo [13] interpreted the deficit as evidence for neutrino oscillations.

In the early 1970's, with the idea of quark-lepton symmetries [14,15] suggests that the proton could be unstable. This brings about the construction of underground detectors, large enough to monitor many protons, and instrumentalized to detect the Čerenkov light emitted by the products of proton decay. By the middle 1980's, several such detectors are in place. They fail to detect proton decay, but in a remarkable serendipitous turn of events, 150,000 years earlier, a supernova erupted in the large Magellanic Cloud, and in 1987, its burst of neutrinos was detected in these detectors! All of a sudden, proton decay detectors turn their attention to neutrinos. Today, these detectors have shown great success in measuring the effects of solar and atmospheric neutrinos. They continue their unheralded watch for signs of proton decay, reassured in the knowledge that lepton number and baryon number violations are connected in most theories, leading to correlations between neutrino masses and proton decay rates. Proton decay, in the mind of most theorists, is just around the corner!

II STANDARD MODEL NEUTRINOS

The standard model of electro-weak and strong interactions contains three left-handed neutrinos, represented by two-components Weyl spinors, ν_i, $i = e, \mu, \tau$, each describing a left-handed fermion (right-handed antifermion). As the upper components of weak isodoublets L_i, they have $I_{3W} = 1/2$, and a unit of the global ith lepton number.

These standard model neutrinos are strictly massless. The only Lorentz scalar made out of these neutrinos is the Majorana mass, of the form $\nu_i^t \nu_j$; it has the quantum numbers of a weak isotriplet, with third component $I_{3W} = 1$, and two units of total lepton number. In the absence of Higgs isotriplet in the Standard Model, there are no tree-level neutrino masses.

Quantum corrections, however, are not limited to renormalizable couplings, and it is easy to make a weak isotriplet out of two isodoublets, yielding the $SU(2) \times U(1)$ invariant

$$\frac{1}{M} L_i^t \vec{\tau} L_j \cdot H^t \vec{\tau} H \ ,$$

where H is the Higgs doublet, and M is an unknown mass scale. But this term is not invariant under lepton number, it is not be generated in perturbation theory, leading to the important conclusion: *The standard model neutrinos are kept massless by global chiral lepton number symmetry.* The detection of neutrino masses is therefore *a tangible indication of physics beyond the standard model.* It also means that observation of a neutrino Majorana masses will give important indications as to the value of M, indicating perhaps the next scale for physics beyond the Standard Model.

III EXPERIMENTAL ISSUES

From the solar neutrino deficit to the spectacular result from SuperKamiokande, experiments suggest neutrino masses, providing the first credible evidence for physics beyond the standard model. Still, there remains several burning issues in neutrino physics that can be settled by future experiments:

- The origin of the Solar Neutrino Deficit

  This is currently being addressed by SuperK, in their measurements of the shape of the ^8B spectrum, of day-night asymmetry and of the seasonal variation of the neutrino flux. Their reach is now being improved by lowering the threshold energy.

  SNO, joining the hunt, is expected to provide a more accurate measurement of the Boron flux. Its *raison d'être*, however, is the ability to measure neutral current interactions. Barring sterile neutrinos, we might have a flavor-independent measurement of the solar neutrino flux, while measuring at the same time the electron neutrino flux!

  KamLand, which uses the flux from many Japanese nuclear reactors, will monitor the MSW effect for solar neutrinos if the mixing angle is large.

  This experiment will be joined by BOREXINO, designed to measure neutrinos from the ^7Be capture. These neutrinos are suppressed in the MSW solutions, which could explain the results from the $p-p$ solar neutrino experiments and those that measure the Boron neutrinos.

- Atmospheric Neutrinos

  Here, there are several long baseline experiments to monitor muon neutrino beams and corroborate the SuperK results. The first, called K2K, already in progress, sends a beam from KEK to SuperK. Another, called MINOS, will monitor a FermiLab neutrino beam at the Soudan mine, 730 km away. A third experiment will send a CERN beam towards the Gran Sasso laboratory (also about 730 km away!). Eventually, these experiments hope to detect the appearance of a τ neutrino.

It is also possible that intense neutrino beams, originating in a neutrino factory [16], through the earth will prove valuable tools in mapping out neutrino masses and mixings, even possibly CP violation.

This brief survey of upcoming experiments in neutrino physics is intended to give a flavor of things to come. These experiments will not only measure neutrino parameters (masses and mixing angles), but will help us answer fundamental questions about the nature of neutrinos, and someday, this new generation of neutrino detectors may detect proton decay, thus realizing the kinship between leptons and quarks.

IV NEUTRINO MASSES

Neutrinos must be extraordinarily light: experiments indicate $m_{\nu_e} < 10$ eV, $m_{\nu_\mu} < 170$ keV, $m_{\nu_\tau} < 18$ MeV [17], and any model of neutrino masses must explain these small values.

Neutrinos masses are easily generated by supplementing each neutrino with its electroweak-singlet Dirac partner, $\overline{N}_i$. These appear naturally in the Grand Unified group $SO(10)$ where they complete each family into its spinor representation. Neutrino Dirac masses stem from the couplings $L_i \overline{N}_j H$ after electroweak breaking. Unless there are extraordinary suppressions, these couplings generate masses that are way too big, of the same order of magnitude as the masses of the charged elementary particles $m \sim \Delta I_w = 1/2$.

Based on recent ideas from string theory, it has been proposed [18] that the world of four dimensions is in fact a "brane" immersed in a higher dimensional space. In this view, all fields with electroweak quantum numbers live on the brane, while standard model-singlet fields can live in the "bulk" as well. One such field is the graviton, others could be the right-handed neutrinos. Their couplings to the brane are reduced by geometrical factors, and the smallness of neutrino masses is due to the naturally small coupling between brane and bulk fields. At present, no natural mechanism for small neutrino masses emerges from this picture, unless of course dictated in hitherto unknown way by the physics in higher dimensions.

There is already one indication of a large mass scale, at which the gauge couplings unify. The value of neutrino masses implies another scale, which happens to be commensurate with the former!

To see this, we introduce Majorana mass terms $\overline{N}_i \overline{N}_j$ for the right-handed neutrinos. The masses of these new degrees of freedom are arbitrary, as they have no electroweak quantum numbers, $M \sim \Delta I_w = 0$. If much larger than the electroweak scale, the neutrino masses are suppressed relative to that of their charged counterparts by the ratio of the electroweak scale to that new scale: the mass matrix (in 3×3 block form) is

$$\begin{pmatrix} 0 & m \\ m & M \end{pmatrix}, \tag{1}$$

leading, for each family, to one small and one large eigenvalue

$$m_\nu \sim m \cdot \frac{m}{M} \sim \left(\Delta I_w = \frac{1}{2} \right) \cdot \left(\frac{\Delta I_w = \frac{1}{2}}{\Delta I_w = 0} \right). \tag{2}$$

This seesaw mechanism [19] provides a natural explanation for small neutrino masses as long as lepton number is broken at a large scale M. With M around the energy at which the gauge couplings unify, this yields neutrino masses at or below tenths of eVs, consistent with the SuperK results.

The 6×6 seesaw Majorana matrix is written in 3×3 block form

$$\mathcal{M} = \mathcal{V}_\nu^t \, \mathcal{D} \mathcal{V}_\nu \sim \begin{pmatrix} \mathcal{U}_{\nu\nu} & \epsilon \mathcal{U}_{\nu N} \\ \epsilon \mathcal{U}_{N\nu}^t & \mathcal{U}_{NN} \end{pmatrix}, \tag{3}$$

where ϵ is the tiny ratio of the electroweak to lepton number violating scales, and $\mathcal{D} = \mathrm{diag}(\epsilon^2 \mathcal{D}_\nu, \mathcal{D}_N)$, is a diagonal matrix. $\mathcal{D}_\nu$ contains the three neutrino masses, and ϵ^2 is the seesaw suppression. The weak charged current is then given by

$$j_\mu^+ = e_i^\dagger \sigma_\mu \mathcal{U}_{MNS}^{ij} \nu_j , \qquad (4)$$

where

$$\mathcal{U}_{MNS} = \mathcal{U}_e \mathcal{U}_\nu^\dagger , \qquad (5)$$

is the Maki-Nakagawa-Sakata [11] (MNS) flavor mixing matrix, the analog of the CKM matrix in the quark sector, written in terms of $\mathcal{U}_e$, the unitary matrix that rotates the lepton doublets L_i, and $\mathcal{U}_\nu$, the matrix that diagonalizes the Majorana mass 3×3 submatrix. It contains three rotation angles, and one CP-violating phase, all new parameters of the Standard Model which await measurement. In the seesaw scenario, it also contains two additional CP-violating phases which cannot be absorbed in a redefinition of the neutrino fields, because of their Majorana masses (these extra phases can be measured only in $\Delta \mathcal{L} = 2$ processes).

Theoretical predictions of lepton hierarchies and mixings depend very much on hitherto untested theoretical assumptions about Yukawa couplings. One can summarize the issues in terms of questions:

- Do the right handed neutrinos have quantum numbers beyond the standard model?

- Are quarks and leptons related by grand unified theories?

- Are quarks and leptons related by anomalies?

- Are there family symmetries for quarks and leptons?

The measured neutrino mass difference at SuperK (barring any fortuitous degeneracies), then suggests a mass for the right-handed neutrinos that is consistent with the scale at which the gauge couplings unify. Is this just a numerical coincidence, or should we view this as a hint for grand unification?

Grand unified theories imply symmetries much larger than the standard model's, necessitating a desert and supersymmetry, but also a carefully designed set of Higgs particles to achieve the desired symmetry breaking. Fortunately, we have learned [20] in string theory that the best features of grand-unified theories can be preserved, because much of the symmetry-breaking is achieved by geometric compactification from higher dimensions.

Since the vanishing of chiral anomalies also leads (mostly) to anomaly-free groups, this author believes that anomaly cancellation is more fundamental than group structure, and one should only retain this feature of grand-unified theories.

A grand-unified theory based on $SO(10)$, proposed in 1982 [21,22], *predicted* large $\nu_\mu - \nu_\tau$ mixing, albeit with a small value for the top quark, and also small $\nu_e - \nu_\mu$ mixings. Here, we will outline a more modern approach [23] to neutrino masses which predicted no Cabibbo-suppression for $\nu_\mu - \nu_\tau$ mixing.

A Theory of Cabibbo Suppression

Quark mass and mixing hierarchies can be parametrized in terms of powers of the Cabibbo angle. In this section, we present a theoretical framework which explains the Cabibbo exponents in terms of extra Abelian flavor symmetries, following a suggestion of Froggatt and Nielsen [24]. Not as predictive as specific grand unified models, such a simple model nevertheless yields [23] no Cabibbo suppression between the muon and tau neutrinos.

Suppression of some Yukawa couplings is explained by the fact that symmetries can forbid tree-level couplings of some standard model-invariant operator, such as $\mathbf{Q}_i \bar{\mathbf{d}}_j H_d$, if there are additional symmetries under which the operator is not invariant. Simplest is to assume an Abelian symmetry, with an electroweak singlet field θ, as its order parameter. Then the interaction

$$\mathbf{Q}_i \bar{\mathbf{d}}_j H_d \left(\frac{\theta}{M}\right)^{n_{ij}} \tag{6}$$

can appear in the potential as long as the family charges balance under the new symmetry. As θ acquires a *vev*, this leads to a suppression of the Yukawa couplings of the order of $\lambda^{n_{ij}}$ for each matrix element, with $\lambda = \theta/M$ identified with the Cabibbo angle, and M is the natural cut-off of the effective low energy theory. As a consequence of the charge balance equation

$$X_{if}^{[d]} + n_{ij} X_\theta = 0 , \tag{7}$$

the exponents of the suppression are related to the charge of the standard model-invariant operator, the sum of the charges of the fields that make up the the invariant.

Each charged lepton Yukawa coupling $L_i \overline{N}_j H_u$, has an extra charge $X_{L_i} + X_{N_j} + X_H$, which gives the Cabibbo suppression of the ij matrix element. Hence, the orders of magnitude of these couplings can be expressed as

$$\begin{pmatrix} \lambda^{l_1} & 0 & 0 \\ 0 & \lambda^{l_2} & 0 \\ 0 & 0 & \lambda^{l_3} \end{pmatrix} \hat{Y} \begin{pmatrix} \lambda^{p_1} & 0 & 0 \\ 0 & \lambda^{p_2} & 0 \\ 0 & 0 & \lambda^{p_3} \end{pmatrix} , \tag{8}$$

where $\hat{Y}$ is a Yukawa matrix with no Cabibbo suppressions, $l_i = X_{L_i}/X_\theta$ are the charges of the left-handed doublets, and $p_i = X_{N_i}/X_\theta$, those of the singlets. The first matrix forms half of the MNS matrix. Similarly, the mass matrix for the right-handed neutrinos, $\overline{N}_i \overline{N}_j$ will be written in the form

$$\begin{pmatrix} \lambda^{p_1} & 0 & 0 \\ 0 & \lambda^{p_2} & 0 \\ 0 & 0 & \lambda^{p_3} \end{pmatrix} \mathcal{M} \begin{pmatrix} \lambda^{p_1} & 0 & 0 \\ 0 & \lambda^{p_2} & 0 \\ 0 & 0 & \lambda^{p_3} \end{pmatrix} . \tag{9}$$

The diagonalization of the seesaw matrix is of the form

$$L_i H_u \overline{N}_j \left(\frac{1}{\overline{N}\,\overline{N}}\right)_{jk} \overline{N}_k H_u L_l , \qquad (10)$$

from which the Cabibbo suppression matrix from the $\overline{N}_i$ fields *cancels*, leaving us with

$$\begin{pmatrix} \lambda^{l_1} & 0 & 0 \\ 0 & \lambda^{l_2} & 0 \\ 0 & 0 & \lambda^{l_3} \end{pmatrix} \hat{\mathcal{M}} \begin{pmatrix} \lambda^{l_1} & 0 & 0 \\ 0 & \lambda^{l_2} & 0 \\ 0 & 0 & \lambda^{l_3} \end{pmatrix} , \qquad (11)$$

where $\hat{\mathcal{M}}$ is a matrix with no Cabibbo suppressions. The Cabibbo structure of the seesaw neutrino matrix is determined solely by the charges of the lepton doublets! The Cabibbo structure of the MNS mixing matrix is also due entirely to the charges of the three lepton doublets. This general conclusion depends on the existence of at least one Abelian family symmetry, which we now argue is implied by the observed structure in the quark sector.

The Wolfenstein parametrization of the CKM matrix [25],

$$\begin{pmatrix} 1 & \lambda & \lambda^3 \\ \lambda & 1 & \lambda^2 \\ \lambda^3 & \lambda^2 & 1 \end{pmatrix} , \qquad (12)$$

and the Cabibbo structure of the quark mass ratios

$$\frac{m_u}{m_t} \sim \lambda^8 \quad \frac{m_c}{m_t} \sim \lambda^4 \; ; \quad \frac{m_d}{m_b} \sim \lambda^4 \quad \frac{m_s}{m_b} \sim \lambda^2 , \qquad (13)$$

can be reproduced [23,26] by a simple *family-traceless* charge assignment for the three quark families, namely

$$X_{\mathbf{Q},\overline{\mathbf{u}},\overline{\mathbf{d}}} = \mathcal{B}(2,-1,-1) + \eta_{\mathbf{Q},\overline{\mathbf{u}},\overline{\mathbf{d}}}(1,0,-1) , \qquad (14)$$

where $\mathcal{B}$ is baryon number, $\eta_{\overline{\mathbf{d}}} = 0$, and $\eta_{\mathbf{Q}} = \eta_{\overline{\mathbf{u}}} = 2$. Two striking facts are evident:

- the charges of the down quarks, $\overline{\mathbf{d}}$, associated with the second and third families are the same,

- $\mathbf{Q}$ and $\overline{\mathbf{u}}$ have the same value for η.

Assume these family-traceless charges are gauged, and not anomalous. To cancel anomalies, the leptons must themselves have family charges.

Anomaly cancellation generically implies group structure. In $SO(10)$, baryon number generalizes to $\mathcal{B} - \mathcal{L}$, where $\mathcal{L}$ is total lepton number, and in $SU(5)$ the fermion assignment is $\overline{\mathbf{5}} = \overline{\mathbf{d}} + L$, and $\mathbf{10} = \mathbf{Q} + \overline{\mathbf{u}} + \overline{e}$, and anomaly cancellation is easily achieved by assigning $\eta = 0$ to the lepton doublet L_i, and $\eta = 2$ to the electron singlet $\overline{e}_i$, and by generalizing baryon number to $\mathcal{B} - \mathcal{L}$, leading to the charges

$$X_{\mathbf{Q},\bar{u},\bar{d},L,\bar{e}} = (\mathcal{B} - \mathcal{L})(2, -1, -1) + \eta_{\mathbf{Q},\bar{u},\bar{d}}(1, 0, -1) , \tag{15}$$

where now $\eta_{\bar{d}} = \eta_L = 0$, and $\eta_\mathbf{Q} = \eta_{\bar{u}} = \eta_{\bar{e}} = 2$. The charges of the lepton doublets are $X_{L_i} = -(2, -1, -1)$, fixing the Cabibbo structure of the MNS lepton mixing matrix

$$\mathcal{U}_{MNS} \sim \begin{pmatrix} 1 & \lambda^3 & \lambda^3 \\ \lambda^3 & 1 & 1 \\ \lambda^3 & 1 & 1 \end{pmatrix} , \tag{16}$$

implying *no Cabibbo suppression in the mixing between ν_μ and ν_τ*. This is consistent with the SuperK discovery and with the small angle MSW [27] solution to the solar neutrino deficit. One also obtains a much lighter electron neutrino, and Cabibbo-comparable masses for the muon and tau neutrinos.

The scale of the neutrino mass values depend on the family trace of the family charge(s). Here we simply quote the results our model [23]. The masses of the right-handed neutrinos are of the orders of magnitude

$$m_{\overline{N}_e} \sim M\lambda^{13} ; \qquad m_{\overline{N}_\mu} \sim m_{\overline{N}_\tau} \sim M\lambda^7 , \tag{17}$$

where M is the scale of the right-handed neutrino mass terms, assumed to be the cut-off. The seesaw mass matrix for the three light neutrinos yields

$$m_0 \begin{pmatrix} a\lambda^6 & b\lambda^3 & c\lambda^3 \\ b\lambda^3 & d & e \\ c\lambda^3 & e & f \end{pmatrix} , \tag{18}$$

where we have added for future reference the prefactors a, b, c, d, e, f, all of order one, and

$$m_0 = \frac{v_u^2}{M\lambda^3} , \tag{19}$$

where v_u is the *vev* of the Higgs doublet. This matrix has one light eigenvalue

$$m_{\nu_e} \sim m_0 \lambda^6 . \tag{20}$$

Without a detailed analysis of the prefactors, the masses of the other two neutrinos come out to be both of order m_0. The mass difference announced by superK [9] cannot be reproduced without going beyond the model, by taking into account the prefactors. The two heavier mass eigenstates and their mixing angle are written in terms of

$$x = \frac{df - e^2}{(d+f)^2} , \qquad y = \frac{d-f}{d+f} , \tag{21}$$

as

$$\frac{m_{\nu_2}}{m_{\nu_3}} = \frac{1 - \sqrt{1-4x}}{1 + \sqrt{1-4x}}, \quad \sin^2 2\theta_{\mu\tau} = 1 - \frac{y^2}{1-4x}. \tag{22}$$

If $4x \sim 1$, the two heaviest neutrinos are nearly degenerate. If $4x \ll 1$, a condition easy to achieve if d and f have the same sign, we can obtain an adequate split between the two mass eigenstates. For illustrative purposes, when $0.03 < x < 0.15$, we find

$$4.4 \times 10^{-6} \leq \Delta m^2_{\nu_e - \nu_\mu} \leq 10^{-5} \text{ eV}^2, \tag{23}$$

which yields the correct non-adiabatic MSW [27] effect, and

$$5 \times 10^{-4} \leq \Delta m^2_{\nu_\mu - \nu_\tau} \leq 5 \times 10^{-3} \text{ eV}^2, \tag{24}$$

for the atmospheric neutrino effect. These were calculated with a cut-off, 10^{16} GeV $< M < 4 \times 10^{17}$ GeV, and a mixing angle, $0.9 < \sin^2 2\theta_{\mu-\tau} < 1$. This value of the cut-off is compatible not only with the data but also with the gauge coupling unification scale, a necessary condition for the consistency of our model, and more generally for the basic ideas of grand unification.

This model is highly falsifiable by upcoming data from the neutrino detectors, as it is consistent with the SMA solar neutrino solutions and contains no sterile neutrinos.

B Brane Neutrinos

Recently, there has been a flurry of theoretical activity promoting the idea that the observed world of three space dimensions is to be viewed as immersed in a larger higher-dimensional world. Two groups have shown how small neutrino masses could be generated in this framework [28,29]. Below we present a general kinematical analysis [30] of the problem, showing that all possibilities depend on unknown bulk-physics.

Consider the simplest case where we live in 3-surface embedded in a four-dimensional space. The relevant invariance group is the Lorentz group in five dimensions $SO(4,1)$, which has only one spinor representation, a complex **4**. The fermion kinetic term is given by

$$S = \int d^4x \int dy \, \overline{\Psi} \gamma^a \partial_A \Psi,$$

where Ψ is a complex four spinor, with dimension 2. It can be written in terms of two Weyl spinors

$$\Psi = \begin{pmatrix} \eta \\ \overline{\xi} \end{pmatrix}(x, y),$$

where both η and $\overline{\xi}$ are left-handed Weyl spinors $\sim (2, 1)$ of the usual Lorentz group, and the indices A, B take on five values.

This kinetic term is invariant under the three global symmetries:

- Lepton number: η with $L = 1$; ξ with $L = -1$.
- Discrete P_5: $y \to -y$, $x \to x$ and $\eta \to \overline{\xi}$, $\xi \to -\overline{\eta}$.
- Discrete Π_5: $y \to -y$, $x \to x$ and $\eta \to \pm\eta$, $\xi \to \mp\xi$.

These symmetries enable us to catalog the possible four-dimensional Lorentz-invariant mass terms. We have four different invariant mass terms:

- Dirac Mass $\Psi^\dagger \gamma_0 \Psi \equiv \overline{\Psi}\Psi = \eta^\dagger \sigma_2 \xi^* + \xi^T \sigma_2 \eta$.
- Majorana Mass: $\Psi^T \sigma_2 \Psi \equiv \overline{\Psi^c}\Psi = \eta^T \sigma_2 \eta - \xi^\dagger \sigma_2 \xi^*$.
- Dirac Vector: $\overline{\Psi}\gamma^5 \Psi = -i\xi^T \sigma_2 \eta + i\eta^\dagger \sigma_2 \xi^*$.
- Majorana Vector: $\overline{\Psi^c}\gamma_5 \Psi = -(i\eta^T \sigma_2 \eta + i\xi^\dagger \sigma_2 \xi^*)$.

The last two, fifth components of five-vectors, can serve as invariant mass terms in the lower-dimensional theory. All of these are distinguished by the three symmetries. Both invariant Dirac mass term and Dirac vector are real expressions. The Majorana vector is not generated by this kinetic term, and we consider it no further.

It follows that bulk physics can generate three different mass terms on the brane, producing very different phenomenologies. With a larger bulk, there are more spinors and more possible mass terms.

To see how possible mass terms are generated, we need to expand the fields in their Kaluza-Klein modes. These in turn depend on the boundary conditions. Either we compactify over the circle (known to sophisticates as toroidal compactification), in which case

$$\eta(x, y + 2\pi R) = \eta(x, y) \qquad \xi(x, y + 2\pi R) = \xi(x, y) ,$$

or over orbifolds, for which

$$\eta(x, y + 2\pi R) = \eta(x, y) \qquad \xi(x, y + 2\pi R) = -\xi(x, y) .$$

The Dirac kinetic term generates Lorentz-invariant contributions coming from the derivative along y. We can expand the fields as

$$\eta(x, y) = c_0 \eta_0(x) + \sum_n [\eta_n^c(x) c_n(y) + \eta_n^s(x) s_n(y)] ,$$

$$\xi(x, y) = c_0 \xi_0(x) + \sum_n [\xi_n^c(x) c_n(y) + \xi_n^s(x) s_n(y)] ,$$

with the expansion coefficients satisfying

$$< c_n \, c_m > = < s_n \, s_m > = \delta_{nm} ; \qquad < c_n \, s_m > = 0 ,$$

where the $< ... >$ denote integration over y. The orbifold case is more restrictive as $\xi_0 = \xi_n^s = 0$. The four-dimensional Lagrangian contains the terms

$$\sum_n \frac{n}{R} [\xi_n^{cT} \sigma_2 \eta_n^s - \xi_n^{sT} \sigma_2 \eta_n^c \eta_n^{ct} \sigma_2 \xi_n^{s*} + \eta_n^{st} \sigma_2 \xi_n^{c*}] \ .$$

A gauged Dirac kinetic term could generate the term

$$-ig A_5 [\xi_0^T \sigma_2 \eta_0 + \sum_n \{\xi_n^{cT} \sigma_2 \eta_n^c + \xi_n^{sT} \sigma_2 \eta_n^s\} - \eta_0^\dagger \sigma_2 \xi_0^* - \sum_n \{\eta_n^{c\dagger} \sigma_2 \xi_n^{c*} - \eta_n^{s\dagger} \sigma_2 \xi_n^{s*}\}] \ .$$

If A_5 acquire a Lorentz-invariant vacuum value, this contributes to the fermion mass terms. The Majorana mass term yields

$$\eta_0^T \sigma_2 \eta_0 - \xi_0^\dagger \sigma_2 \xi_0^* + \sum_n [\eta_n^{cT} \sigma_2 \eta_n^c - \xi_n^{c\dagger} \sigma_2 \xi_n^{c*} + \eta_n^{sT} \sigma_2 \eta_n^s - \xi_n^{s\dagger} \sigma_2 \xi_n^{s*}] \ ,$$

and the Dirac mass term gives

$$\eta_0^\dagger \sigma_2 \xi_0^* + \xi_0^T \sigma_2 \eta_0 + \sum_n [\eta_n^{c\dagger} \sigma_2 \xi_n^{c*} + \xi_n^{cT} \sigma_2 \eta_n^c + \eta_n^{s\dagger} \sigma_2 \xi_n^{s*} + \xi_n^{sT} \sigma_2 \eta_n^s] \ .$$

Call M be the Majorana mass, M_D the Dirac mass, and define

$$D \equiv -iM_D - gA_5 \ , \qquad K_n \equiv -i\frac{n}{R} \ ,$$

$$\begin{pmatrix} M & D & \cdot & 0 & 0 & 0 & 0 & \cdot \\ D & M & \cdot & 0 & 0 & 0 & 0 & \cdot \\ \cdot & \cdot & \cdot & \cdot & \cdot & \cdot & \cdot & \cdot \\ 0 & 0 & \cdot & M+K_n & D & 0 & 0 & \cdot \\ 0 & 0 & \cdot & D & M-K_n & 0 & 0 & \cdot \\ 0 & 0 & \cdot & 0 & 0 & M+K_n & D & \cdot \\ 0 & 0 & \cdot & 0 & 0 & D & M-K_n & \cdot \\ \cdot & \cdot & \cdot & \cdot & \cdot & \cdot & \cdot & \cdot \end{pmatrix} \ .$$

Its eigenvalues are then

$$M_0^\pm = \sqrt{(M \pm gA_5)^2 + M_D^2} \ ,$$

for the lowest modes, and for each block, two copies of

$$M_n^\pm = \sqrt{([M^2 + \frac{n^2}{R^2}]^{1/2} \pm gA_5)^2 + M_D^2} \ , \qquad n = 1, 2 \dots \ .$$

This mass spectrum will be changed by the coupling to the brane fields.

The most naive brane-bulk coupling is that between the standard model invariant $L_i H_u$ and the five-dimensional fermion field Ψ (or its conjugate), evaluated at the brane located at $y = 0$. This yields the interaction term

$$\frac{h_i}{\sqrt{M_s}} \eta^T(x,0) \sigma_2 L_i(x) H(x) + \frac{y_i}{\sqrt{M_s}} \xi^T(x,0) \sigma_2 L_i(x) H(x) \ ,$$

where L_i are the three lepton doublets, H is the standard model Higgs field, and M_s is the some scale, identified with the string scale in string theory compactifications. Note that with both couplings, the lepton number symmetry of the five-dimensional Dirac kinetic term is explicitly broken. Thus if we gauge the five-dimensional theory, it is not broken when A_5 gets a vacuum value, and we should be careful not to take both of the above couplings. Expanding into components, we obtain extra mass terms of the form

$$\nu_i^T \sigma_2 (\mu_1^i (\eta_0 + \sum_n \eta_n^c) + \mu_2^i (\xi_0 + \sum_n \xi_n^c)) ,$$

with the complex parameters μ_1^i, μ_2^i of the order of the electroweak breaking scale. Ignoring for the moment the family indices, we get the new Majorana mass matrix

$$\mathcal{M} = \begin{pmatrix} 0 & \mathbf{v}^T \\ \mathbf{v} & \mathbf{D} \end{pmatrix} ,$$

where the brane-bulk couplings are given by the infinite-dimensional vector $\mathbf{v} \equiv (\alpha_0, \beta_0, \cdots, \alpha_n, \beta_n, \gamma_n, \delta_n, \cdots)$, and $\mathbf{D}$ is the diagonal matrix $(M_0^+, M_0^-, \ldots, M_n^+, M_n^-, M_n^+, M_n^-, \ldots)$. The eigenvalues and eigenvectors of the physical fields are obtained by forming the hermitian matrix $\mathcal{M}\mathcal{M}^\dagger$,

$$\mathcal{M}\mathcal{M}^\dagger = \begin{pmatrix} v^2 & \mathbf{v}^T \mathbf{D} \\ \mathbf{D} \mathbf{v}^* & \mathbf{v} \mathbf{v}^\dagger + \mathbf{D}^2 \end{pmatrix} ,$$

where $v^2 \equiv \mathbf{v}^\dagger \mathbf{v}$. The eigenvalue equations

$$\begin{pmatrix} v^2 & \mathbf{v}^T \mathbf{D} \\ \mathbf{D} \mathbf{v}^* & \mathbf{v}\mathbf{v}^\dagger + \mathbf{D}^2 \end{pmatrix} \begin{pmatrix} \psi_L \\ \Psi_T \end{pmatrix} = x \begin{pmatrix} \psi_L \\ \Psi_T \end{pmatrix} ,$$

are most easily solved in perturbation theory when $v << D$. In that case, we find one small eigenvalue, x_0 which vanishes when $\mathbf{v}$ is set to zero, for which the characteristic equation yields

$$x_0 \approx |\mathbf{v}^T \mathbf{D}^{-1} \mathbf{v}|^2$$

with the corresponding eigenvector

$$\psi_L^{[0]} \approx \frac{1}{\sqrt{1 + \mathbf{v}^\dagger \mathbf{D}^{-2} \mathbf{v}}} , \quad \Psi_T^{[0]} \approx \frac{1}{\sqrt{1 + \mathbf{v}^\dagger \mathbf{D}^{-2} \mathbf{v}}} \frac{1}{\mathbf{D}} \mathbf{v}^* .$$

Generically, all other eigenvalues are large and close to some element of $\mathbf{D}$. Consider the eigenvalue that is close to $M_0^\pm$. In this case the operator $(\mathbf{D} - x_0^\pm)^{-1}$ contains only one large denominator. Neglecting all others, we obtain the eigenvalue and eigenvector

$$x_0^+ \approx M_0^{+2} + 2|\alpha_0|^2 , \quad \psi_L^{[0+]} \approx \frac{|\alpha_0|}{M_0^+} ; \quad \Psi_T^{[0+]} \approx -\frac{\vec{\alpha}_0}{|\alpha_0|} ,$$

where $\vec{\alpha}_0$ is the vector with component along α_0. Similarly, there is another eigenvalue and eigenvector

$$x_0^- \approx M_0^{-2} + 2|\beta_0|^2 \; ; \qquad \psi_L^{[0-]} \approx \frac{|\beta_0|}{M_0^-} \; ; \qquad \Psi_T^{[0-]} \approx -\frac{\vec{\beta}_0}{|\beta_0|} \; ,$$

with $\vec{\beta}_0$ the vector with non-zero component along β_0. These vectors are orthogonal to $\mathcal{O}(v/M)$.

Then for each fourset in each floor of the Kaluza-Klein tower, there are two degenerate sets of eigenvalues, and thus two small denominators for each eigenvalue $x_n^{\pm}$. Their determination is best done by requiring orthogonality with $\Psi^{[0]}$.

The complexity of these equations makes it clear that until there are credible theories for how these "mass" parameters are generated, it is not possible to see how small eigenvalues can be generated. Work is in progress in trying to classify the possibilities in terms of the global symmetries we have introduced [31]. Much work remains to be done.

The hallmark of this type of physics is the appearance of Kaluza-Klein towers of sterile neutrinos. However, outside of a very intriguing idea [32] has been recently introduced, tying such towers to the reported Karmen anomaly, we are not aware of any phenomena that demand their existence in the present data. In the absence of any credible dynamics for bulk-physics, We therefore take the attitude that for the moment *"one neutrino on the brane is worth two in the bulk"*. While it is likely that the bulk opens up, it does so at much shorter length scales.

V OUTLOOK

Theoretical predictions of neutrino masses and mixings depend on developing a credible theory of flavor. The present experimental situation is somewhat unclear: the LSND results [33] imply the presence of a sterile neutrino; and superK favors $\nu_\mu - \nu_\tau$ oscillation over $\nu_\mu - \nu_{\text{sterile}}$. The origin of the solar neutrino deficit remains a puzzle, which several possible explanations. One is the non-adiabatic MSW effect in the Sun, which our theoretical ideas seem to favor, but it is an experimental question which is soon to be answered by the continuing monitoring of the ^8B spectrum by SuperK, and the advent of the SNO detector. If neutrino masses reflect (through the seesaw) the value of the ultraviolet cut-off, they set the scale for the strength of proton decay interactions, implying that observation may not be far in the future. Neutrino physics has given us a first glimpse of physics at very short distances, and proton decay cannot be too far behind.

VI ACKNOWLEDGEMENTS

I wish to thank Professors A. Halprin and J. Nieves for their kind invitation to this very pleasant, stimulating and informative workshop. I also wish to acknowledge a

Dr Lee visiting Fellowship at Christ Church College in Oxford, where some of this work was done.

REFERENCES

1. O. Von Bayer, O. Hahn, and L. Meitner, Phys. Zeitschr. **12**, 273(1911); *ibid* **13**, 273(1911); **13** 264(1912).
2. J. Chadwick, Verh. d. D. Phys. Ges., **16**, 383(1914).
3. C. D. Ellis and W. A. Wooster, Proc. Royal Soc. **A117**, 109(1927).
4. B. Pontecorvo, Chalk river Report PD-205, November 1946, unpublished.
5. C.L. Cowan, F. Reines, F.B. Harrison, H.W. Kruse, A.D. McGuire, Science **124**, 103(1956).
6. B. Pontecorvo, JETP (USSR) **34**, 247(1958).
7. Raymond Davis Jr., Phys Rev **97**, 766(1955).
8. M. Goldhaber, L. Grodzins, A.W. Sunyar, Phys.Rev. **109**, 1015(1958).
9. Super-Kamiokande Collaboration, Phys. Rev. Lett. **81**, 1562(1998)
10. G. Danby, J.M. Gaillard, K. Goulianos, L.M. Lederman, N. Mistry, M. Schwartz, J. Steinberger, Phys.Rev.Lett. **9**, 36(1962).
11. Z. Maki, M. Nakagawa and S. Sakata, Prog. Theo. Physics, **28**, 247(1962). B. Pontecorvo, Zh. Eksp. Teor. Fiz. **53**, 1717(1967).
12. Raymond Davis Jr., D. Harmer and K. Hoffman, Phys. Rev. Lett. **20**, 1205(1968).
13. V. Gribov and B. Pontecorvo, Phys. Lett. **B28**, 493(1969).
14. J. Pati and A. Salam, Phys. Rev. Lett. **31**, 661(1973)
15. H. Georgi and S. L. Glashow, Phys. Rev. Lett. **32**, ; H. Fritzsch and P. Minkowski, Annals Phys. **93**, 193(1975); H. Georgi, in AIP Conference Proceedings no 23, Williamsburg, Va, 1975; F. Gürsey, P. Ramond and P. Sikivie, Phys. Lett. **60B**, 177(1976)
16. See C. Albright et al, FermiLab Report, FN-692, April 2000.
17. Particle Data Group, R. M. Barnett *et al.*, Phys Rev **D54**, 1(1996).
18. N. Arkani-Hamed, S. Dimopoulos, Gia Dvali *Phys.Lett.* **B429**, 263(1998); E. Dudas, K. Dienes, and T. Gherghetta, *Phys.Lett.* **B436**, 55(1998)
19. M. Gell-Mann, P. Ramond, and R. Slansky, in Sanibel Talk, CALT-68-709, Feb 1979 (unpublished), and in *Supergravity* (North Holland, Amsterdam 1979). T. Yanagida, in *Proceedings of the Workshop on Unified Theory and Baryon Number of the Universe*, KEK, Japan, 1979.
20. P. Candelas, G. Horowitz, A. Strominger and E. Witten, Nucl. Phys. **B258**, 46(1985)
21. J. A. Harvey, P. Ramond and D. B. Reiss, Nucl. Phys. **B199**, 223(1982)
22. The Case for Neutrino Oscillations, P. Ramond, Proceedings of the Los Alamos Neutrino Workshop, LA-9358-C, June 1981.
23. N. Irges, S. Lavignac and P. Ramond, Phys. Rev. **D58**, 035003(1998).
24. C. Froggatt and H. B. Nielsen Nucl. Phys. B147 (1979) 277; P. Ramond, R.G. Roberts and G.G. Ross, Nucl. Phys. B406 (1993)
25. L. Wolfenstein, Phys. Rev. Lett. **51**, 1945(1983).
26. J. Elwood, N., Irges, and P. Ramond, Phys. Rev. Lett. **81**, 5064(1998)

27. L. Wolfenstein, Phys. Rev. D17, 2369 (1978); V. Barger, K. Wishnant, S. Pakvasa, and R. J. N. Phillips, Phys Rev D22, 2718(1980); S. Mikheyev and A. Yu Smirnov, Nuovo Cim. **9C**, 17 (1986). (1994), 221-228.
28. E. Dudas, K. Dienes, and T. Gherghetta, hep-ph/9811428.
29. Arkani-Hamed, S. Dimipoulos, G. Dvali, and J. March-Russell, hep-ph/9811448.
30. P. Ramond, G. G. Ross, unpublished.
31. , A. Lukas, P. Ramond, A. Romanino, and G.G. Ross, in preparation.
32. A. Lukas and A Romanino, hep-ph/0004130
33. C. Athanassopoulos *et al.*, Phys. Rev. Lett. **75**, 2560(1995); **77**, 3082(1996); nucl-ex/9706006.

NEUTRINO EXPERIMENTS

Final Neutrino Oscillation Results from LSND

W. C. Louis
representing the LSND Collaboration [1]

*Physics Division, Los Alamos National Laboratory,
Los Alamos, NM 87545, USA
E-mail: louis@lanl.gov*

Abstract. The LSND experiment at Los Alamos has conducted searches for $\bar{\nu}_\mu \to \bar{\nu}_e$ oscillations using $\bar{\nu}_\mu$ from μ^+ decay at rest and for $\nu_\mu \to \nu_e$ oscillations using ν_μ from π^+ decay in flight. For the $\bar{\nu}_\mu \to \bar{\nu}_e$ search, a total excess of $83.3 \pm 21.2 \pm 12.0$ events is observed with e^+ energy between 20 and 60 MeV, while for the $\nu_\mu \to \nu_e$ search, a total excess of $18.1 \pm 6.6 \pm 4.0$ events is observed with e^- energy between 60 and 200 MeV. If attributed to neutrino oscillations, the most favored allowed region from a fit to the entire data sample is a band from 0.2 to 2.0 eV^2. This result implies that at least one neutrino has a mass greater than 0.4 eV/c^2 and that neutrinos contribute more than 1% to the mass of the universe.

I INTRODUCTION

The LSND experiment collected data for six years, from 1993 to 1998, during which time the LAMPF/LANSCE accelerator operated for 17 months of calendar time and delivered 28,898 C (~ 0.3 g) of protons on target. Using partial data samples, evidence for neutrino oscillations has been published previously for both $\bar{\nu}_\mu \to \bar{\nu}_e$ [1,2] and $\nu_\mu \to \nu_e$ [3] oscillations. In this report we present the final LSND oscillation results that include the entire 1993-1998 data sample, that combines the two oscillation searches in a global analysis, and that makes use of a new event reconstruction that has greatly improved the event spatial and angular resolutions. An excess of events consistent with neutrino oscillations is observed and implies

---

[1] The LSND Collaboration presently consists of the following people and institutions: E. D. Church, I. Stancu, W. Strossman, G.J. VanDalen (Univ. of California, Riverside); W. Vernon (Univ. of California, San Diego); D.O. Caldwell, S. Yellin (Univ. of California, Santa Barbara); D. Smith (Embry-Riddle Aeronautical Univ.); R.L. Burman, J.B. Donahue, G.T. Garvey, W.C. Louis, G.B. Mills, V. Sandberg, R. Tayloe, D.H. White (Los Alamos National Laboratory); R. Imlay, H.J. Kim, A. Malik, W. Metcalf, M. Sung, N. Wadia (Louisiana State Univ.); K. Johnston (Louisiana Tech Univ.); A. Fazely, R.M. Gunasingha (Southern Univ); L.B. Auerbach, R. Majkic (Temple Univ.)

that at least one neutrino has a mass greater than 0.4 eV/c² and that neutrinos contribute more than 1% to the mass of the universe.

The old event position reconstruction was hampered due to the charge response of the 8" phototubes used in LSND (Hamamatsu R1408). For these phototubes, the single photoelectron distribution is essentially a broad Gaussian peak followed by an exponential charge tail that extends to arbitrarily high values. As the position and angle fits weight the hit phototubes by their charge, this charge tail has the effect of smearing the reconstructed event positions and angles. (It also has the effect of smearing the energy resolution. At 50 MeV the electron energy resolution is $\sim 7\%$, much worse than the $\sim 3\%$ resolution that would be expected from photon statistics alone.) To ameliorate this effect, a new reconstruction was developed that weights the hit phototubes by a function of their expected charge and not by their actual charge. (The new reconstruction also has other improvements, such as the inclusion of timing information.) This has resulted in an improvement in the correlated positions for muon decay events. The mean reconstructed distance between the muon and decay electron has improved from 22 cm with the old reconstruction to 14 cm with the new reconstruction. For 2.2 MeV γ from neutron capture, the most likely distance has improved from 74 cm to 55 cm.

II DETECTOR

The Liquid Scintillator Neutrino Detector (LSND) experiment at Los Alamos [4] was designed to search with high sensitivity for $\bar{\nu}_\mu \to \bar{\nu}_e$ oscillations from μ^+ decay at rest. The LANSCE accelerator is an intense source of low energy neutrinos due to its 1 mA proton intensity and 800 MeV energy. For the 1993-1995 running period the beam stop consisted of a 30-cm long water target (20-cm in 1993) followed by a water-cooled Cu Beam dump, while for the 1996-1998 running period the beam stop was reconfigured with the water target replaced by a close-packed high-z target for testing tritium production. The muon decay-at-rest neutrino flux with this new configuration is only 2/3 of the neutrino flux with the old beam stop; however, the pion decay-in-flight neutrino flux has been reduced to 1/2 of the original flux, so that the 1996-1998 data serve as a systematic check. The neutrino source is well understood because almost all neutrinos arise from π^+ or μ^+ decay; π^- and μ^- are readily captured in the Fe of the shielding and Cu of the beam stop. [5] The production of kaons and heavier mesons is negligible at these energies. The $\bar{\nu}_e$ rate is calculated to be only 4×10^{-4} relative to $\bar{\nu}_\mu$ in the $36 < E_\nu < 52.8$ MeV energy range, so that the observation of a significant $\bar{\nu}_e$ rate would be evidence for $\bar{\nu}_\mu \to \bar{\nu}_e$ oscillations.

The LSND detector consists of an approximately cylindrical tank 8.3 m long by 5.7 m in diameter. The center of the detector is 30 m from the neutrino source. On the inside surface of the tank 1220 8-inch Hamamatsu phototubes provide 25% photocathode coverage. The tank is filled with 167 metric tons of liquid scintillator consisting of mineral oil and 0.031 g/l of b-PBD. This low scintillator concentration allows the detection of both Čerenkov light and scintillation light and yields a relatively long attenuation length of more than 20 m for wavelengths greater than 400 nm. [6] A typical 45 MeV electron created in the detector produces a total of ~ 1500 photoelectrons, of which ~ 280 photoelectrons are in the Čerenkov cone. The phototube time and pulse height signals are used to reconstruct the track with an average r.m.s. position resolution of ~ 14 cm, an angular resolution of ~ 12 degrees, and an energy resolution of $\sim 7\%$. The Čerenkov cone for relativistic particles and the time distribution of the light, which is broader for non-relativistic particles, give excellent particle identification. Surrounding the detector is a veto shield [7] which tags cosmic ray muons going through the detector.

III DATA ANALYSIS

The primary oscillation search in LSND is the search for $\bar{\nu}_\mu \to \bar{\nu}_e$ oscillations, where the $\bar{\nu}_\mu$ arise from μ^+ decay at rest in the beam stop and the $\bar{\nu}_e$ are identified through the reaction $\bar{\nu}_e p \to e^+ n$. This reaction allows a two-fold signature of a positron with a 52 MeV endpoint and a correlated 2.2 MeV γ from neutron capture on a free proton. The positron/electron selection criteria (LSND is unable to determine charge) for this primary oscillation search is the following. First, in order to eliminate muon decay events, it is required that there be no event within 8 μs in the future or within 12 μs in the past. Second, the event particle identification parameter, χ_p, is required to lie in the range $-1.5 < \chi_p < 0.5$, where the precise range values are determined by maximizing the acceptance divided by the square root of the beam-off background. Third, there must be less than 4 veto hits associated with the event and the time of the nearest veto hit must be more than 30 ns from the event time. Fourth, the positron energy is required to be in the range $20 < E_e < 60$ MeV. Fifth, the event reconstructed position must be more than 35 cm from the nearest phototube surface. Finally, it is required that there be no more than one correlated γ with $R_\gamma > 10$ (see below) in order to reduce the background from cosmic-ray neutrons, which will typically knock-out additional neutrons.

The correlated 2.2 MeV γ selection criteria makes use of the likelihood ratio, R_γ, which is defined to be the likelihood that the γ is correlated divided by the likelihood that the γ is accidental. R_γ depends on three quantities: the number of hit phototubes associated with the γ (the multiplicity is proportional to the γ energy), the distance between the reconstructed γ and positron positions, and the time between the γ and positron. As checks of the likelihood distributions, Figs. 1 and 2 show the R_γ distributions for $\nu_e C \to e^- N_{gs}$ exclusive events, [8] where the N_{gs} beta decays, and $\nu_\mu C \to \mu^- X$ and $\bar{\nu}_\mu C \to \mu^+ X$ and $\bar{\nu}_\mu p \to \mu^+ n$ inclusive events. [9] By definition, the former reaction has no recoil neutron, so that its R_γ distribution should be consistent with a purely accidental γ distribution; indeed, a fit to the R_γ distribution finds that the fraction of events with a correlated γ, f_c, is $f_c = -0.004 \pm 0.007$ ($\chi^2 = 4.6/9$ DOF). For the latter reactions, correlated γ

TABLE 1. Numbers of beam-on events that satisfy the selection criteria for the primary $\bar{\nu}_\mu \to \bar{\nu}_e$ oscillation search with $R_\gamma > 1$, $R_\gamma > 10$, and $R_\gamma > 100$. Also shown are the correlated γ efficiencies, the beam-off background, the estimated neutrino background, and the excess of events that is consistent with neutrino oscillations.

Selection & Efficiency	Beam-On Events	Beam-Off Background	ν Background	Event Excess
$R_\gamma > 1$ (51.15%)	195	98.1 ± 2.4	37.7	59.2 ± 14.2
$R_\gamma > 10$ (39.29%)	83	33.7 ± 1.4	16.6	32.7 ± 9.2
$R_\gamma > 100$ (16.86%)	25	7.9 ± 0.7	5.4	11.7 ± 5.0

are expected for $\sim 14\%$ of the events. [10] A fit to the R_γ distribution gives $f_c = 0.129 \pm 0.013$ ($\chi^2 = 8.2/9$ DOF), in agreement with expectations. Note that with the new reconstruction, the correlated γ efficiency has increased while the accidental γ efficiency has decreased. For $R_\gamma > 10$, the correlated and accidental efficiencies are 0.393 and 0.003, respectively. With the old reconstruction the correlated and accidental efficiencies were 0.230 and 0.006, respectively.

The secondary oscillation search in LSND is the search for $\nu_\mu \to \nu_e$ oscillations, where the ν_μ arise from π^+ decay in flight in the beam stop and the ν_e are identified through the reaction $\nu_e C \to e^- X$. The electron selection criteria for this primary oscillation search is almost the same as for the primary search, except that the electron energy is required to be in the range $60 < E_e < 200$ MeV and there must be no associated 2.2 MeV γ.

IV NEUTRINO OSCILLATION RESULTS

Table 1 gives the statistics for events that satisfy the selection criteria for the primary $\bar{\nu}_\mu \to \bar{\nu}_e$ oscillation search. An excess of events is observed over what is expected from beam-off and neutrino background that is consistent with neutrino oscillations. A fit to the R_γ distribution, as shown in Fig. 3, gives $f_c = 0.0578 \pm 0.0108$ ($\chi^2 = 9.2/9$ DOF), which leads to a beam on-off excess of 113.3 ± 21.2 events with a correlated neutron. Subtracting the neutrino background from μ^- decay at rest followed by $\bar{\nu}_e p \to e^+ n$ scattering (21.6 events) and π^- decay in flight followed by $\bar{\nu}_\mu p \to \mu^+ n$ scattering (8.4 events) [11] leads to a total excess of 83.3 ± 21.2 events or an oscillation probability of $(0.25 \pm 0.06 \pm 0.04)\%$. (Note that with the old reconstruction the oscillation probability was determined to be $(0.33 \pm 0.09 \pm 0.05)\%$.)

A fairly clean sample of oscillation candidate events can be obtained by requiring $R_\gamma > 10$, where as shown in Table 1, the beam on-off excess is 49.3 ± 9.2 events while the estimated neutrino background is only 16.6 events. Fig. 4 displays the energy distribution of events with $R_\gamma > 10$. The shaded region shows the estimated neutrino background while the curves show the expected distributions

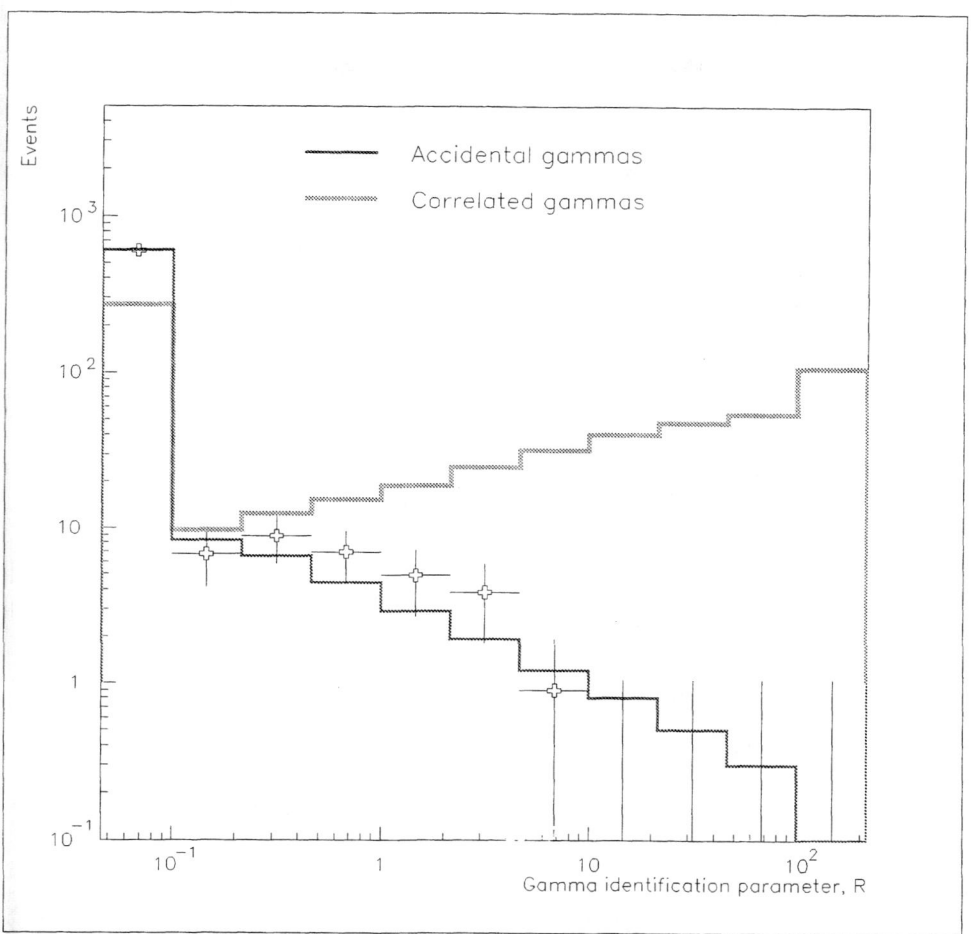

FIGURE 1. The R_γ distribution for $\nu_e C \to e^- N_{gs}$ exclusive events, where the N_{gs} beta decays.

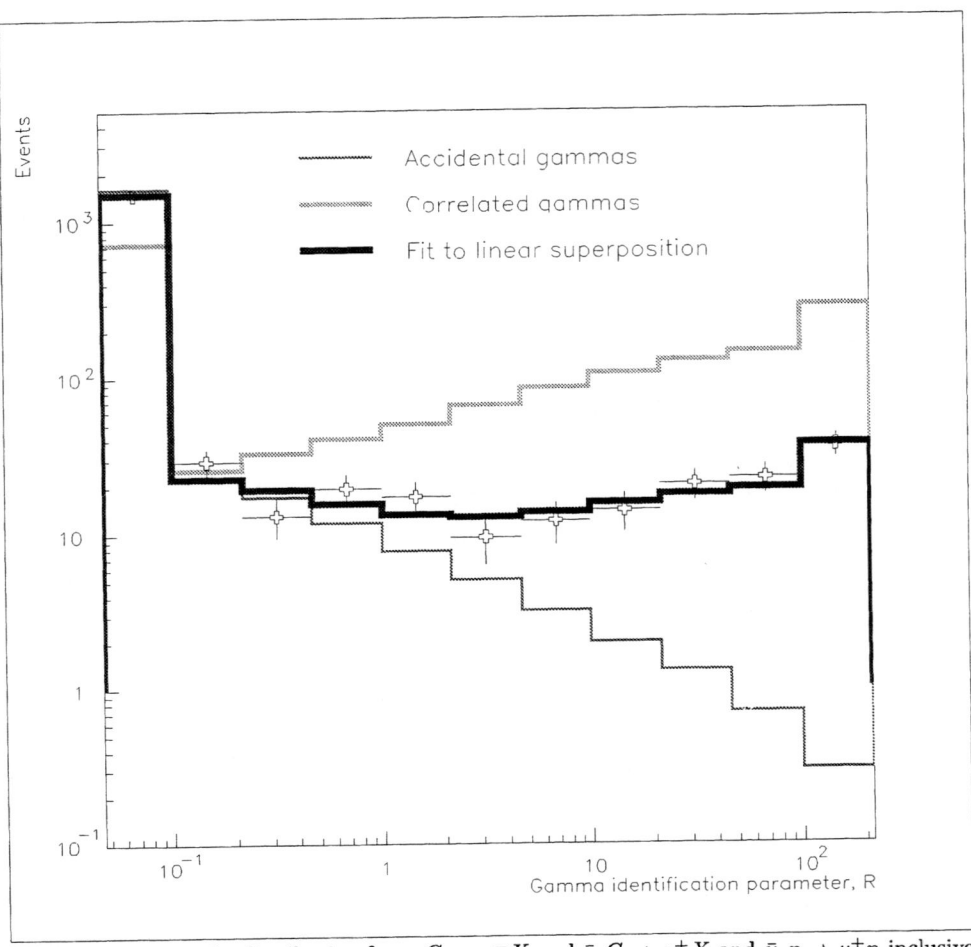

FIGURE 2. The R_γ distribution for $\nu_\mu C \to \mu^- X$ and $\bar\nu_\mu C \to \mu^+ X$ and $\bar\nu_\mu p \to \mu^+ n$ inclusive events.

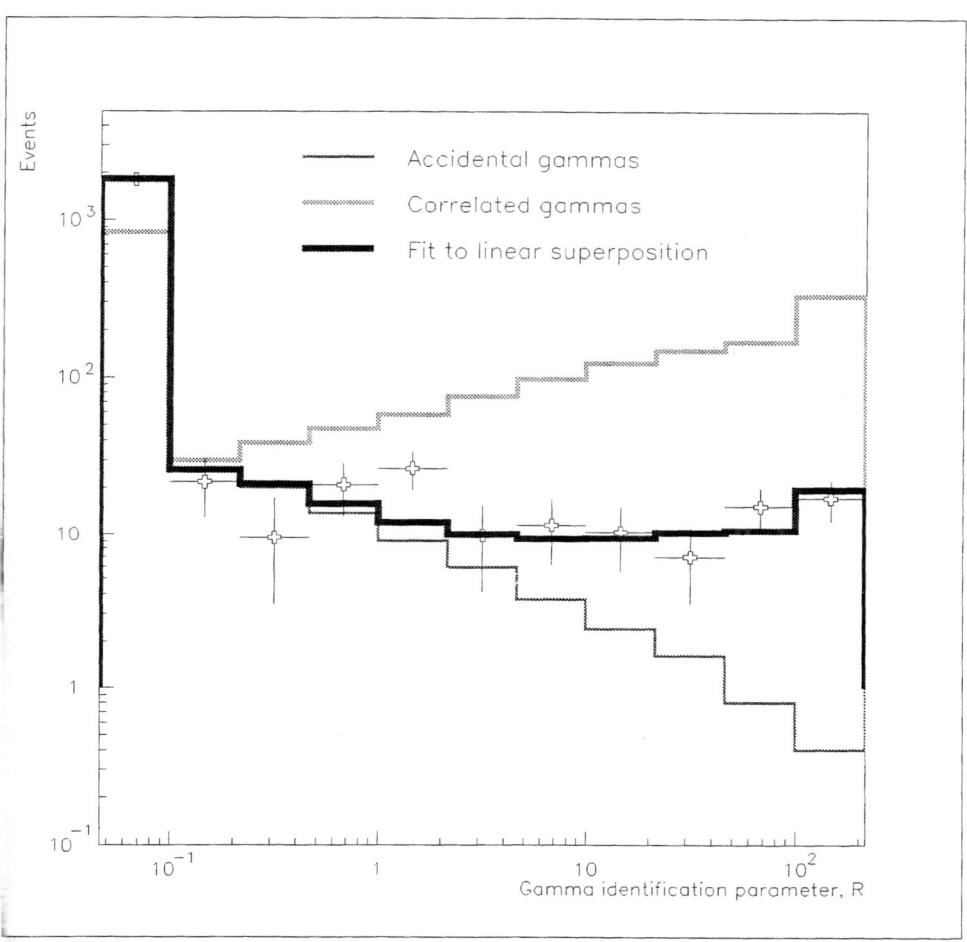

FIGURE 3. The R_γ distribution for events that satisfy the selection criteria for the primary $\bar{\nu}_\mu \to \bar{\nu}_e$ oscillation search.

TABLE 2. Number of beam on-off excess events that satisfy the selection criteria for the primary $\bar{\nu}_\mu \to \bar{\nu}_e$ oscillation search with one $R_\gamma > 10$ associated γ and with > 1 $R_\gamma > 10$ associated γ. The excess of events with > 1 correlated γ is approximately zero, which is what is expected for the reaction $\bar{\nu}_e p \to e^+ n$.

Energy Selection	1 Associated γ	> 1 Associated γ
$20 < E_e < 60$ MeV	49.2 ± 9.1	-2.8 ± 1.7
$36 < E_e < 60$ MeV	20.8 ± 5.8	-2.8 ± 1.0

from a combination of neutrino background plus neutrino oscillations at high or low Δm^2. The data agree well with the oscillation hypothesis. Fig. 5 shows the spatial distribution for events with $R_\gamma > 10$, where z is along the axis of the tank (and approximately along the beam direction), y is vertical, and x is transverse. The data agree well with the distributions from $\nu_e C \to e^- N_{gs}$ scattering (solid histogram), where the reaction is identified by the N_{gs} beta decay.

A test of the oscillation hypothesis is to check whether there is an excess of events with > 1 correlated γ. If the excess of events is indeed due to the reaction $\bar{\nu}_e p \to e^+ n$, then there should be no excess with > 1 correlated γ because the recoil n is too low in energy (< 5 MeV) to knock out additional neutrons. If, on the other hand, the excess involves higher energy neutrons (> 20 MeV), then one would expect a comparable excess with > 1 correlated γ. However, as shown in Table 2, the excess of events with > 1 correlated γ is approximately zero, as expected for the reaction $\bar{\nu}_e p \to e^+ n$.

A Δm^2 vs. $\sin^2 2\theta$ oscillation parameter fit for the entire data sample, $20 < E_e < 200$ MeV, is shown in Fig. 6. The fit includes both $\bar{\nu}_\mu \to \bar{\nu}_e$ and $\nu_\mu \to \nu_e$ oscillations, as well as all known neutrino backgrounds. The inner and outer regions correspond to 90% and 99% CL allowed regions, while the curves are 90% CL limits from the Bugey reactor experiment, [12] the CCFR experiment at Fermilab, [13] the NOMAD experiment at CERN, [14] and the KARMEN experiment at ISIS. [15] The most favored allowed region is the band from 0.2 to 2.0 eV^2, although a region around 7 eV^2 is also possible. Applying the above analysis to the $60 < E_e < 200$ MeV data sample, involving secondary $\nu_\mu \to \nu_e$ oscillations only, results in a total excess of $11.4 \pm 12.5 \pm 4.0$ oscillation events or an oscillation probability of $(0.13 \pm 0.14 \pm 0.05)$%. This result is consistent with our higher precision analysis of the 1993-1995 data sample, [3] which gave a total excess of $18.1 \pm 6.6 \pm 4.0$ oscillation events, corresponding to an oscillation probability of $(0.26 \pm 0.10 \pm 0.05)$%. (Note that the 1996-1998 data sample had reduced flux and higher beam-off background compared to the 1993-1995 data.)

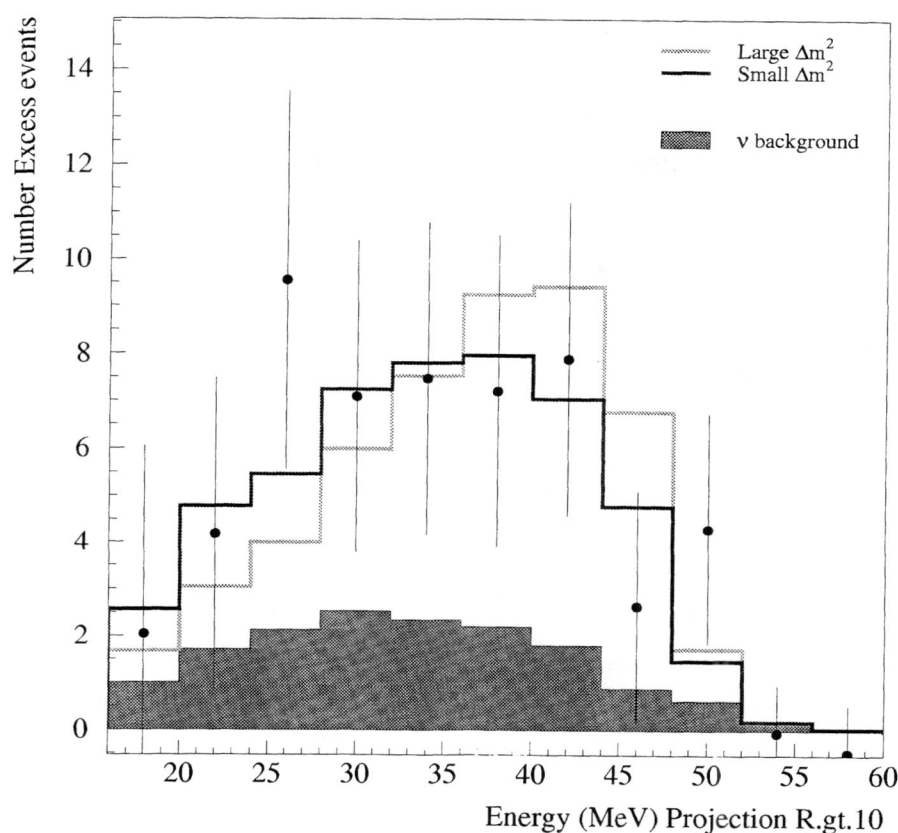

FIGURE 4. The energy distribution of events with $R_\gamma > 10$. The shaded region shows the estimated neutrino background while the curves show the expected distributions from a combination of neutrino background plus neutrino oscillations at high or low Δm^2.

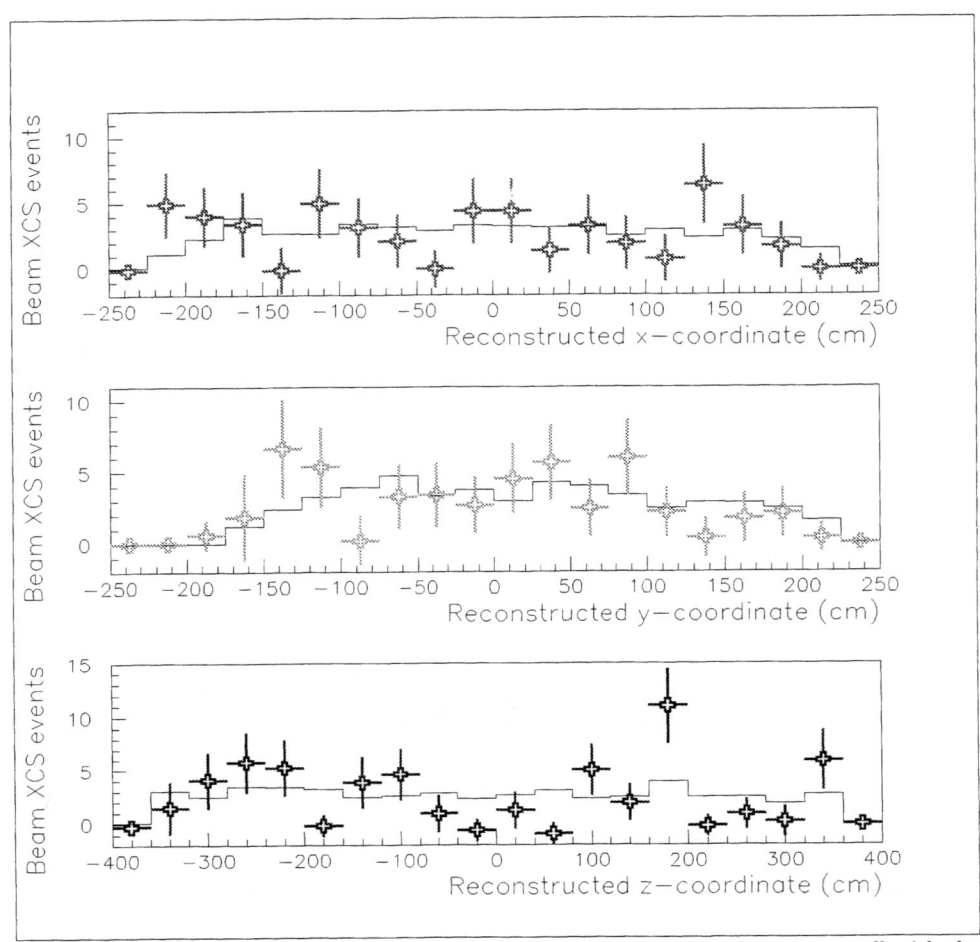

FIGURE 5. The spatial distributions for events with $R_\gamma > 10$. The data agree well with the distributions from $\nu_e C \to e^- N_{gs}$ scattering (solid histogram), where the reaction is identified by the N_{gs} beta decay.

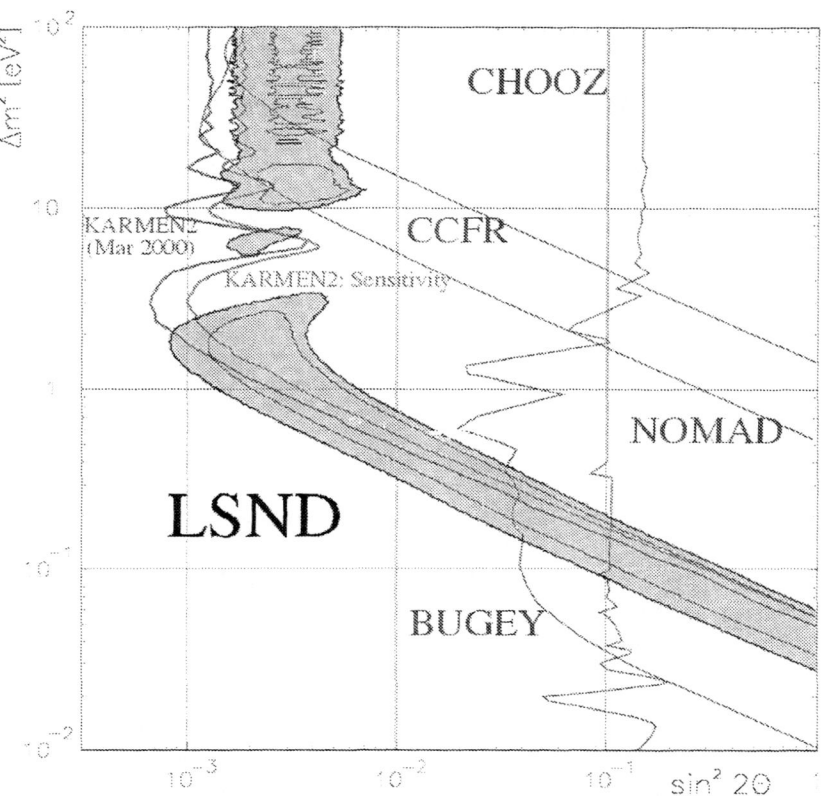

FIGURE 6. A Δm^2 vs. $\sin^2 2\theta$ oscillation parameter fit for the entire data sample, $20 < E_e < 200$ MeV. The fit includes primary $\bar{\nu}_\mu \to \bar{\nu}_e$ oscillations and secondary $\nu_\mu \to \nu_e$ oscillations, as well as all known neutrino backgrounds. The inner and outer regions correspond to 90% and 99% CL allowed regions, while the curves are 90% CL limits from the Bugey reactor experiment, the CCFR experiment at Fermilab, the NOMAD experiment at CERN, and the KARMEN experiment at ISIS.

V CONCLUSIONS

The LSND experiment provides evidence for neutrino oscillations from both the primary $\bar{\nu}_\mu \to \bar{\nu}_e$ oscillation search and the secondary $\nu_\mu \to \nu_e$ oscillation search. At present, this remains the only evidence for appearance neutrino oscillations and implies that at least one neutrino has a mass greater than 0.4 eV/c^2 and that neutrinos comprise more than 1% of the mass of the universe. The MiniBooNE experiment at Fermilab, [16] which is presently under construction, will provide a definitive test of the LSND results, and if the neutrino oscillation results are confirmed, will make a precision measurement of the oscillation parameters.

REFERENCES

1. C. Athanassopoulos et al., Phys. Rev. Lett. **75**, 2650 (1995).
2. C. Athanassopoulos et al., Phys. Rev. C **54**, 2685 (1996); C. Athanassopoulos et al., Phys. Rev. Lett. **77**, 3082 (1996).
3. C. Athanassopoulos et al., Phys. Rev. Lett. **81**, 1774 (1998); C. Athanassopoulos et al., Phys. Rev. C **58**, 2489 (1998).
4. C. Athanassopoulos et al., Nucl. Instrum. Methods A **388**, 149 (1997).
5. R.L. Burman, M.E. Potter, and E.S. Smith, Nucl. Instrum. Methods A **291**, 621 (1990); R.L. Burman, A.C. Dodd, and P. Plischke, Nucl. Instrum. Methods in Phys. Res. A **368**, 416 (1996).
6. R.A. Reeder et al., Nucl. Instrum. Methods A **334**, 353 (1993).
7. J.J. Napolitano et al., Nucl. Instrum. Methods A **274**, 152 (1989).
8. C. Athanassopoulos et al., Phys. Rev. C **55**, 2078 (1997).
9. C. Athanassopoulos et al., Phys. Rev. C **56**, 2806 (1997).
10. E. Kolbe, K. Langanke, F.-K. Thielmann, and P. Vogel, Phys. Rev. C **52**, 3437 (1995).
11. This background estimate also includes contributions from $\bar{\nu}_\mu C \to \mu^+ n X$ and $\nu_\mu C \to \mu^- n X$.
12. B. Achkar et al., Nucl. Phys. **B434**, 503 (1995).
13. A. Romosan et al., Phys. Rev. Lett. **78**, 2912 (1997).
14. D. Autiero, talk presented at the 1998 International Conference on High Energy Physics in Vancouver, Canada.
15. K. Eitel, talk presented at the 2000 International Neutrino Conference in Sudbury, Canada.
16. E. Church et al., "A proposal for an experiment to measure $\nu_\mu \to \nu_e$ oscillations and ν_μ disappearance at the Fermilab Booster: BooNE", LA-UR-98-352, Fermilab experiment 898.

The Fermilab Neutrino Oscillation Program

Michael H. Shaevitz

Fermilab
Batavia, Illinois 60510
Columbia University
New York, New York 10027

Abstract. There are presently three experimental indications of neutrino oscillations from solar, atmospheric, and the LSND neutrino experiments. In the near future, these indications will be checked using terrestrial neutrino beams. At Fermilab, the Minos and MiniBooNE experiments will make definitive explorations of the atmospheric and LSND results respectively. Studies of a future ν-factory program using a muon storage ring show that such a facility would allow measurements of other neutrino mixing parameters, the neutrino mass hierarchy, and possibly CP violations.

INTRODUCTION

Whether neutrinos have mass has been a key question in particle physics since the existence of neutrinos was first postulated. Theoretically, neutrinos are expected to be massive since they are the Standard Model partners of the massive charged leptons. Experimentally it has been shown that neutrino masses are very small. Some models, for example the See-Saw Model, have been developed that explain the smallness of neutrino masses through a mixing process with new isosinglet, heavy neutrinos. In this way, the measurement of a finite neutrino mass would give insights into new physics at the scale of these new heavy particle.

Direct neutrino mass measurements set limits many orders of magnitude smaller than the charged lepton masses and are now reaching their limits of sensitivity. To probe neutrino masses significantly below 1 eV requires indirect methods such as neutrino oscillation experiments. If the weak or flavor neutrino eigenstates are mixtures of several mass eigenstates with different masses, then one flavor of neutrino can oscillate to another as it propagates over a finite distance. With three generations of neutrinos, the flavor and mass eigenstates are related by a (3×3) unitary mixing matrix similar to the CKM mixing matrix of quarks.

$$\begin{pmatrix} \nu_1 \\ \nu_2 \\ \nu_3 \end{pmatrix} = \begin{pmatrix} c_{12}c_{13} & s_{12}c_{13} & s_{13}e^{-i\delta} \\ -s_{12}c_{23} - c_{12}s_{23}s_{13}e^{i\delta} & c_{12}c_{23} - s_{12}s_{23}s_{13}e^{i\delta} & s_{23}c_{13} \\ s_{12}s_{23} - c_{12}c_{23}s_{13}e^{i\delta} & -c_{12}s_{23} - s_{12}c_{23}s_{13}e^{i\delta} & c_{23}c_{13} \end{pmatrix} \begin{pmatrix} \nu_1 \\ \nu_2 \\ \nu_3 \end{pmatrix} \quad (1)$$

With three mass eigenstates, one can form three values for Δm^2, $\Delta m_{12}^2 = m_1^2 - m_2^2$, $\Delta m_{23}^2 = m_2^2 - m_3^2$, $\Delta m_{31}^2 = m_3^2 - m_1^2$, but only two are independent. At each Δm^2, there can be oscillations between all the neutrino flavors with different mixing angle combinations. For example with $\Delta m_{23}^2 \gg \Delta m_{12}^2$, the dominant oscillation probability equations are

$$P(\nu_\mu \to \nu_\tau) = \cos^4 \theta_{13} \sin^2 2\theta_{23} \sin^2 \left(1.27 \Delta m_{32}^2 L/E_\nu\right) \quad (2)$$
$$P(\nu_\mu \to \nu_e) = \sin^2 \theta_{23} \sin^2 2\theta_{13} \sin^2 \left(1.27 \Delta m_{32}^2 L/E_\nu\right)$$
$$P(\nu_e \to \nu_\tau) = \cos^2 \theta_{23} \sin^2 2\theta_{13} \sin^2 \left(1.27 \Delta m_{32}^2 L/E_\nu\right)$$

indicating finite probabilities for $\nu_\mu \to \nu_\tau$, $\nu_\mu \to \nu_e$, and $\nu_e \to \nu_\tau$ oscillations.

With active mixing between three or more flavors, the phase δ can lead to CP violation effects. One popular example is a measurement of the difference between $\nu_\mu \to \nu_e$ and $\bar{\nu}_\mu \to \bar{\nu}_e$ oscillations given by

$$P(\nu_\mu \to \nu_e) - P(\bar{\nu}_\mu \to \bar{\nu}_e) = 4 \left(U_{\mu 1} U_{e1}^* U_{\mu 1}^* U_{e1}\right)(s_{12} + s_{23} + s_{31})$$

where $s_{ij} = \sin\left(\delta m_{ij}^2 L/E\right)$ and $\delta m_{ij}^2 = m_i^2 - m_j^2$

For this difference to be sizable, all the terms, s_{12}, s_{23}, and s_{31} must be $\gg 1$ or the process is effectively a two component oscillation. For example, if $s_{12} \approx 0$, then $s_{13} \approx -s_{23}$ which implies that $s_{12} + s_{23} + s_{31} \approx 0$ and small CP violation effects. In general, CP violation measurements can only be made if an experiment is sensitive to oscillations to at least three neutrino flavors. For the current situation, this implies experiments with oscillation sensitivity either to solar and atmospheric or atmospheric and LSND-like Δm^2 values.

There are currently three experimental indications of neutrino oscillations, the solar neutrino deficit, the atmospheric muon neutrino deficit, and the LSND indication of $\bar{\nu}_\mu \to \bar{\nu}_e$ oscillations. These three results would correspond to oscillations at three different Δm^2 scales as shown in Figure 1. Experimentally, the next step is to answer the questions: 1) Are all three indications really neutrino oscillations? 2) Which flavors are responsible for the oscillations? 3) What are the precise values for the oscillation parameters, Δm^2 and $\sin^2 2\theta$? The solar neutrino experiments are covered in several other talks at this conference with several new results expected over the next several years from Super-Kamiokande, SNO, Borexino, and KamLand.

I   ATMOSPHERIC NEUTRINO OSCILLATIONS

The Super-Kamiokande experiment has reported strong evidence for the neutrino oscillations from their measurements of interactions of neutrinos produced in the

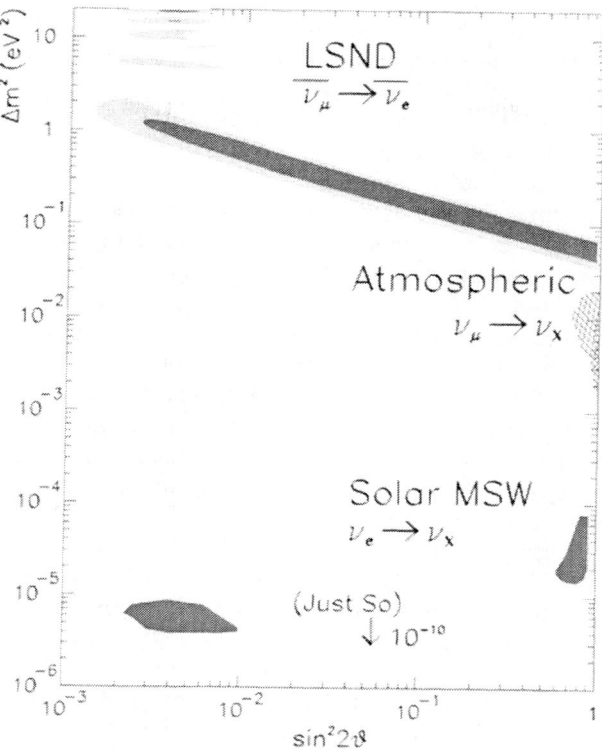

FIGURE 1. Three experimental indications of neutrino oscillations from solar, atmospheric and the LSND experiments.

atmosphere [1]. They observe a significant depletion of muon-like events relative to electron-like events. The depletion is dependent on energy and zenith angle which strongly supports an oscillation hypothesis. The neutrino flux spans a range between 300 and 2000 MeV and the zenith angle variation corresponds to distances 20 to 10^4 km. As shown in Figure 1, the result corresponds to oscillations with $\Delta m^2 \sim 3 \times 10^{-3}$ eV^2 and $\sin^2 2\theta$ near 1.0.

Most analyses assume the the atmospheric result corresponds to $\nu_\mu \to \nu_\tau$ oscillations but this remains an experimental question. Currently, several experimental measurements provide restrictions on the possible flavors contributing to the atmospheric deficit. The CHOOZ, Bugey, and Palo Verde reactor experiments have set limits on oscillations involving electron neutrino disappearance, $\nu_e \to \nu_e$, that rule out a dominant contribution of $\nu_\mu \to \nu_e$ oscillations in the atmospheric region. A sub-dominant component is possible for mixing angles, $\sin^2 2\theta$, below 0.10.

Another possibility is that muon neutrinos are oscillating to sterile neutrinos which are predicted in many models for neutrino mass. The Super-Kamiokande experiment has put restrictions on this possibility in two ways. First, Super-K has measured the ratio of π^0-like neutral-current (NC) to charge-current (CC) like events as compared to Monte Carlo (MC) predictions. They find a ratio $(NC(Data/MC)/CC(Data/MC)) = 1.11\pm0.06\pm0.26$ where the second systematic error dominates due to uncertainties in the $\nu N \to \nu N \pi^0$ cross section. Comparison to the MC expectation, 1.0 for oscillations to ν_τ and ~ 0.8 for $\nu_{sterile}$, is inconclusive at this time but may be improved in the future when the K2K experiment reduces the π^0 cross section uncertainty. Second, the high- energy Super-K data from large zenith angles are sensitive to matter effects in the earth. Interactions with matter in the earth (especially at $\cos\theta_{zenith}$ near 1.0) changes the $\nu_\mu \to \nu_{sterile}$ oscillation probability since the $\nu_{sterile}$ has no NC interactions with the quarks in the material. Using model predictions for these matter effects, Super-K has restricted oscillations to sterile neutrinos at the 1-2σ for most of the allowed parameter space.

II LONGBASELINE EXPERIMENTS

Several experiments are being set up to probe neutrino oscillations in the atmospheric parameter region using accelerator produced neutrino beams. For the K2K experiment, a low energy neutrino beam ($\langle E_\nu \rangle = 1.4$ GeV) produced at KEK is directed towards the Super-Kamiokande detector 250 km away. The experiment has several near detectors at 100 m to measure the beam flux and characteristics. K2K will look for a disappearance signal by measuring the rate of muon neutrino events in the near and far detectors. The experiment is currently running and expects about 170 events with no oscillations in the 22.5 kton Super-K fiducial volume after 3 years of data. With this data sample, K2K can probe most of the atmospheric region at the 90% CL and the upper half in Δm^2 at the 2-3σ level. As of June, 2000, the experiment reported an observation of 17 (9) fully-contained (μ-like) events with an no oscillation expectation of 29.2 ± 3.5 (15.8) events.

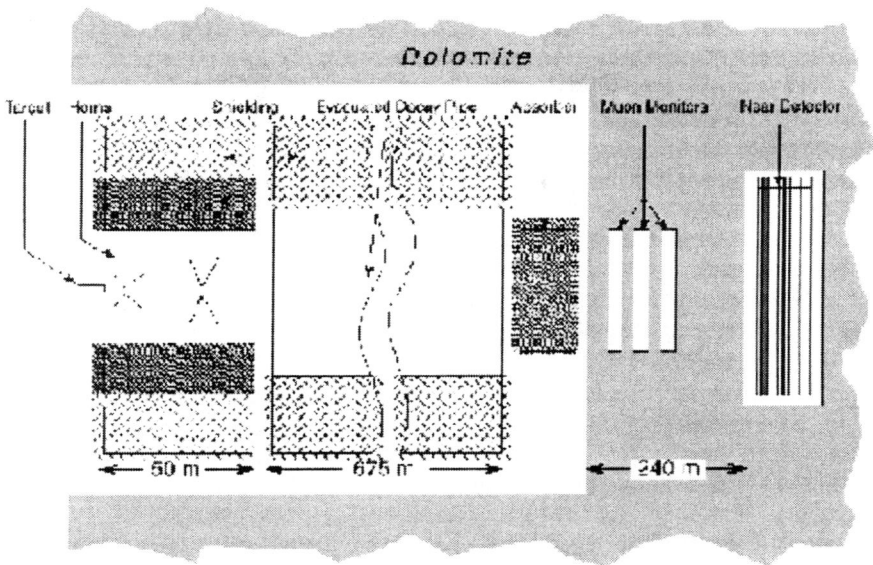

FIGURE 2. Schematic layout of the NuMI beamline from the primary proton target to the near detector hall. (Horizontal scale is compressed.)

A The NuMI/Minos Experiment at Fermilab

The new, high-intensity NuMI neutrino beam [3] is being constructed at Fermilab to illuminate the Minos neutrino detector in Soudan, Minnesota starting at the end of 2003. The 120 GeV protons from the Main Injector will hit a BeO target producing secondary pions and kaons which will be directed with a magnetic horn down a 675 m evacuated decay pipe as shown in Figure 2. The beam will be monitored with muon detectors after the hadron absorber and by a 980 ton near detector placed 240 m downstream of the decay pipe.

The beamline is directed towards the Soudan mine in Minnesota 730 km from Fermilab. The 5400 ton Minos far detector is located in the mine about 2400 feet below the surface. The detector is an 8 m diameter fine-grained iron/scintillator sampling calorimeter with 486 1 inch steel absorber plates magnetized to create a ~ 1.3 T toroidal field. Solid scintillator planes, made up of strips 1 cm thick and 4 cm wide, are interspersed after each steel plate and read out by wavelength shifter fibers into multi-anode photomultiplier tubes. The fine-grained nature of the detector will allow Minos to separate ν_μ CC events with muons from ν_e and NC events that will not penetrate very far through the steel. The experiment can

TABLE 1. NuMI Beam Options

Beam	$\langle E_\nu^{peak} \rangle$ GeV	CC events/year
High Energy	14	10,000
Medium Energy	7	5,000
Low Energy	3	700

also try to isolate ν_e from NC events due to the collimated nature of an electron shower in steel.

The experiment is designed to be flexible and cover the full atmospheric oscillation parameter space. With both a near and far detector, the experiment will be much less sensitive to the difficult modeling of the neutrino beam than a single detector experiment. Oscillations can be searched for using several techniques for comparing the near and far detector including the NC to CC ratio, the energy spectrum and rate for ν_μ CC events, and the rate for ν_e-like events. The physics goals include obtaining firm evidence if oscillations are present, measuring the oscillation parameters with good precision, and determining the flavor composition for any observed oscillation signal. To accomplish these goals, the beamline has been designed with the capability to produce three different energy distributions with the characteristics shown in Table 1.

With this flexibility, the Minos experiment can be tuned to the best energy spectrum for investigating oscillation signals for the full range of Δm^2 values from 10^{-3} to 10^{-2} eV^2. For example, with the "low energy" beam, Minos can cover the full Super-K atmospheric region at the 3.5 σ level including systematic uncertainties as shown in Figure 3. For $\Delta m^2 \gtrsim few \times 10^{-3}$ eV^2, $\sin^2 2\theta$ can be measured to 10% with the "medium energy" beam. The "high energy" beam will give large event samples for $\Delta m^2 \gtrsim 5 \times 10^{-3}$ eV^2 and possibly allow ν_τ events to be isolated.

Minos can also discriminate between $\nu_\mu \to \nu_\tau$ and $\nu_\mu \to \nu_{sterile}$ from the measured CC to NC ratio. For $\nu_\mu \to \nu_\tau$ oscillations, the ν_τ interactions will produce τ-leptons that 80% of time will look like NC events without a penetrating muon and will make the CC to NC event ratio go down relative to no oscillation expectation. On the other hand, for a $\nu_\mu \to \nu_{sterile}$ oscillation, both CC and NC events will be suppressed keeping the ratio constant. Estimates of the sensitivity for this technique for two year of Minos data with the "low energy" beam yields a 4σ separation for Δm^2 values above 2×10^{-3} eV^2 as shown in Figure 4.

B The CERN to Gran Sasso (CNGS) ν Oscillation Program

Cern has approved a program to send a ν_μ neutrino beam starting some time after 2005 from CERN to the Gran Sasso underground laboratory 750 km away [4]. The beam is similar to the Minos high-energy beam but about a factor of two less in rate. The program is to emphasize appearance measurements with ν_τ and

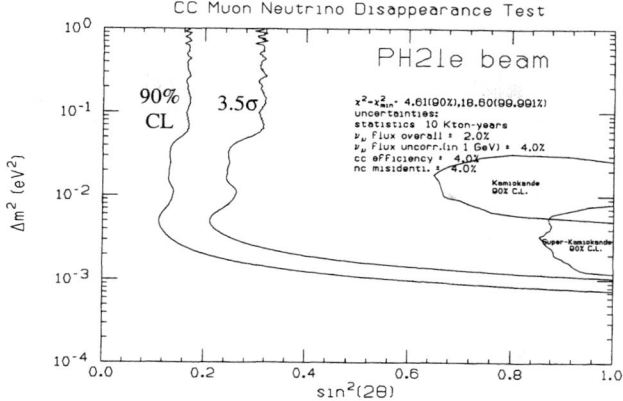

FIGURE 3. Sensitivity for Minos experiment at 90% CL and 3.5σ as compared to the Super-K and Kamiokande allowed regions.

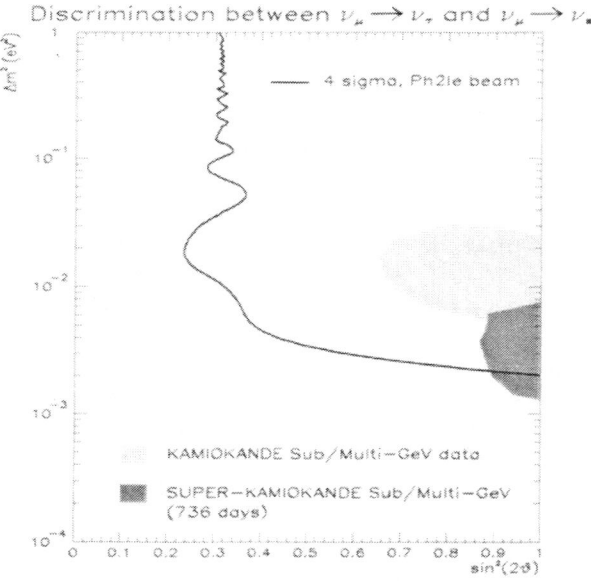

FIGURE 4. Minos discrimination between oscillations to ν_τ and $\nu_{sterile}$. The curve indicates the region with 4σ separation for two years of Minos data.

ν_e identification but no near detector.

Two experiments for the Gran Sasso laboratory have been proposed, the Opera emulsion experiment and the ICANOE hybrid liquid argon experiment. The Opera experiment will use emulsion bricks interspersed with electronic trackers to identify which bricks have neutrino interactions and should be developed and scanned. The goal is a 1.5 kton detector with a event reconstruction efficiency of about 10%. The experiment will have very low background so only a few events can confirm an oscillations of $\nu_\mu \to \nu_\tau$. For a five year exposure with 2.2×10^{20} protons on target, the experiment would expect 20 $\nu_\mu \to \nu_\tau$ detected oscillation events over a 0.5 event background for $\Delta m^2 = 3.5 \times 10^{-3}$ eV^2.

The ICANOE experiment proposes to use long drift length liquid argon calorimeter modules interspersed with magnetized fine grain solid calorimeter modules. The plan is to have four 1250 ton liquid argon and four 625 ton solid calorimeter modules for a total target mass of 7500 tons. The goal of the experiment is to detect and identify all neutrino species allowing both a ν_τ and ν_e appearance search. Tau neutrinos will be identified using fine grain nature of the detector to make kinematical selections of events with missing momentum as done in the NOMAD experiment. Electron neutrino interactions will be isolated using tracking combined with electromagnetic shower identification. With four years of data assuming 1.8×10^{20} protons on target, the experiment claims to cover at 90% CL $\nu_\mu \to \nu_\tau$ oscillations with $\Delta m^2 > 1.3 \times 10^{-3}$ eV^2 and $\nu_\mu \to \nu_e$ oscillations for $\sin^2 2\theta$ down to 1.5×10^{-3}.

III THE BOOSTER NEUTRINO OSCILLATION EXPERIMENT (BOONE) AT FERMILAB

The third indication of neutrino oscillations comes from the LSND experiment which has reported evidence for $\bar\nu_\mu \to \bar\nu_e$ oscillations with an oscillation probability of $\sim 0.3\%$ [2]. If this result is proven to be correct, then it will change many assumptions about the phenomenology of neutrino oscillations. Sterile neutrinos may be needed since there would be three different Δm^2 values; the LSND Δm^2 range would require at least one neutrino to have a mass near 1 eV^2; and CP violation studies would become much more possible. Restrictions from other experiments (CCFR, Nomad, Bugey, and Karmen) limit the range of a possible oscillation interpretation of the LSND results to the region with $0.2 < \Delta m^2 < 2.0$ eV^2 and $0.001 < \sin^2 2\theta < 0.03$. None of these experiments will cover the full region of the LSND result.

The BooNE neutrino oscillation experiment at Fermilab is designed to make a definitive check of the LSND result with the capability to measure the oscillation parameters with good precision [5]. The experiment will use the 8 GeV Booster to produce a low energy ($\langle E_\nu \rangle \approx 1$ GeV) neutrino beam and search for oscillations in the $\nu_\mu \to \nu_e$ appearance and ν_μ disappearance channels. The BooNE program has been approved to start with a single detector (MiniBooNE) with the goal of covering the entire LSND mass region and establish definitively whether there are neutrino oscillations. If a positive signal is observed, a two detector experiment would be proposed where a second detector at a different distance is added to the MiniBooNE setup.

The MiniBooNE experiment is designed to have a detector with a large fiducial mass and good particle identification for neutrino events in the $0.10 < E_\nu < 2.0$ GeV energy region. At these low energies, a totally active detector is necessary. A detector based on a large volume of mineral oil is both cost effective and very powerful for particle identification. Many of the critical detector components are available from the LSND experiment including the 1220 eight-inch photomultiplier tubes (PMTs) with readout and data acquisition system. The mineral oil will be contained in a spherical tank 12 m in diameter leading to a fiducial volume corresponding to $\sim$ 445 tons. The phototubes will be located at a radius of 5.5 m, and the 50 cm veto region between the outer wall and phototubes will be optically isolated from the main volume and viewed by an additional 292 phototubes facing outward.

MiniBooNE is on schedule to start data taking in December, 2001. The detector tank has been completed and PMT installation will start in the fall of 2000. Detector commissioning is to begin in the summer of 2001. The contract for the target station and beamline fabrication has been awarded and construction is beginning. One year of data during 2002 will be needed to refute or confirm the LSND signal at the $> 5\sigma$ level.

With the proposed beam intensity, MiniBooNE will record ~500,000 ν_μ CC quasi-elastic events per year, and if the LSND signal is due to neutrino oscillations, there should between 800 and 1000 excess ν_e events per year, depending on the Δm^2 and $\sin^2 2\theta$ of the oscillation. The oscillation signal has a different energy distribution as compared to the intrinsic ν_e backgrounds and experimental methods have been developed to determine the level of the backgrounds directly from observed data. Particle misidentification can be reduced to the $\approx 10^{-3}$ level while keeping the ν_e and ν_μ efficiency at $\sim$60%. The identification techniques use the spatial and time correlation of the detected Čerenkov and scintillation light by the PMTs. The level of these backgrounds can be measured directly from the data using the preponderance of events which are identified unambiguously. Systematic uncertainties on the intrinsic beam and misidentification backgrounds are expected to be at the 5-10% level.

The sensitivity of the MiniBooNE experiment for one year of data including statistical and systematic errors is shown in Figure 5. The ability of the experiment to isolate a ν_e appearance oscillation signal and measure the parameters for the signal can be estimated from energy dependent fits to simulated data generated with given input parameters. For one year of running with 5×10^{20} protons on target, Table 2 shows the results for two sets of parameters that span the allowed edge of the LSND 90% CL region. In both cases, MiniBooNE will establish an oscillation signal at greater than 15 standard deviations and measure the parameters with good accuracy.

As shown above, if a signal is observed in MiniBooNE, Δm^2 will be determined

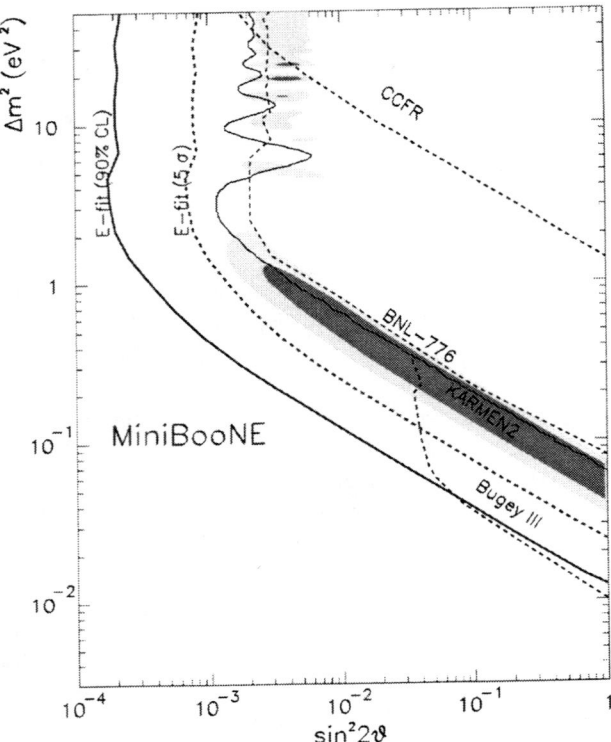

FIGURE 5. Sensitivity for the MiniBooNE experiment. The outer solid curve is th 90% CL region and the inner dashed curve is the 5σ region.

TABLE 2. Energy dependent fits for two example oscillation signals. The statistical sample corresponds to one year of MiniBooNE running with 5×10^{20} protons on target.

Δm_0^2	$\sin^2 2\theta_0$	$\delta\left(\Delta m^2\right)$	$\delta\left(\sin^2 2\theta\right)$	Signal Significance
0.3 eV^2	0.03	0.10 eV^2	0.02	44 σ
2.0 eV^2	0.002	0.10 eV^2	0.0002	15 σ

with sufficient accuracy (≈ 0.1 eV^2) to indicate where a second detector should be placed for the follow-up two detector BooNE experiment. For example, if $\Delta m^2 \approx 0.3$ eV^2 (2.0 eV^2) then the second detector should be placed at 2 km (0.25 km) to best determine the oscillation parameters as indicated in Fig. 6. As an example of the expectations for BooNE (the full 2 detector experiment), consider the case of low Δm^2 where the second detector is placed at 2 km. For this example, BooNE will measure Δm^2 to ± 0.014 eV^2 and $\sin^2 2\theta$ to ± 0.002 for one year of running with 5×10^{20} protons on target.

IV MUON STORAGE RING ν-FACTORY

A high intensity muon storage ring could provide a unique neutrino beam facility for pursuing neutrino oscillation measurements. The composition and spectra of the intense neutrino beam from a muon storage ring is selectable by the charge, momentum, and polarization of the stored muons, through the decays $\mu^- \to e^- \nu_\mu \bar{\nu}_e$ or $\mu^+ \to e^+ \nu_e \bar{\nu}_\mu$. There is no other comparable source of electron neutrinos and antineutrinos or neutrino beam with such well-understood fluxes. For muon energies above 20 GeV, the beam intensity from a muon storage ring is much greater than conventional hadron focused neutrino beams. The combination of almost equal numbers of electron and muon neutrinos allows the study of all types of oscillation processes including $\nu_e \to \nu_{\mu \text{ or } \tau}$. Furthermore, the beam is highly collimated and very long baseline experiments from 2000 to 10,000 km are possible.

Two studies of a muon storage ring facility have recently been performed. The first was an accelerator study [6] charged to develop a design concept for a muon storage ring and associated support facility that could support a compelling neutrino based research program. The goal of the study was to look at a design for a 50 GeV muon storage facility that would provide 2×10^{20} μ decays per year directed to a site 2900 km away such as Fermilab to SLAC or LBNL. The study came up with the conceptual design shown in Figure 7 and identified the areas needing R&D. Further, the study showed that a ν-Factory would be a unique facility that could be accomplished through a staged program of increasing energy and intensity. The machine and detector parameters can be tuned together since the oscillation sensitivity is proportional to the product, *Storage ring energy* × *Storage ring intensity* × *Mass detector*. The conclusion of the study was: "The result of this study clearly indicates that a neutrino source based on the concepts, which are

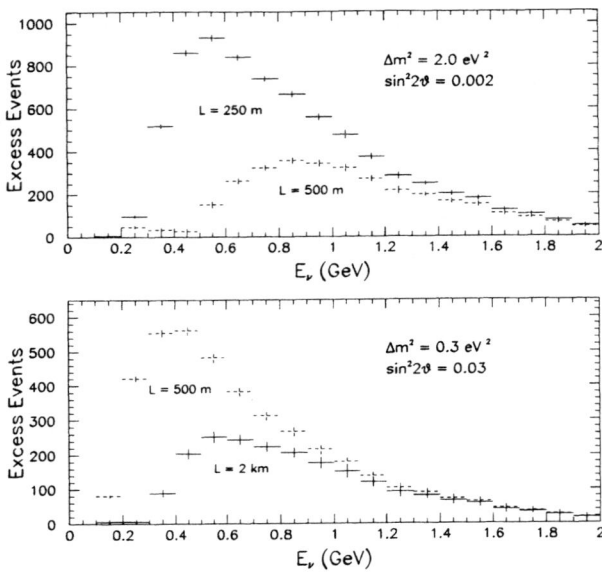

FIGURE 6. The energy distribution of excess events over background for two example oscillation signals and several distances for the detector. The solid (dashed) points in the top plot are for $\Delta m^2 = 2\ eV^2$, $\sin^2 2\theta = 0.002$ at a distance of 250 m (500 m). The solid (dashed) points in the bottom plot are for $\Delta m^2 = 0.3\ eV^2$, $\sin^2 2\theta = 0.03$ at a distance of 2 km (500 m). The statistical sample corresponds to an increased BooNE sample corresponding to 2×10^{21} protons on target.

presented here, is technically feasible."

The second study [7] looked at the physics motivation for a neutrino source based on a muon storage ring and how such a facility would allow the study of neutrino oscillations beyond the current set of experiments. The study was to focus on the physics capability as a function of stored muon energy, intensity, and polarization as well as possible baseline lengths.

A ν-factory could possibly be brought to fruition near the end of this decade. At that time, the current round of oscillation experiments and some possible upgrades will have been completed. Each of the three indications of oscillations should be either confirmed or refuted and the mixing parameters for the atmospheric, solar, and LSND regions should be measured to 10%. The dominant flavors involved in the oscillation for each region should also be know as well as if sterile neutrinos are needed.

In this context, the motivation for a ν-factory would be to extend the oscillation measurements to higher precision and other parameters associated with the neutrino mixing matrix given in Eq. 1. These measurements will allow tests of

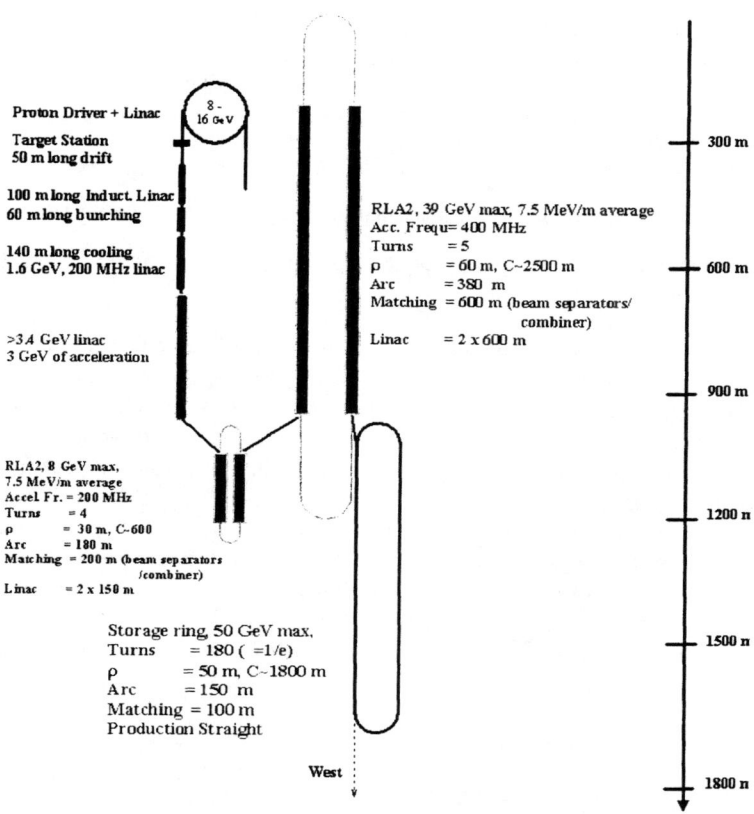

FIGURE 7. Layout of the muon storage ring facility from the ν-factory machine study.

the mixing and CP violation phenomenology predicted similar to studies of CKM quark mixing matrix. Specifically, a ν-factory could provide:

1. Unique access to ν_e oscillations to ν_μ or ν_τ

2. Measurements of Δm_{23}^2, θ_{23} to 1% and θ_{13} to the 10^{-4} level

3. A determination of the neutrino mass hierarchy by measuring the sign of Δm_{23}^2

4. Measurements of δ, the CP violation parameter, if solar or LSND parameters are favorable

A ν-factory has many advantages over a ν beam program. The CC ν event rate for a 20 GeV μ storage ring with 2×10^{20} μ decays per year is eight time higher than the NuMI beam. The angular divergence of the beam is much better especially for $E_\mu > 30$ GeV allowing much longer baseline experiments. Conventional beams only produce ν_μ with a small ν_e background that is hard to model; a ν-factory beam almost equal ν_μ and ν_e rates. The flux versus energy from a ν-factory beam is easily understood from the rate and energy of the stored muons whereas a conventional beam relies heavily on knowing the production rates and spectrum for hadronic production of pions and kaons. The oscillation mode $\nu_e \to \nu_\mu$ is inherently much more precise than $\nu_\mu \to \nu_e$ for measuring $\sin^2 2\theta_{13}$.

The measurement of $\sin^2 2\theta_{13}$ can be made using the second process listed in Eq. 2 but specifically $\nu_e \to \nu_\mu$

$$P(\nu_e \to \nu_\mu) = \sin^2 \theta_{23} \sin^2 2\theta_{13} \sin^2 \left(1.27 \Delta m_{32}^2 L/E_\nu\right)$$

By using the $\nu_e \to \nu_\mu$ signal, the measurement becomes a search for wrong-sign muons instead of a ν_e signal which has much higher background. For example, stored μ^+'s normally give CC $\bar\nu_\mu$ interactions with an outgoing μ^+ but an oscillation an $\nu_e \to \nu_\mu$ would lead to an ν_μ interaction with a outgoing μ^- as shown below:

$$\mu^+ \to \nu_e + \bar\nu_\mu + e^+$$
$$\hookrightarrow \nu_\mu \to \mu^- + hadrons$$
(with $\bar\nu_\mu \to \mu^+ + hadrons$)

Studies [7] have shown that the wrong-sign muon background can be reduced to the 10^{-4} level with requirements that $P_\mu > 4$ GeV and $P_T^2 > 2$ GeV^2. With the wrong-sign muon technique sensitivities as shown in Figure 8 can be reached.

For $\nu_e \to \nu_\mu$ oscillations, matter effects can change the oscillation probability as the neutrino propagates through the earth. This happens since the $\nu_e + e \to \nu_e + e$ process has both NC and CC contributions whereas the $\nu_\mu + e \to \nu_\mu + e$ has only a NC contribution. Through this mechanism, the oscillation probability becomes sensitive to the sign of $\Delta m_{32}^2 = m_3^2 - m_2^2$ and therefore allows a measurement of the mass hierarchy, whether $m_3 > m_2$ or not. Furthermore, the difference between $\nu_e \to \nu_\mu$ and $\bar\nu_e \to \bar\nu_\mu$ oscillations is sensitive to CP violations (δ in Eq. 2). Figure 9

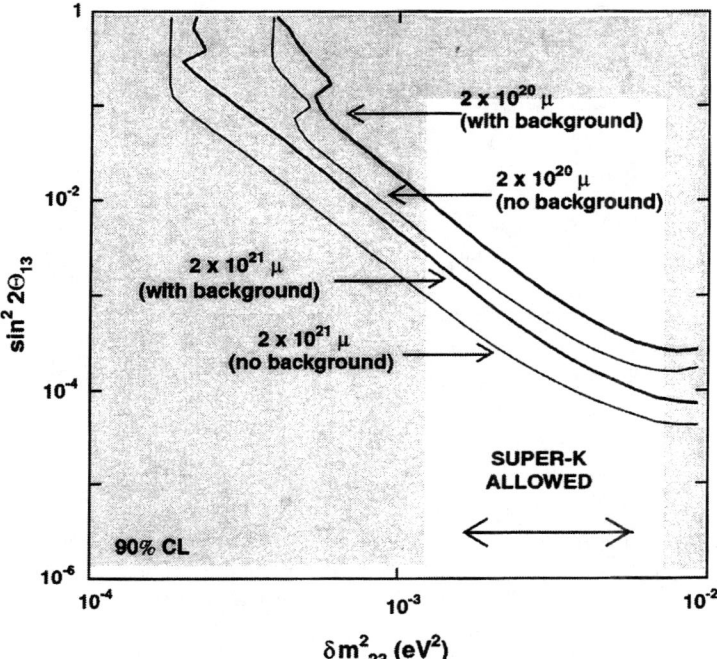

FIGURE 8. Limits on $\sin^2 2\theta_{13}$ that would result from the absence of a $\nu_e \to \nu_\mu$ signal in a 10 kt. detector 7400 km downstream of a 30 GeV neutrino factory in which there are 10^{20} and 10^{21} μ^+ decays, followed by the same number of μ^- decays. The limits are shown as a function of Δm_{32}^2 with and without background effects.

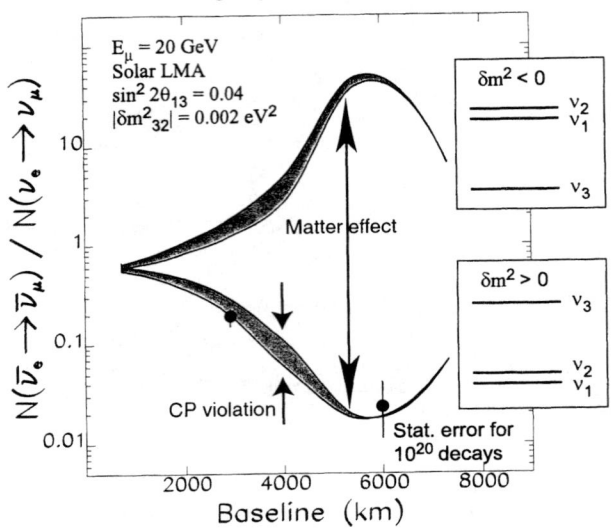

FIGURE 9. Predicted ratios of $\bar{\nu}_e \to \bar{\nu}_\mu$ to $\nu_e \to \nu_\mu$ rates at a 20 GeV neutrino factory. The upper(lower) band is for $\Delta m_{32}^2 < 0$ ($\Delta m_{32}^2 < 0$). The range of possible CP violation determines the widths of the bands. The statistical errors shown correspond to 10^{20} muon decays of each sign and a 50kt detector [7]

shows the ratio of number of $\bar{\nu}_e \to \bar{\nu}_\mu$ to $\nu_e \to \nu_\mu$ oscillations as function of baseline length for a 20 GeV muon storage ring with 10^{20} μ decays. There is a clear difference between the case with $\Delta m_{32}^2 < 0$ (top) and $\Delta m_{32}^2 > 0$ (bottom). The width of the curves represent the difference in rate for $\delta = 0$ and $\delta = \pi/2$. The black points with error bars represent measurements with the above statistics and clearly show that CP violation can be measured with a low-energy, modest-intesity ν-factory with the above parameters.

The physics study [7] looked at the physics reach for a ν-factory as a function of muon energy and the number of muon decays per year. Figure 10 shows a summary of these results indicating that an entry level machine with modest energy and intensity could make probe the new oscillation modes and measure the mass hierarchy leading to a high intensity machine that could measure CP violation. In summary, experiments at a ν-factory would be able to simultaneously measure, or limit, all of the appearance modes associated with ν_μ, ν_e, and ν_τ oscillations. The unique, well-understood ν_e component of the beam offers a way to probe ν_e oscillation channels that offer the best methods for detecting MSW matter and CP violation effects. A ν-factory is a unique, stageable facility for neutrino oscillation studies that could provide the mechanism for discoveries in the latter part of this decade.

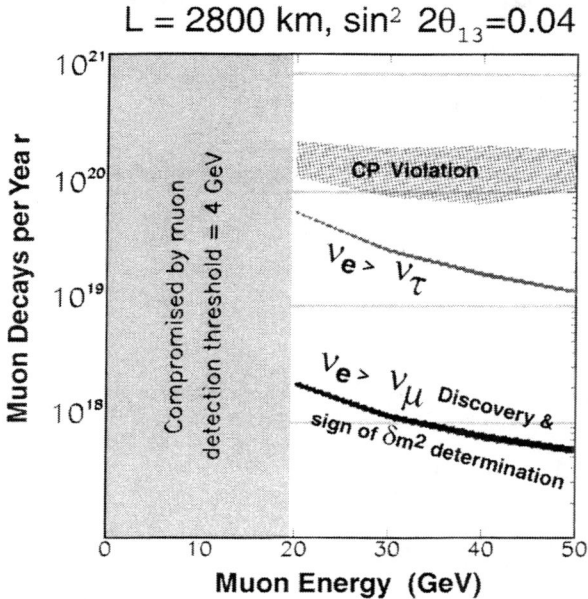

FIGURE 10. The required number of muon decays needed in a neutrino factory to observe $\nu_e \to \nu_\mu$ oscillations in a 50 kton detector and determine the sign of Δm^2 and the number of decays needed to observe $\nu_e \to \nu_\tau$ oscillations in a few kton detector, and ultimately put stringent limits on (or observe) CP violation in the lepton sector with a 50 kton detector [7].

REFERENCES

1. Y. Fukuda *et. al.*, Phys. Rev. Lett. **81**, 1562 (1998).
2. C. Athanassopoulos *et. al.* , Phys. Rev. Lett. **81**, 1774 (1998); C. Athanassopoulos *et. al.* , Phys. Rev. Lett. **77**, 3082 (1996).
3. http://www.hep.anl.gov/ndk/hypertext/numi.html
4. http://ngs.web.cern.ch/NGS/
5. http://www-boone.fnal.gov/
6. http://www.fnal.gov/projects/muon_collider/nu-factory
7. http://www.fnal.gov/projects/muon_collider/nu/study/study.html

SUPER–KAMIOKANDE'S PAST, PRESENT, AND FUTURE

MARK R. VAGINS FOR THE SUPER–KAMIOKANDE COLLABORATION

Department of Physics and Astronomy, University of California, Irvine
4129 Reines Hall, Irvine, CA 92697, USA
E-mail: mvagins@uci.edu

Abstract. Results from the first 1117 days of Super–Kamiokande's solar neutrino data are presented, including the absolute flux, energy spectrum, day/night and seasonal variation. The possibility of MSW and vacuum oscillations is discussed in light of these results. Results from the first 1144 days of Super-K's atmospheric neutrino analysis are also presented, including the evidence for $\nu_\mu \to \nu_\tau$ oscillations, against $\nu_\mu \to \nu_{sterile}$ oscillations, and the current limits on proton decay.

I INTRODUCTION

Super–Kamiokande is the product of the collaborative effort of approximately 110 Japanese and American astrophysicists, many of whom previously worked on either the Kamiokande or IMB water Cherenkov experiments. The Super–Kamiokande site is about 300 kilometers west–northwest of Tokyo near the small town of Mozumi in the Japanese Alps. Located under 1 kilometer of rock (2,700 meters water equivalent) in the same ancient zinc mine as the recently decommissioned Kamiokande, Super-K shares the same basic design as its former neighbor and namesake: a cylinder of ultra–pure water surrounded with inward–facing photomultiplier tubes [PMT's], a light barrier, a layer of outward–facing PMT's, and a veto region of water, all contained within a stainless steel tank.

Roughly an order of magnitude larger than its predecessors, Super-K has been designed to be a premier facility for studying solar neutrinos, atmospheric neutrinos, nucleon decay, and neutrinos from galactic supernovae. Weighing in at 50,000 tons of water, and holding over 11,000 fifty-centimeter diameter PMT's and 1,850 twenty-centimeter PMT's, Super-K is the world's largest underground water Cherenkov detector. Figure 1 shows a cutaway view of the detector.

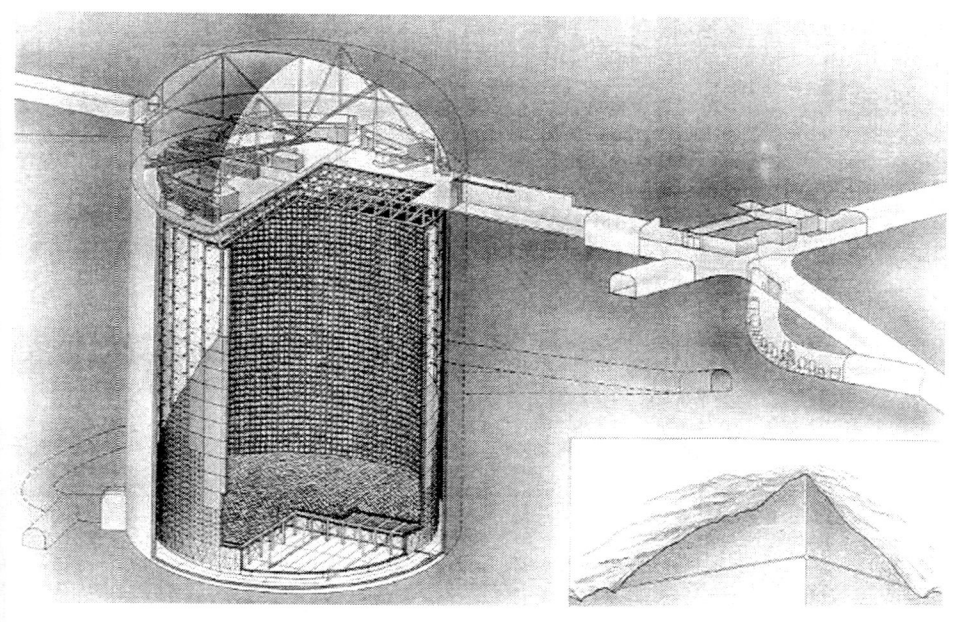

FIGURE 1. Cutaway view of the Super–Kamiokande detector. The detector itself stands some 42 meters tall — to get a sense of the scale, consider that the tunnels shown in the drawing are large enough for full-sized trucks to drive through. The curving tunnel on the right side of the sketch contains our water purification plant; just above that is our main control room.

II LOW ENERGY ANALYSIS — SOLAR NEUTRINOS AT SUPER–KAMIOKANDE

A Solar Intro

When we talk about Super–Kamiokande's "low-energy" analysis, we generally mean events with less than 100 MeV of visible energy deposited in the detector. With an endpoint energy of about 15.0 MeV, Super–Kamiokande's primary source of solar neutrinos is the following nuclear reaction in the Sun:

$$^8B \to\, ^8Be^* + e^+ + \nu_e \tag{1}$$

These ^{8}B neutrinos are seen in Super–K via elastic scattering:

$$\nu_e + e^- \to \nu_e + e^- \tag{2}$$

We have obtained what is by far the largest single sample of solar neutrino events in the world. Our most recent results, representing 1117 live days of analyzed low-energy data, spanning the period of May 31, 1996 to April 24, 2000, will be presented below.

B Calibration of Super–Kamiokande's Absolute Energy Scale

1 The Need for Precise Calibration

Super–Kamiokande observes ^{8}B neutrinos originating from the sun by elastic scattering of electrons in water. The energy of the neutrino cannot be reconstructed kinematically, so only the recoil electron spectrum of the ^{8}B neutrinos can be observed. The recoil electron's energy being only a lower limit for the neutrino energy, this spectrum falls steeply with increasing energy and small uncertainties in the calibration of the energy scale lead to large uncertainties in the measured flux and energy spectrum. Therefore, a precise calibration of Super–Kamiokande's energy spectrum is necessary for us to obtain useful solar neutrino results, especially if we wish to be able to study the possibility of solar neutrino oscillations.

The two most powerful calibration tools presently at our disposal are an electron linear accelerator [LINAC], which has been used to great effect in the past, and a new device introduced in 1999, a Deuterium–Tritium [D–T] neutron generator.

2 LINAC Calibration

Super–Kamiokande's most precise calibration has been achieved with a Mitsubishi ML-14MIII electron linear accelerator. It injects single electron pulses of

Position	A	B	C	D	E	F	G	H	I
X (m)	-3.88	-3.88	-8.13	-8.13	-12.37	-12.37	-3.88	-12.37	-8.13
Y (m)	-0.71	-0.71	-0.71	-0.71	-0.71	-0.71	-0.71	-0.71	-0.71
Z (m)	11.97	-0.06	11.97	-0.06	11.97	-0.06	-12.09	-12.09	-12.09

TABLE 1. Linear accelerator beam-pipe end positions.

beam momentum (MeV/c)	5.08	6.03	7.00	8.86	10.99	13.65	16.31
Ge energy (MeV)	4.25	5.21	6.17	8.03	10.14	12.80	15.44
in-tank energy (MeV)	4.89	5.84	6.79	8.67	10.78	13.44	16.09

TABLE 2. Linear accelerator beam momentum and associated energy. The second line is the energy as measured by the Ge detector, while the third line is the total energy of the electrons after leaving the beam pipe.

energies ranging between 5.0 MeV and 16.0 MeV into Super-Kamiokande's water tank.

The details of this calibration are described in Reference 6. The experimental setup can be seen in Figure 2. At eight positions in the water tank the energy spectrum of the downward-going monochromatic beam was studied at seven fixed beam momenta (see Tables 1 and 2). For each beam momentum the energy of the outgoing electrons is determined by a Germanium detector.

This method of calibration has the advantage that electrons of a well-known energy and direction are used. Also, the end of the beam pipe contains a scintillation counter, allowing the study of detector response to electrons on an event-by-event basis. However, for each change of position the beam pipe has to be disassembled at the old position and reassembled at the new position. Each energy requires tuning of the beam. Therefore, a comparatively large number of people (about six people for assembly and about four people to operate the beam in two shifts) need to devote most of their time to operate the accelerator while it is running. A scan through several positions and all energies together with frequent measurements of the accelerator energy with the Ge detector is referred to as "LINAC festival." Such festivals typically last between two to four weeks.

In January, 1999, a permanent magnet was installed at the end of the beam pipe to bend the beam into different (non-vertical) directions. This data has been used to study the directional dependence of the energy scale.

3 D-T Generator Calibration

A suitable energy calibration source in the solar neutrino energy range is ^{16}N which decays into ^{16}O with a half-life of 7.13 sec. The endpoint of the spectrum is 10.4 MeV, and for 72% of the decays a 6.1 MeV (66%) or 7.1 MeV (5%) photon is produced in addition to an electron and a neutrino. Muon capture on ^{16}O produces ^{16}N "naturally." It can be produced "artificially" by a reaction of neutrons with

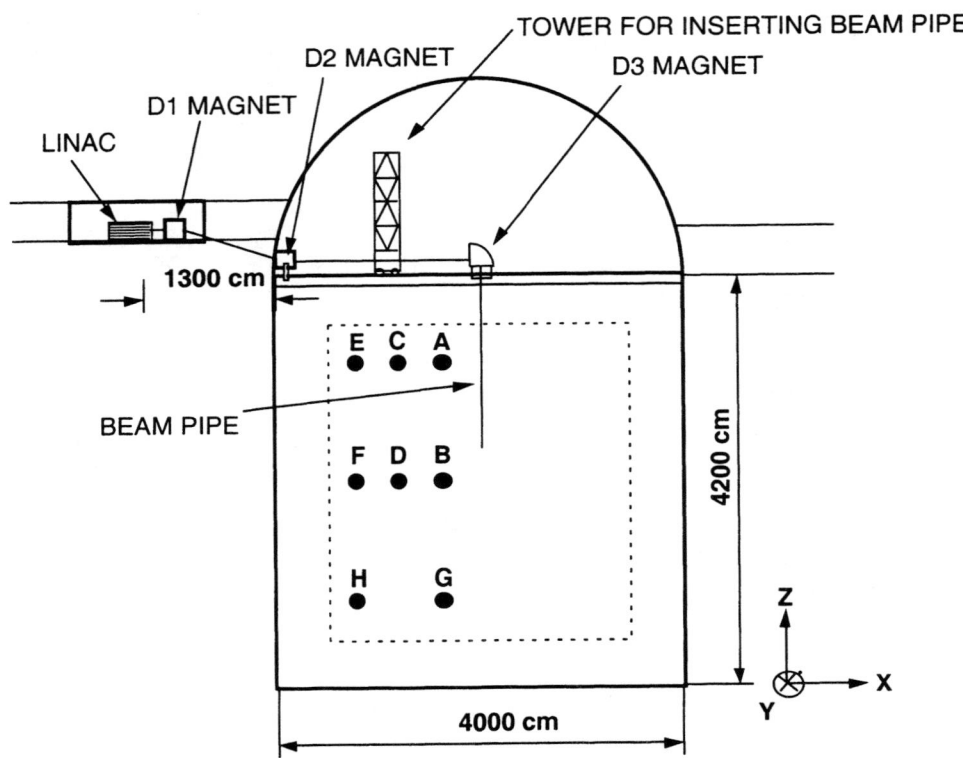

FIGURE 2. Schematic view of the linear accelerator setup and the Super–Kamiokande tank. The eight positions where data was taken are marked A through H. Data was also recently taken at a ninth position I between G and H.

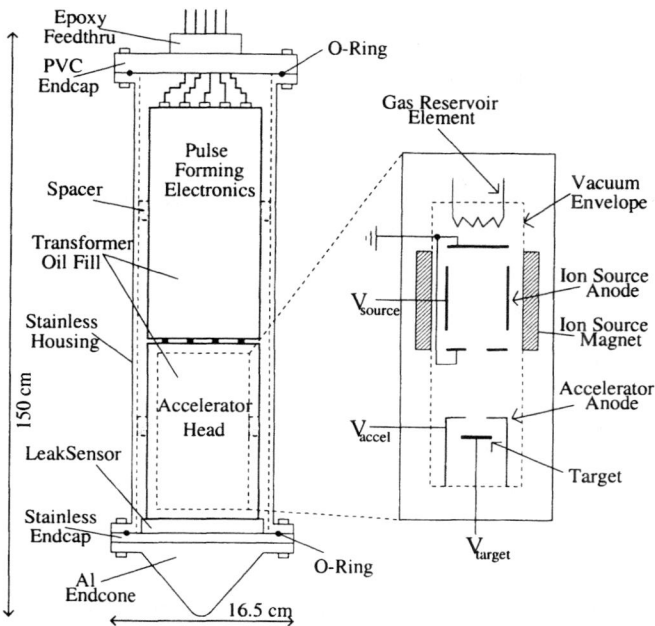

FIGURE 3. Schematic view of the Deuterium-Tritium neutron generator (D-T generator). A Penning ion source generates Deuterium ions which are accelerated by a voltage between 80 and 180 kV into a Tritium target. The resulting nuclear fusion reaction yields 14.2 MeV neutrons.

^{16}O. Super–Kamiokande uses a deuterium accelerator with a tritium target (see Figure 3) to produce large numbers of ^{16}N.

The D-T generator is small enough to be immersed in the Super–Kamiokande tank (see Figure 4); the depth is controlled with a computer-operated crane. A high voltage pulse accelerates a bunch of deuterium ions into the tritium target and produces roughly a million 14.2 MeV neutrons. The D-T generator is automatically pulled upwards leaving behind a cloud of approximately 20,000 ^{16}N atoms. During the next 45 seconds, the ^{16}N decay events are collected, each pulse leading to about 5,000 usable ^{16}N calibration events. After the events are collected, the crane automatically lowers the D-T generator and the next pulse is fired. The process is continued until sufficient statistics have been collected. Note that in just two or three firings of the D-T generator (less than three minutes) more ^{16}N events are observed than all the naturally-occurring ^{16}N events seen during the entire four years Super-K has been running!

At the end of March, 1999, the first D-T generator data was taken. A D-T generator energy calibration with high statistics can be done quickly. In July of 1999, a combined LINAC and D-T festival took LINAC and D-T data at the same positions to directly compare the two calibrations and study the directional dependence of

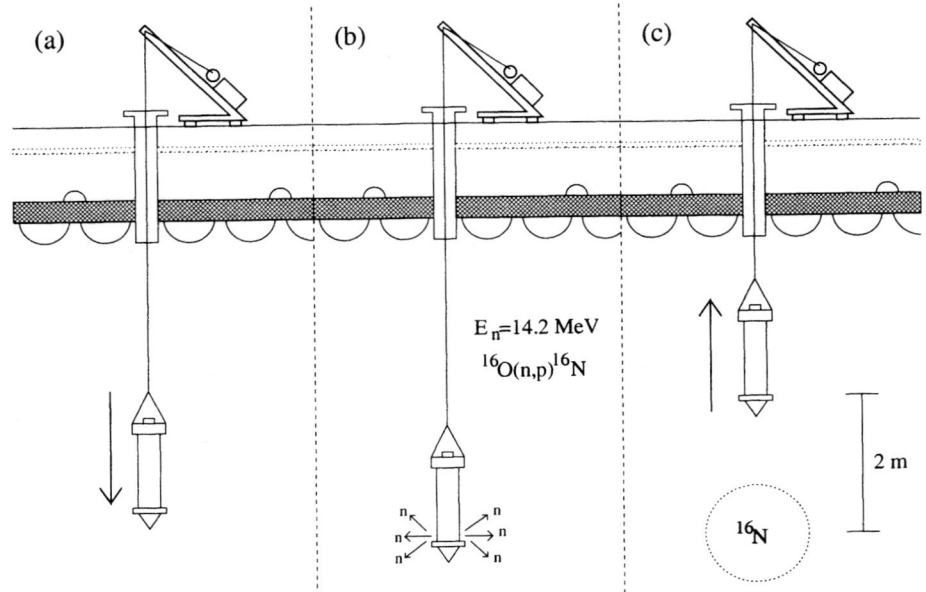

FIGURE 4. D-T data taking: the generator is suspended from a computer-controlled crane, and for each data-taking cycle (a) the generator is lowered to the desired position, (b) neutrons are generated, and (c) the generator is retracted, leaving behind a cloud of ^{16}N atoms.

the energy scale. It was expected that, once a well-understood correlation was established between the LINAC and D-T calibration data, periodic D-T data-taking would largely replace the slow, cumbersome, and manpower-intensive LINAC festivals of the past. Happily, the LINAC and D-T data agreed very closely, and now D-T data is now taken regularly (about once per month) to monitor the time stability of the energy scale and the trigger efficiency as a function of energy.

4 Results of the Calibration

The goal of the calibration was to tune the energy scale of Super–Kamiokande's Monte Carlo simulation with linear accelerator data to a precision of better than 1%. The deviation of Monte Carlo reconstructed energy to linear accelerator data can be seen in Figure 5. It is clear that a 1% calibration has been achieved. It has been shown that the D-T generator data agrees well with the linear accelerator data, and that the absolute energy scale does not depend strongly on the event direction (see Figure 6).

C Recent Results of the Solar Neutrino Analysis

1 Solar Neutrino Flux

Our standard method of displaying the solar neutrino signal is through the use of $\cos\theta_{sun}$ plots, where $\cos\theta_{sun}$ is the angle between a reconstructed low-energy event's direction and the direction defined by a line drawn between the Sun's current position and the vertex position.

The solar neutrino signal for the energy range 5.5 MeV to 20.0 MeV is shown in Figure 7. The peak above the background in the direction of $\cos\theta_{sun} = 1$ (i.e., originating from the direction of the Sun) are our solar neutrinos. There are some 15,000 events under the peak and above the background. This plot is one of the recent official results of the ongoing Super-K solar neutrino analysis, and represents the end result of all reduction and background suppression for 1117 live days of unified low-energy (6.5 MeV and above) and super low-energy (below 6.5 MeV) data [LE+SLE] and 22.5 ktons of fiducial volume.

Note that, unlike atmospheric neutrinos, one can only identify solar neutrinos in a statistical fashion. No one has yet devised a way to prove that any given event in our detector actually originated from the Sun. For this reason, reducing the sea of background events under the solar peak is of central importance in all low-energy investigations.

The best fit to the data points is given by the flux predicted by the BP98 version of the Standard Solar Model [1] multiplied by a factor of 46.5%. More specifically, we measure:

---

[1] J. Bahcall, S. Basu, and M. Pinsonneault, *Phys. Lett. B.*, **433**, 1 (1998)

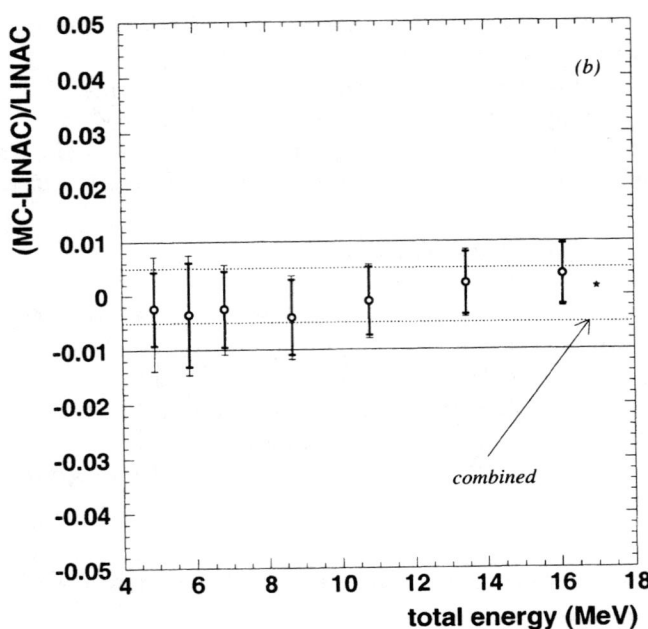

FIGURE 5. Difference of the energy scale between accelerator data and Monte Carlo. For each energy point, the position average is taken. The inner error bars are the RMS of the spread over position, the outer one is the systematic error. Dotted and solid lines show 0.5% and 1% difference between accelerator data and Monte Carlo. The last point is the average over all beam energies.

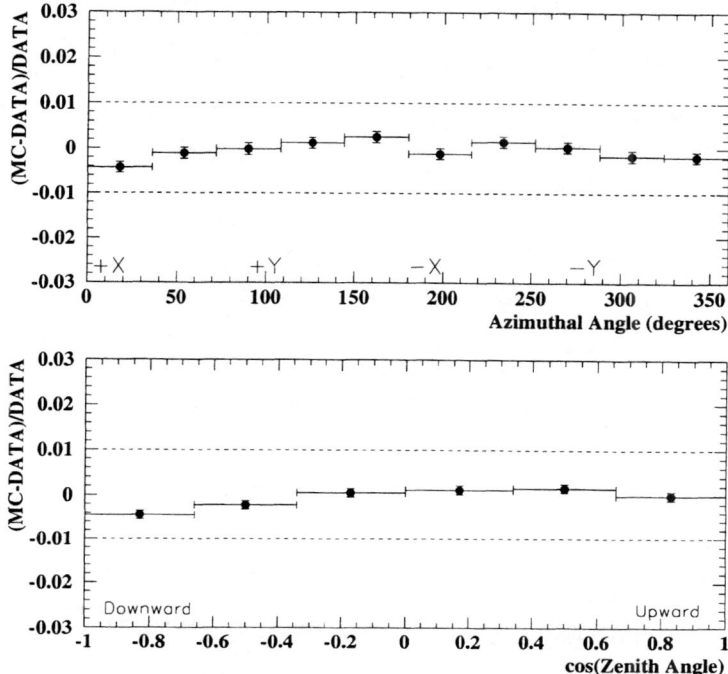

FIGURE 6. Difference of the energy scale between DT data and Monte Carlo as a function of azimuth and zenith angle. Each point is the result of a position average. The dependence of the energy scale on the direction is within 0.5%.

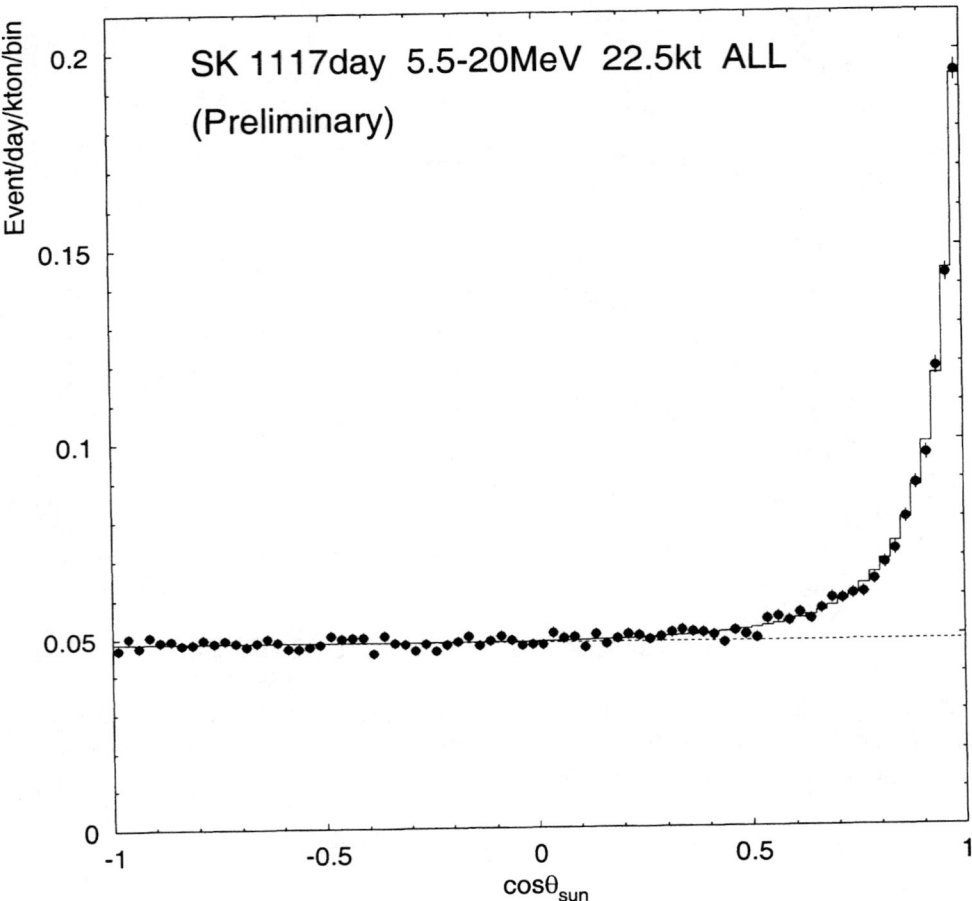

FIGURE 7. Solar neutrino signal between 5.5 MeV and 20.0 MeV. This is the result of 1117 live days of data and a 22.5 kton fiducial volume. The line is a fit to 46.5% of the BP98 SSM.

$$\text{flux} = 2.40 \pm 0.03(stat.) \pm 0.08(syst.) X 10^6/cm^2/sec \qquad (3)$$

and

$$\frac{\text{Data}}{\text{SSM}_{BP98}} = 0.465 \pm 0.005(stat.) \pm 0.014(syst.) \qquad (4)$$

This measurement of flux well below the predicted value is one of the manifestations of the so-called "solar neutrino problem," and in fact a result near 50% is rather suggestive. If electron neutrinos are in fact oscillating to a single (unseen) species and back again, and if they have fully oscillated many times by the time they reach the Earth, then we would expect to measure exactly 50% of the predicted flux. Since it is becoming widely accepted that atmospheric muon neutrinos are in fact oscillating [2], the possibility that neutrino oscillations will be the answer to the solar neutrino problem seems much more likely than it did just two years ago.

Figure 7 is a one-dimensional presentation of our data, but of course we know the angle between the Sun and the recoil electrons in two dimensions. A rather more dramatic visualization of our solar neutrino data is shown in Figure 8. This image can be thought of as a view of the Sun seen in "neutrino light." The graded grey scale is related to the number of neutrinos in each pixel, where the width of each pixel is approximately 0.5° across in the sky. Hence, the actual sun would just fit within a single pixel. This gives a powerful example of how difficult (impossible!) it is to show that any given event in our detector actually originated from the Sun; unlike the background-free atmospheric neutrinos, one can only identify solar neutrinos in a statistical fashion.

2 Low-Energy Neutrino Sky Map

In some sense one can invert the preceding flux analysis, and instead of requiring that events under consideration point back at the Sun one can ask "Where do *all* of our solar neutrino candidates point to in the sky?" In this way one can search for other cosmic sources of low-energy neutrinos than the nearest star.

Figure 9 shows a projection back into the sky of all of our low-energy data (and not just the data in the peak pointing back to the Sun) from the first 825 days of Super–K's running period. The heavens are divided into 768 equal area pixels, and the number of low-energy events emanating from each pixel is plotted. The path of the Sun is clearly seen tracing out the ecliptic, giving us confidence that the projection has been done correctly.

Next, in Figure 10 our solar neutrino signal events are subtracted from Figure 9. The path of the Sun dutifully disappears, but unfortunately no other sources are seen; the distribution of events per pixel histogrammed in the upper right-hand

---

[2] Y. Fukuda et al., *Phys. Rev. Lett.*, **81**, 1562 (1998).

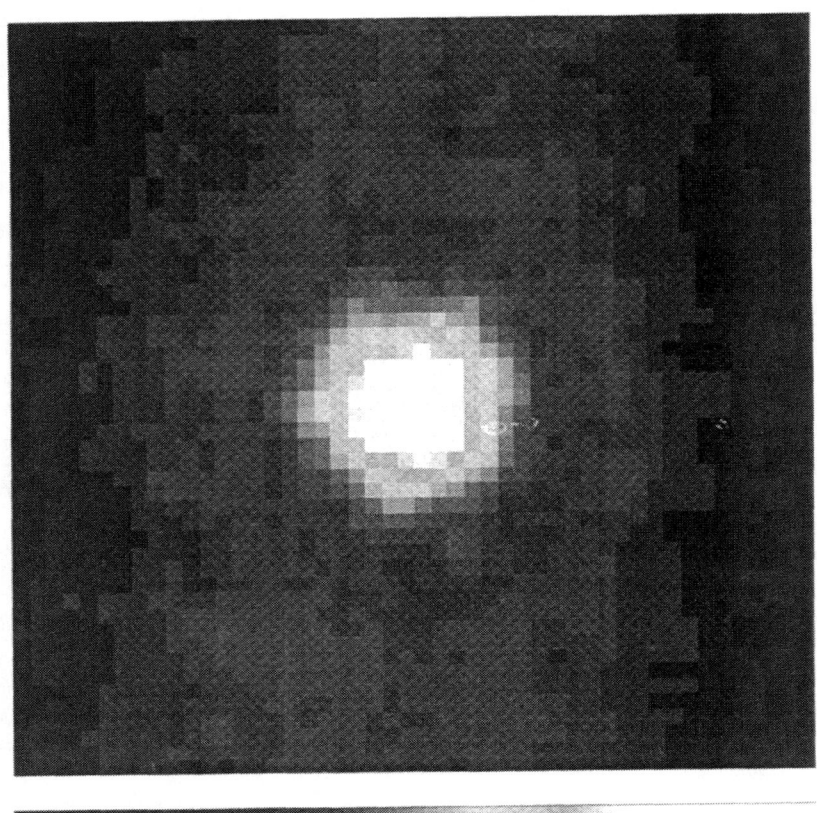

FIGURE 8. A two-dimensional plot of our solar flux data, this is the Sun as imaged in "neutrino light." The grey scale is related to the number of neutrinos in each pixel, where each pixel is approximately 0.5° across.

corner looks perfectly random. We will continue looking for point sources of low-energy neutrinos as more data are collected, however, since, like gamma ray bursts or supernovae, such sources could very likely turn on and then fade rapidly.

3 Day/Night Effect

Figure 7 contains all the LE and SLE data – if the data are broken down into bins based on where the Sun was in relation to the horizon at the time the signal was received we get Figure 11. The bins on the right side of the plot are defined in the upper right-hand corner of the figure.

Although it certainly seems plausible to conclude from Figure 11 that there is little variation between our daytime and our nighttime solar neutrino signals, in fact the maximum possible difference predicted due to matter-enhanced (MSW) neutrino oscillations [3] is similar to the size of the errors on the combined day and night points. In fact, it is still (barely) possible that in this plot we are seeing hints of a real day/night effect. At present our value for the overall day/night difference is as follows:

$$\frac{D - N}{(D + N)/2} = -0.034 \pm 0.022(stat.) \pm 0.013(syst.) \qquad (5)$$

If instead of equal-area bins we divide the night data into three equal-area bins (N1, N2, N3), one slightly oversized bin (N4+), and one bin containing only the core of the Earth, we arrive at Figure 12. Interestingly, the events which have passed through the core of the Earth look slightly suppressed, but the other night bins do not. This is quite difficult to explain via any possible matter-enhanced oscillation hypothesis, though the statistical errors on the bins are still rather large.

We have observed over 15,000 solar neutrinos between 5.5 MeV and 20.0 MeV since the start of the experiment, and are now recording a 1.3σ difference in the day vs. night signals. Continued acquisition and analysis of Super–K's data should allow us to resolve conclusively whether we are seeing any day/night effect or not.

4 Seasonal Variation of Flux

Another interesting study which can be performed by breaking up the data is the search for seasonal variations in the flux. Such variations would be due to vacuum oscillations as the Earth moves around the Sun. Our results are shown in Figure 13. The wavy line represents the expected $\frac{1}{r^2}$ variation in the flux due to eccentricity of the Earth's orbit. Figure 14 shows what the previous plot looks like when the data are combined in 1.5 month bins. In either plot it can be seen that the data lie along the expected line with no strong deviations, though once the data are combined the fit to the expected no-vacuum-oscillation line is rather

---

[3] S. P. Mikheyev and A. Y. Smirinov, *Nuov. Cim.*, **9**, 17 (1986)

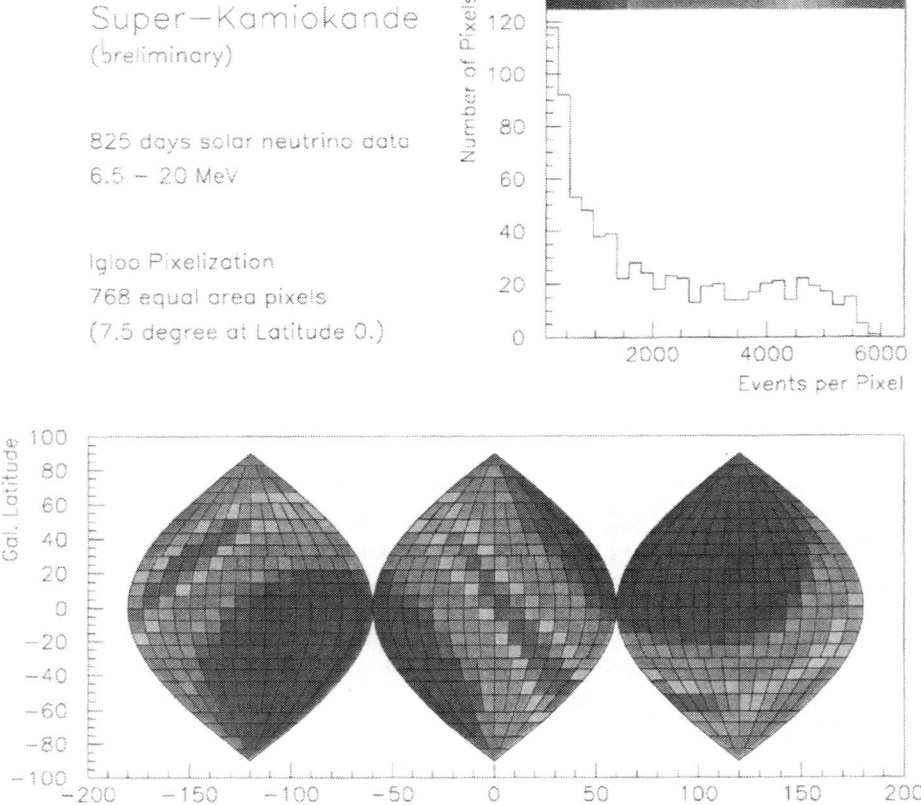

FIGURE 9. The entire first 825 days worth of low-energy data between 6.5 MeV and 20.0 MeV, not just those events pointing back to the Sun, are projected onto the sky. The path of the Sun (the ecliptic) is clearly visible.

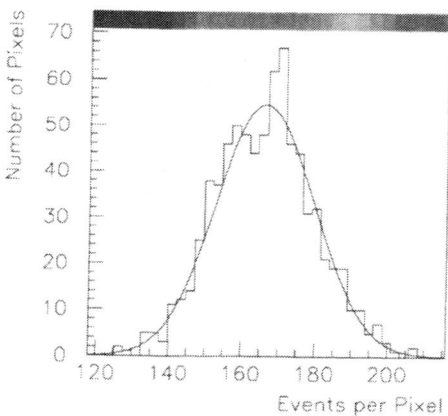
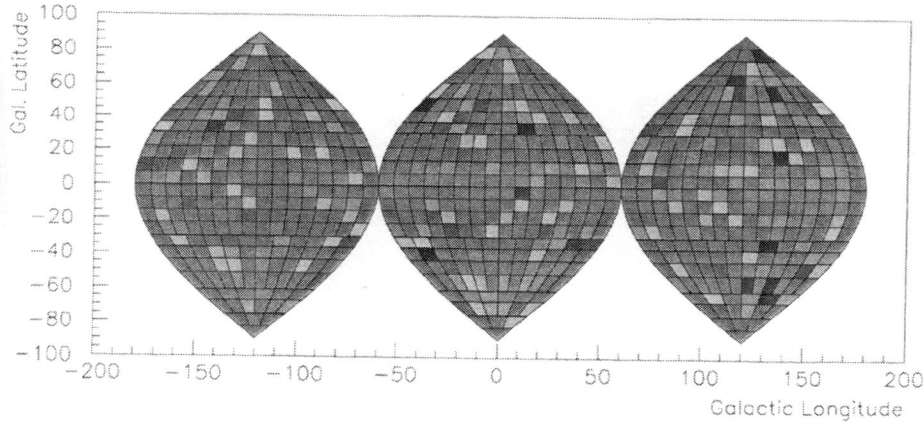

FIGURE 10. After subtraction of our solar neutrino signal, all remaining events from 825 days' worth of low-energy data between 6.5 MeV and 20.0 MeV are projected onto the sky. No point sources are seen — the distribution of events per pixel seen in the upper right-hand corner looks quite random.

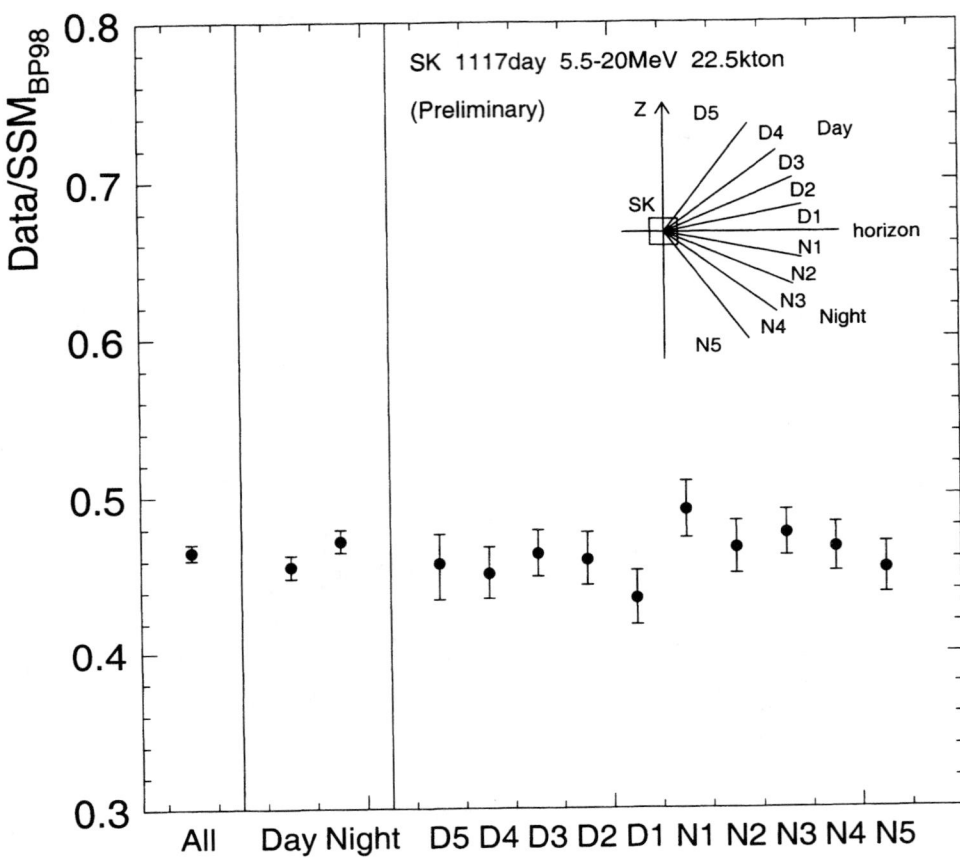

FIGURE 11. Day/night variation in solar neutrino signal between 5.5 MeV and 20.0 MeV. This is the result of 1117 live days of data and a 22.5 kton fiducial volume. The angular divisions D1 through D5 and N1 through N5 are defined in the upper right-hand corner.

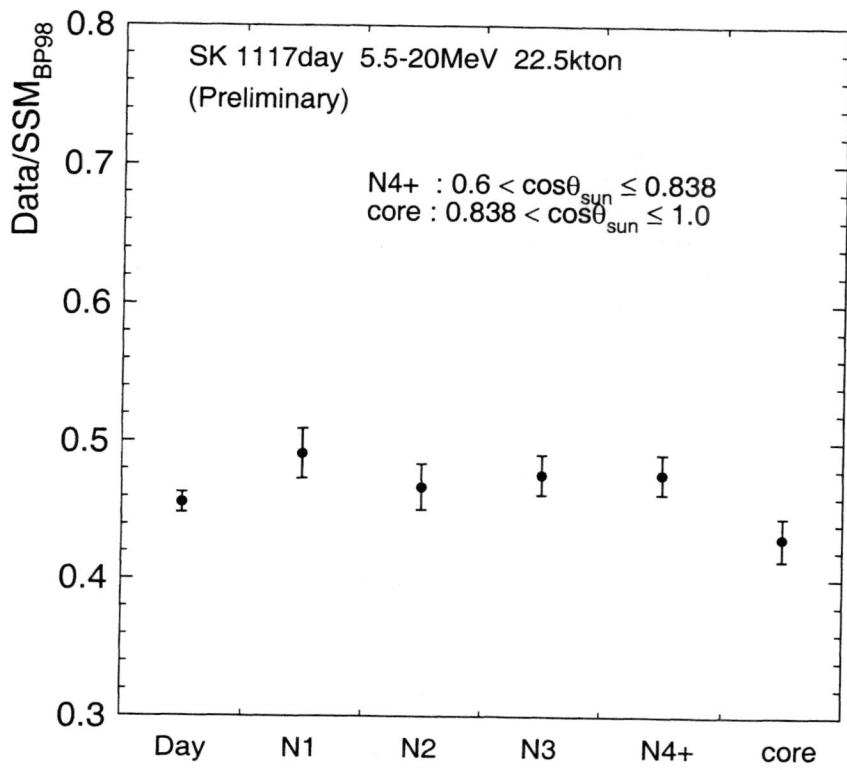

FIGURE 12. Day/night variation in solar neutrino signal between 5.5 MeV and 20.0 MeV. This is the result of 1117 live days of data and a 22.5 kton fiducial volume. The bins N4 and N5 (defined in the previous Figure) have been replaced with an expanded N4+ and a narrowed final bin containing only those events passing through the core of the Earth.

good. It will nevertheless be necessary to collect more data and shrink the error bars some more before we will be able to state conclusively whether or not there is any unexpected behavior going on as the Earth swings around the Sun.

There has recently been some theoretical interest as to whether the upcoming Solar Maximum will have any effect on the solar neutrino flux via correlations with more-frequent sunspots, solar flares, and other signs of increased solar activity. Figure 15 shows the data in Figure 13 compensated for the eccentricity of the Earth's orbit. This fairly stable rate of events is to be compared with Figure 16, which is a record of the number of observed sunspots during the same period. Even though the increase in the number of sunspots as Solar Max is approached is quite dramatic, there is no corresponding change in solar neutrino activity observed.

5 Energy Spectrum

Perhaps the most powerful test of oscillations, however, is made by looking at the energy spectrum of the recoil electrons from the ^{8}B solar neutrinos. Assuming that neutrinos are massive, neutrinos of a given energy will have an opportunity to execute a given number (or fractional number) of oscillations before reaching Super–Kamiokande. Therefore, deviations from the predicted spectral shape would constitute rather strong evidence of oscillations, since neutrinos of certain energies would then be more (or less) likely to be seen in our detector than neutrinos of other energies.

The data and the value of the predicted flux in each energy bin is shown in Figure 17. Note that, for the first time, new SLE data at 5.0 MeV is shown. We have a total of 837 live days of SLE data versus 1117 days of LE data because (as will be described in a coming section) the collection of SLE data was first implemented about one year after the experiment began.

The results of our energy spectrum analysis can be seen in Figure 18, where the data points have been divided by the non-oscillating prediction for each bin. If these points fell in a straight, flat line then they would be consistent with an unoscillated spectrum. In fact, the present shape seen in Figure 18 has a very good fit to flat. This lack of deviations will allow us to rule out certain oscillation hypotheses in a coming section.

6 hep Analysis

Since they seem to be a bit elevated compared to the rest of the (very flat) spectrum, there has been much attention given to the four highest-energy points in Figure 18, though their statistics are still rather poor. Some speculation has arisen that an excess at the upper end of our nominally pure ^{8}B spectrum could be the result of contamination from solar neutrinos produced in the much rarer so-called "hep" process:

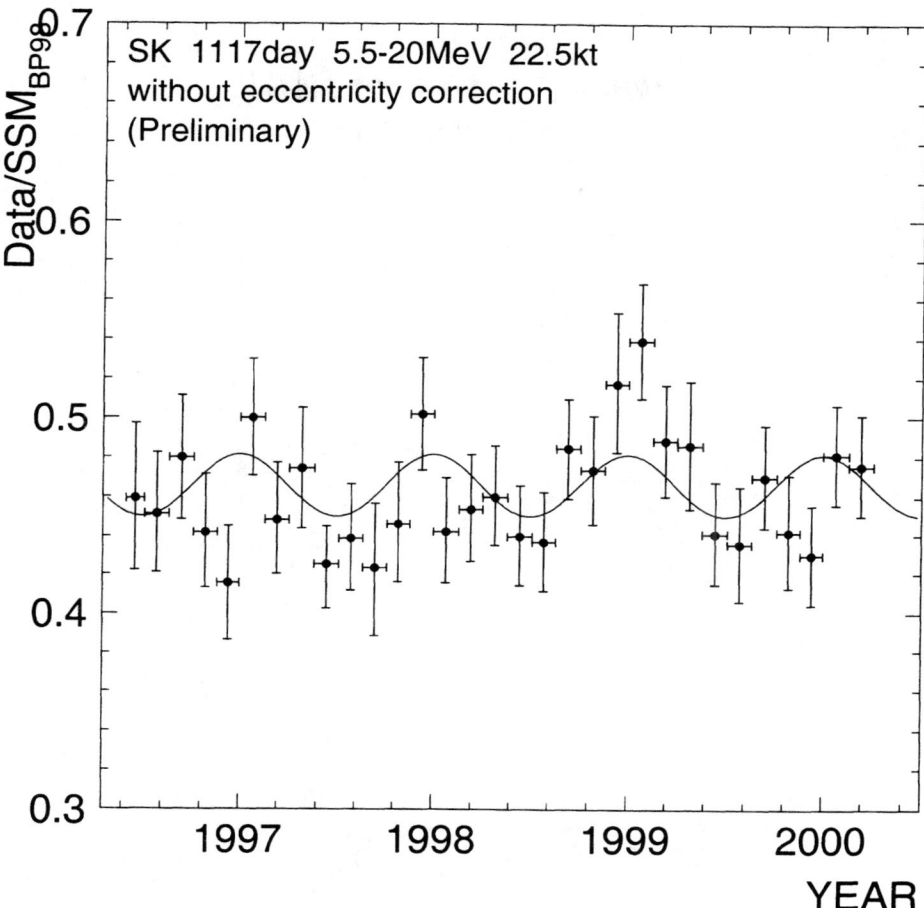

FIGURE 13. Seasonal variation in solar neutrino signal between 5.5 MeV and 20.0 MeV. This is the result of 1117 live days of data and a 22.5 kton fiducial volume. The wavy line represents the expected $\frac{1}{r^2}$ variation in the flux due to the eccentricity of Earth's orbit around the Sun. Deviations from this line could constitute evidence of vacuum oscillations.

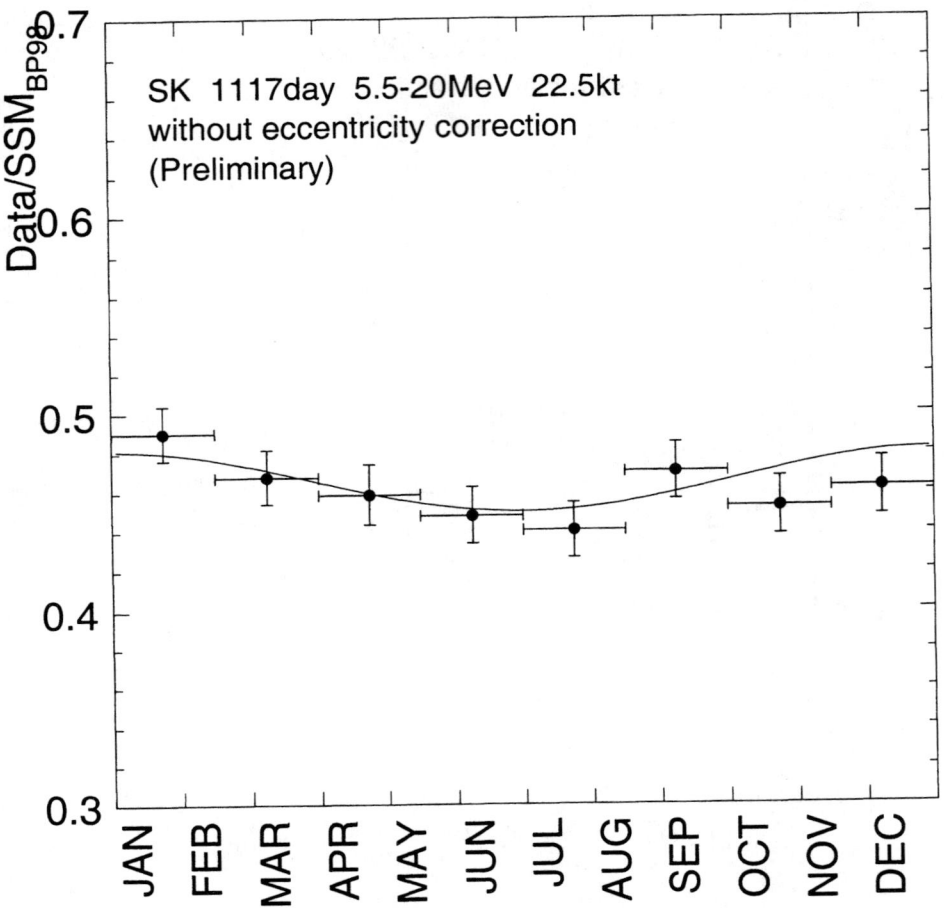

FIGURE 14. Seasonal variation in solar neutrino signal between 5.5 MeV and 20.0 MeV plotted in 1.5 month bins. This is the the same data as shown in the previous plot, but events which occurred at similar times during different years have been combined. The wavy line represents the expected $\frac{1}{r^2}$ variation in the flux due to the eccentricity of Earth's orbit around the Sun. While deviations from this line could constitute evidence of vacuum oscillations, the fit to the line is quite good.

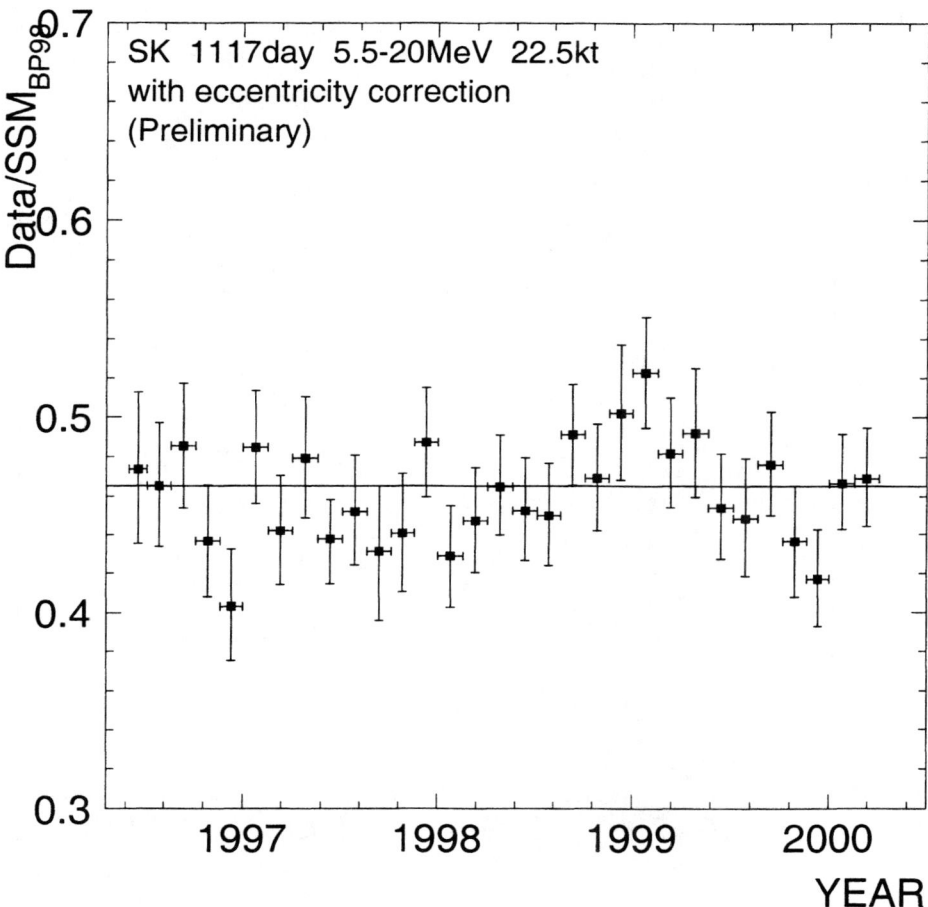

FIGURE 15. Time variation in solar neutrino signal between 5.5 MeV and 20.0 MeV. This is the result of 1117 live days of data and a 22.5 kton fiducial volume. The expected $\frac{1}{r^2}$ variation in the flux due to the eccentricity of Earth's orbit around the Sun has been corrected.

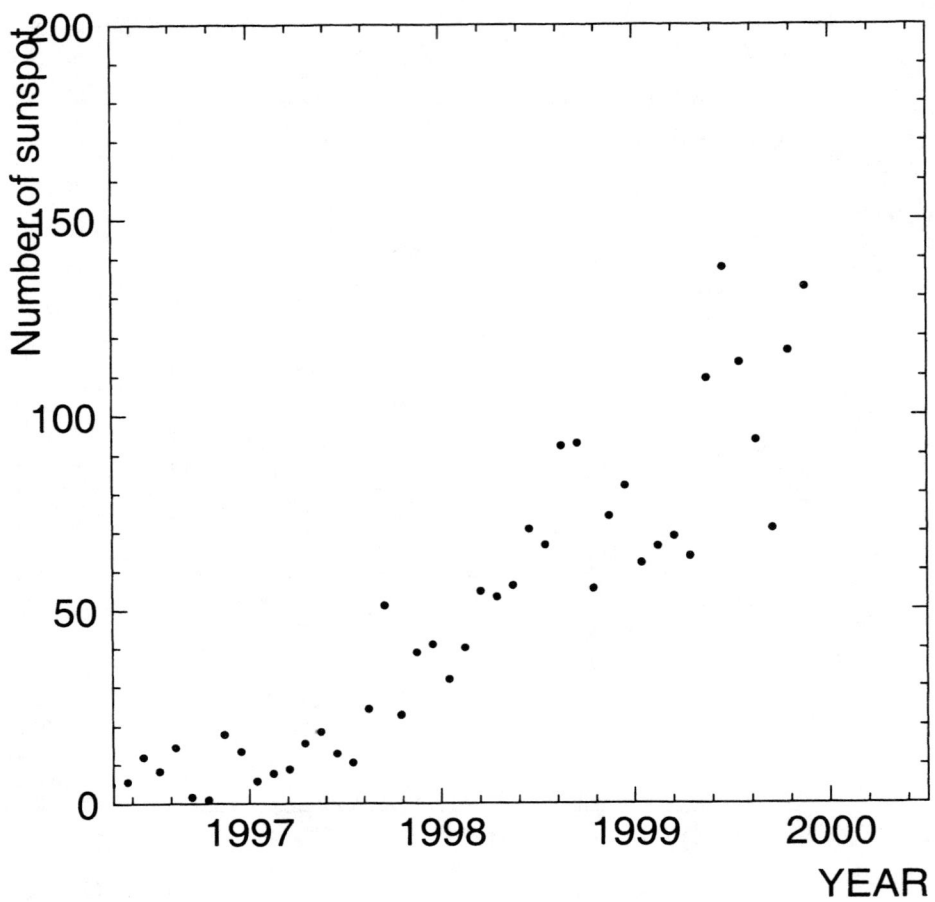

FIGURE 16. The number of observed sunspots is plotted as Solar Maximum is approached. This dramatically increasing rate is to be compared with the steady solar neutrino flux of Figure 15.

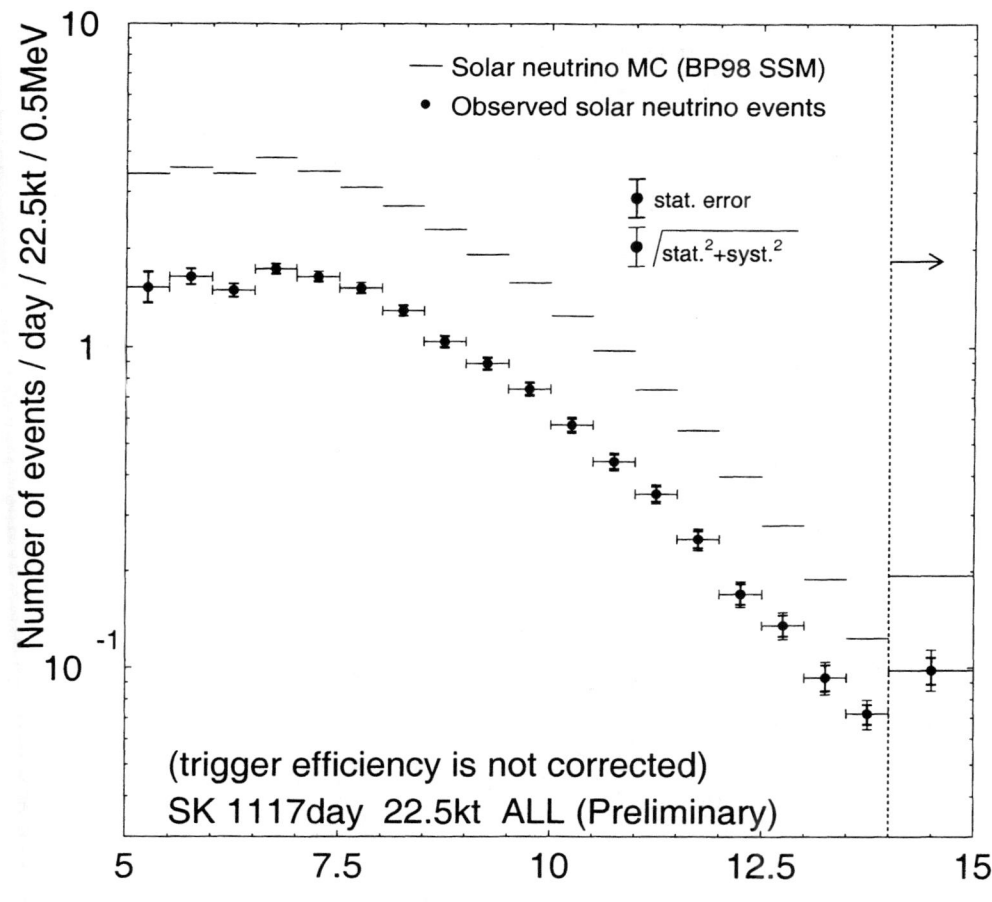

FIGURE 17. Energy spectrum of solar neutrino recoil electrons between 5.0 MeV and 20.0 MeV. This plot contains 1117 days of LE data and 837 days of SLE data, both within our usual 22.5 kton fiducial volume, as well as the theoretical ^{8}B flux predicted for each bin.

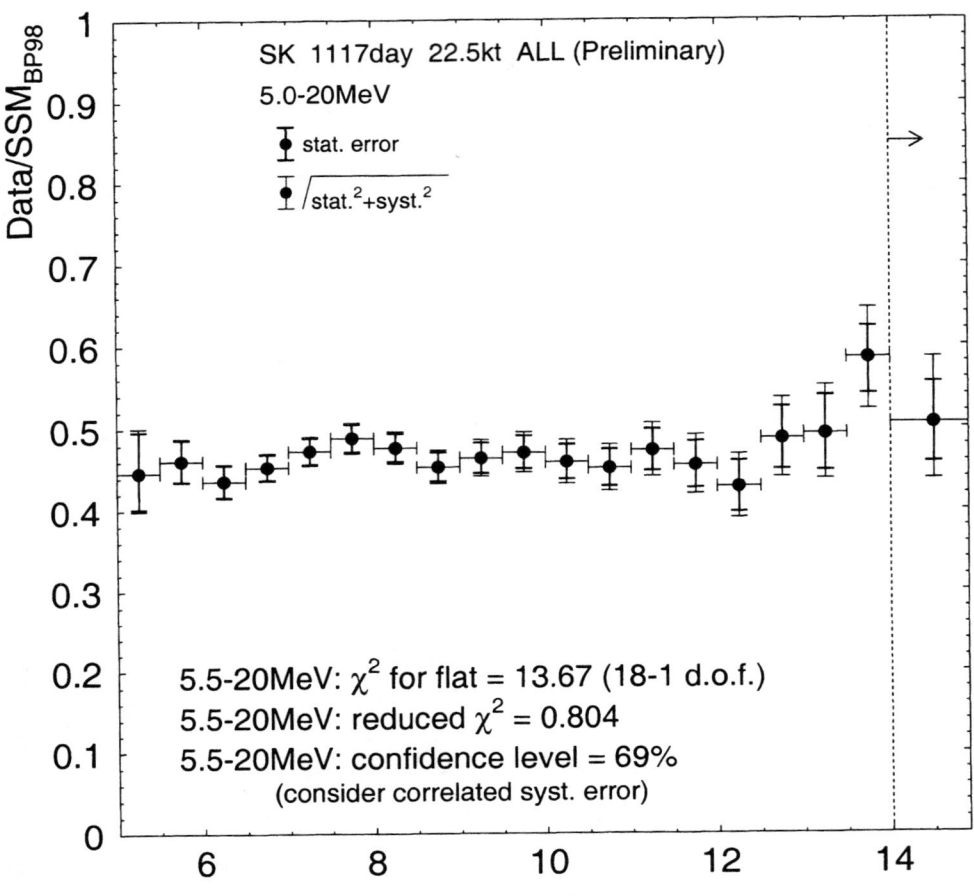

FIGURE 18. Energy spectrum of solar neutrino recoil electrons, divided by theoretical predictions, between 5.0 MeV and 20.0 MeV. This plot contains 1117 days of LE data and 837 days of SLE data, both within our usual 22.5 kton fiducial volume. Deviations from a flat distribution would have constituted evidence of MSW neutrino oscillations. However, the fit to flat is a good one.

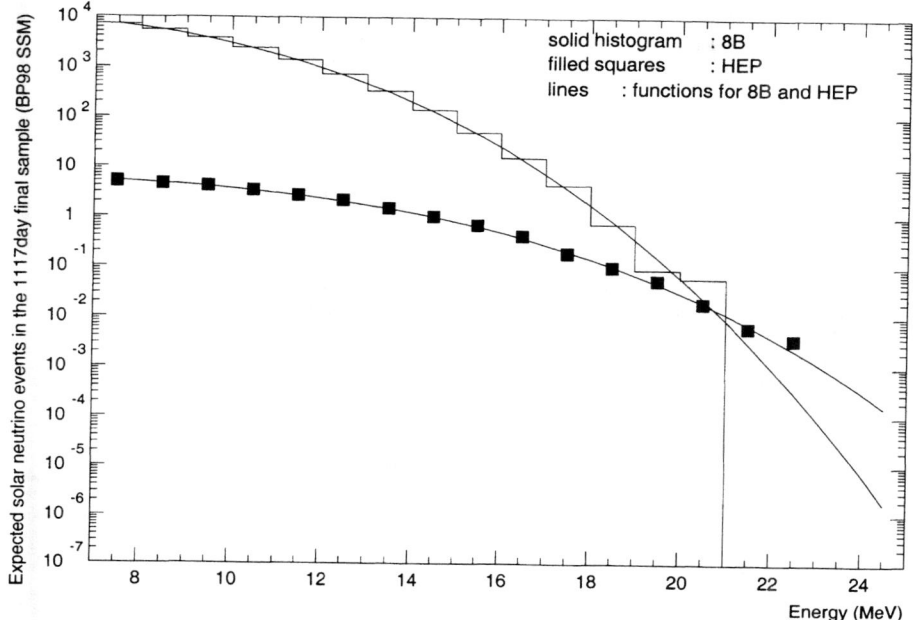

FIGURE 19. Fluxes predicted for ^{8}B and hep solar neutrinos by the BP98 Standard Solar Model [SSM] between 7.0 MeV and 25.0 MeV. The absolute normalization of the hep flux is very uncertain.

$$^3He + p \rightarrow \alpha + e^+ + \nu_e \tag{6}$$

Unfortunately, the standard theoretical predictions (BP98) of the hep flux are still uncertain to an order of magnitude or more.

Figure 19 shows, on a log scale, the relative BP98 predictions for the ^{8}B and hep solar neutrino fluxes as a function of energy. However, since the absolute normalizations of these lines is so uncertain, we may treat them as free parameters in a fit to our data. That is what is done in Figure 20 on both a log (top) and a linear (bottom) scale. We find the best fit occurs for 0.464 times the BP98 value for the ^{8}B flux plus 5.4 ($\pm$4.6) times the BP98 value for the hep flux.

The official spectrum overlaid with the ^{8}B plus hep best fit line is shown in Figure 21. Given the statistical weakness of the hep contribution (just one sigma above zero) it is not particularly surprising that this best fit line is not so different from flat.

In fact, due to the weakness of the potential hep contribution we generally choose to be a bit more conservative than this "free fit" method. Figure 22 shows our "aperture" method: a 90% upper-limit integrated Monte Carlo spectrum for ^{8}B

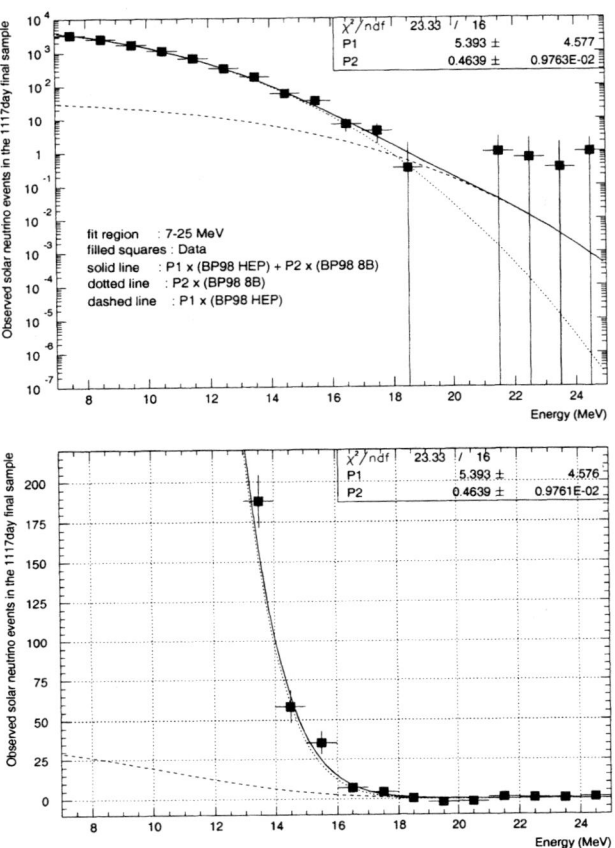

FIGURE 20. The ^{8}B and hep fluxes predicted by the BP98 SSM between 7.0 MeV and 25.0 MeV are fit to 1117 days of Super–K solar neutrino data. The same data are shown on a log scale (top) and a linear scale (bottom). Bins in the log plot with no data are the result of negative values produced by our background subtraction.

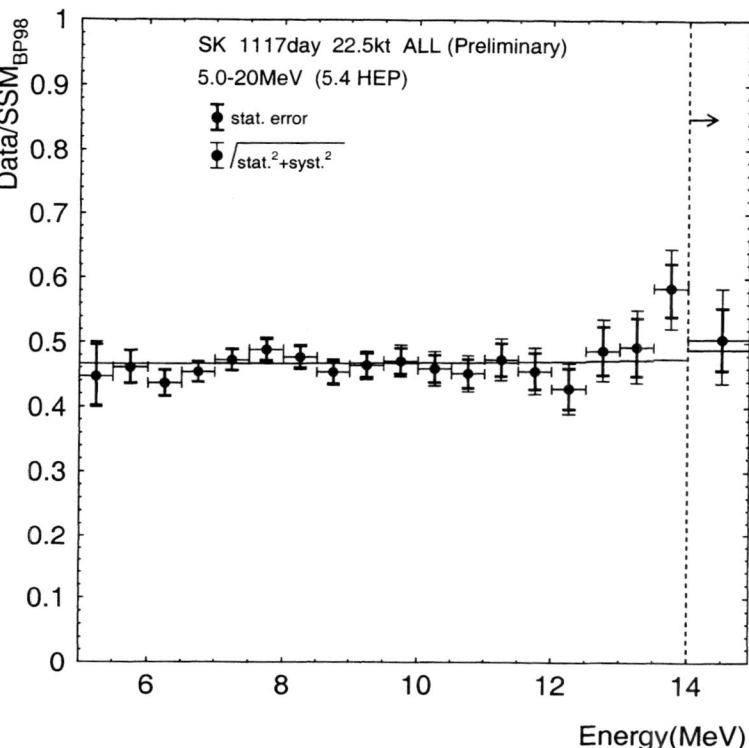

FIGURE 21. Energy spectrum of solar neutrino recoil electrons, divided by theoretical predictions, between 5.0 MeV and 20.0 MeV. This plot contains 1117 days of LE data and 837 days of SLE data, both within our usual 22.5 kton fiducial volume. Here the best fit for the ^{8}B and hep contributions treated as free parameters is plotted over the data as a solid line.

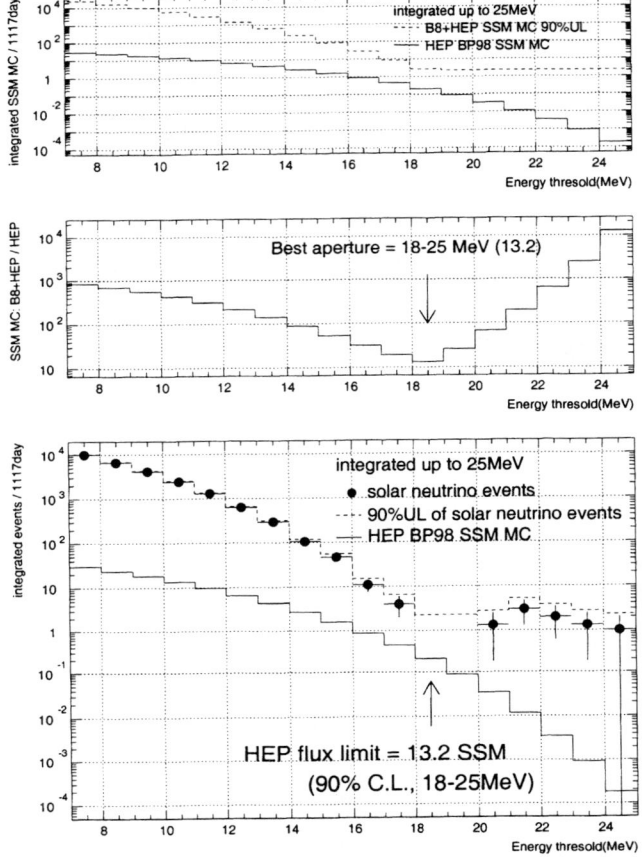

FIGURE 22. The top plot shows the ^{8}B plus hep Monte Carlo expectations vs. hep alone. The point of closest approach is found in the second plot. The third plot confirms with data that the upper limit of the hep flux is 13.2 times the BP98 SSM prediction.

plus hep is compared with the theoretical MC prediction for hep alone. The point of closest approach is found — it is 13.2 times the BP98 hep flux for the integrated flux between 18.0 MeV and 25.0 MeV. We then see that in our data this is indeed the point of closest approach and take, at the 90% confidence level, 13.2 times the SSM prediction of the hep flux as our official upper limit.

The hep analysis just presented demonstrates the extent to which we may extend our reach to higher energies. Continued running is the only way to reduce the errors on these points — however, there really is very little solar neutrino signal left above about 20.0 MeV.

The next part of this writeup will deal with the push towards the other end of the energy spectrum: the quest for Super Low Energy data.

D The Quest for Super Low Energy Data

1 The Challenge of SLE Data

When Super–Kamiokande officially turned on at the stroke of midnight on April 1, 1996, the low-energy threshold at which we triggered the detector stood at 5.5 MeV, while the threshold down to which we could reliably analyze the resulting data stood at 6.5 MeV.

One of the original design goals of Super–K called for a 5.0 MeV analysis threshold. As can be seen in Figure 23, the ^{8}B flux is still near its peak around 6.5 MeV. Since this flux remains near its peak value for another MeV or so below 6.5 MeV, in principle there should be plenty of solar neutrino signal at these energies. [4]

Since the inception of Super–K it has been considered highly desirable to extend the spectrum down to energies below 6.5 MeV in order to give our studies more statistical power. This Super Low Energy [SLE] region of the spectrum is also quite sensitive to vacuum oscillations, as well as to MSW oscillations (particularly the "small angle solution" MSW region).

As it turns out, the rate of background events rises sharply as the threshold is dropped. This is the result of a number of factors, including gammas from the rock surrounding the detector, radioactive decay in the PMT glass itself, and Radon contamination in the water. All become more pernicious as the trigger threshold is lowered, and in fact the hardware trigger rate of Super–K increases by approximately an order of magnitude for each MeV by which this threshold is reduced. This is why we originally set our analysis threshold to 6.5 MeV: our original off-line data acquisition [DAQ] system could only handle a maximum steady-state trigger rate of about 30 Hz.

Fortunately, when vertex fits are performed on super low-energy background events, the resulting vertex positions tend to be strongly clustered at the walls of the detector, outside of the nominal 22.5 kton fiducial volume. This can be easily

---

[4] J. Bahcall's website, http://www.sns.ias.edu/~jnb

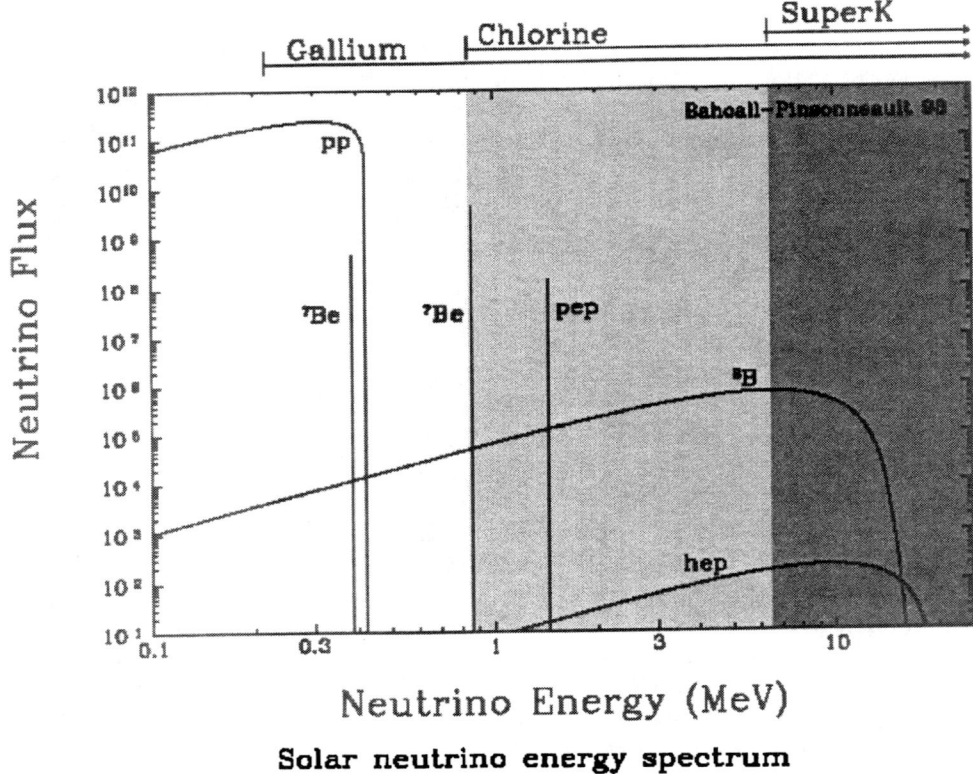

FIGURE 23. Solar energy spectra given by BP98. Super–K's lower threshold is still shown here at 6.5 MeV... looks like John needs to update his website!

seen in Figure 24. Therefore, it was realized that if it were possible to reconstruct SLE events on-line and reject those with unacceptable vertices then it should be possible to drastically reduce the flood of background events which contaminate the data stream.

2 The Intelligent Trigger

The challenge of lowering the low-energy threshold has been taken up in what is interchangeably referred to as the "SLE trigger" or "Intelligent Trigger" [IT] project. An ever-growing cluster of powerful computers running a suite of custom, real-time software filters the SLE data on-line by performing a double vertex fit (using two different fitter algorithms with rather different systematics), keeping only those events which fall within the 22.5 kton fiducial volume according to both fitting routines. This filtering drastically reduces the amount of background events which are passed out of the Super–K main control room and written to tape.

Figure 25 dramatically demonstrates the effect of the Intelligent Trigger on Super–Kamiokande's data collection rate. This automatically-generated plot (the few points off the main line are due to data glitches) strikingly shows, through abrupt changes in the slope of the line, the point in mid 1997 when the first IT machine began operations, the point in early 1999 when the threshold was lowered a second time after the addition of the second IT machine, and the point towards the end of 1999 when four more IT computers were introduced. With the addition of *six* new IT machines in July, 2000, we are now 100% efficient for triggering down to 4.5 MeV. Super–Kamiokande's on-line trigger rate has been raised over 15,000% since we first began taking data in 1996, but thanks to the power of our Intelligent Trigger system the amount of data written to tape has been increased by less than a factor of two!

The combination of hardware and software known as the Intelligent Trigger has worked to successfully and significantly lower our trigger threshold. Almost three years of SLE data have been accumulated since this system became operational, and results making use of these data are now an integral part of our official plots. Figure 26 shows $\cos\theta_{sun}$ plots for the four lowest-energy bins in the present SLE analysis for 837 days of SLE data. It is clear that the remaining background still rises rapidly with decreasing energy, but the solar peaks continue to stand out above background all the way down to 4.5 MeV. There is real hope that collecting and analyzing data all the way down to 4.0 MeV, a full MeV below Super–K's original design specifications, will be an achievable goal in the very near future.

As it turns out, extending the lower end of the energy spectrum has played an key role in producing some of the low-energy group's most dramatic results to date. The allowed phase space left for solar neutrino oscillations has been sharply cut back, essentially eliminating two classic possible solutions to the solar neutrino problem. We'll get to that analysis a little later in this writeup.

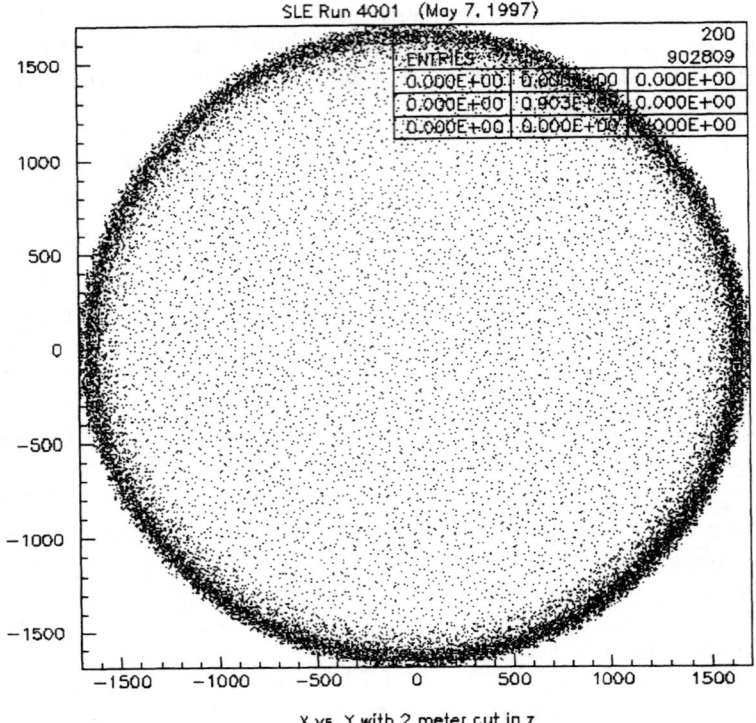

FIGURE 24. Distribution of super low-energy vertices in the Super–Kamiokande detector. View is looking down from above, and axes are labeled in centimeters. Note strong concentration around the walls (which are defined by the inner plane of PMT's) of the detector. A 2 meter cut from the top and bottom of the inner detector volume has been made to make the plot clearer, but no cut has been applied in the x or y directions.

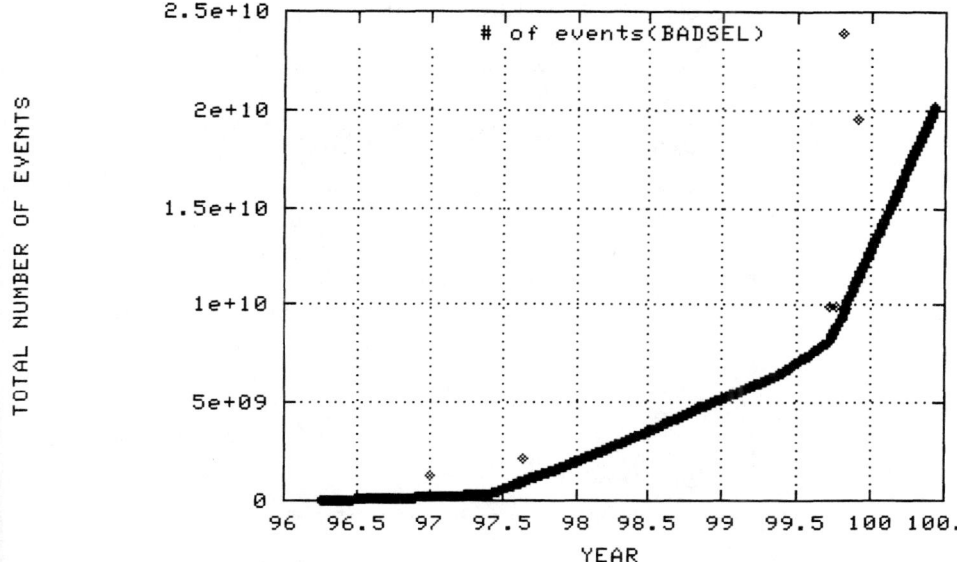

FIGURE 25. The total number of triggers in Super–K since the beginning of the experiment. The points at which new Intelligent Trigger machines were installed and the trigger rate was increased are quite evident at 97.4, 99.3, and 99.7. As this plot is generated automatically by one of the on-line processes, it is thought that points off the main line are due to real-time data glitches.

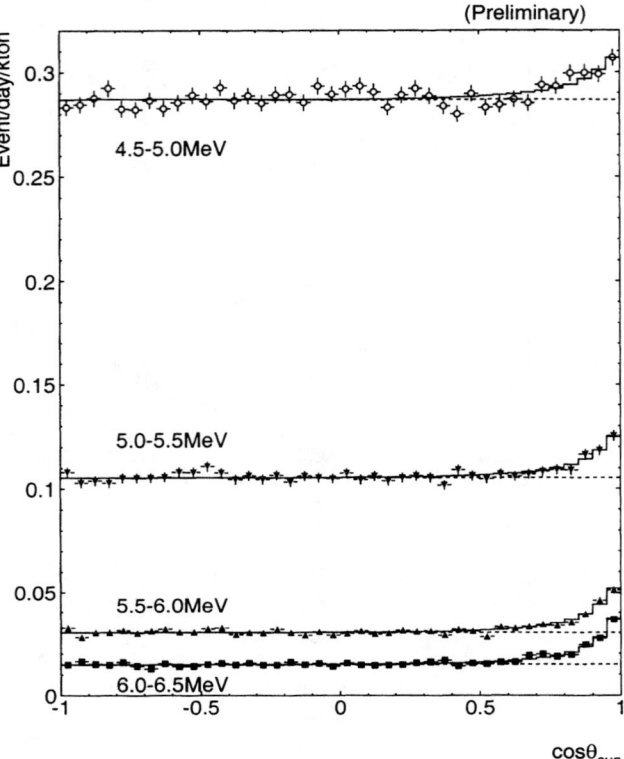

FIGURE 26. Angular distribution of 837 days of SLE data for the four lowest-energy bins. Note the rapid rise of the background level with decreasing energy. However, even in the 4.5 MeV to 5.0 MeV bin a peak pointing at the Sun is easily seen.

E Solar Neutrino Oscillation Analysis

1 Oscillation Signatures

In addition to a simple reduction of the overall solar neutrino flux, the presence of solar neutrino oscillations has the potential to cause three distinct effects in Super–Kamiokande's low-energy data set:

1. a distortion of the energy spectrum
2. a zenith-angle dependent flux (day/night effect)
3. a seasonal dependence of the flux (seasonal variation)

Super–Kamiokande has looked for distortions of the spectrum and time variations of the flux. The results of these studies, which in many ways represents the main conclusions of our solar neutrino analysis, will now be presented.

2 Combined Flux Results — Allowed Regions

The probability of flavor oscillation (in the simplest, two-component case) is given by the well-known expression

$$P(\nu_a \to \nu_b) = \sin^2 2\Theta \sin^2(1.27 \Delta m^2 [\text{eV}^2] L[\text{km}]/E[\text{GeV}]). \tag{7}$$

Because of the Δm^2 term, proof of oscillations would provide evidence of at least one non-zero neutrino mass. Indeed, these two phenomena, oscillations and massive neutrinos, are inextricably linked.

By combining Super–K's total flux with the fluxes measured by solar neutrino experiments using Chlorine and Gallium as their detection media, we arrive at the allowed regions in $\sin^2 2\Theta$ and Δm^2 phase space seen in Figure 27 for oscillations into active neutrino species and Figure 28 for oscillations into sterile neutrinos.

3 Day/Night Spectrum Results — Excluded Regions

The greatest sensitivity to MSW oscillation effects is achieved by combining the zenith-angle flux variation with the spectral distortion. We explore this combination by fitting oscillation predictions to a day and a night spectrum. The extreme flatness of the spectrum over a large range of energies (thanks in part to the addition of SLE data), combined with little day/night variation, leads to the powerful exclusion regions seen in Figures 29 and 30.

These exclusion regions were made assuming the SSM predictions for ^{8}B and hep solar neutrino fluxes as a function of energy. But what if the highly uncertain hep flux is allowed to float freely? Figure 31 compares the excluded MSW regions between our standard analysis and one in which the hep flux is allowed to take on its best-fit value. Happily (and not surprisingly, given the small contribution of possible hep fluxes to the spectral shape), the regions are almost identical.

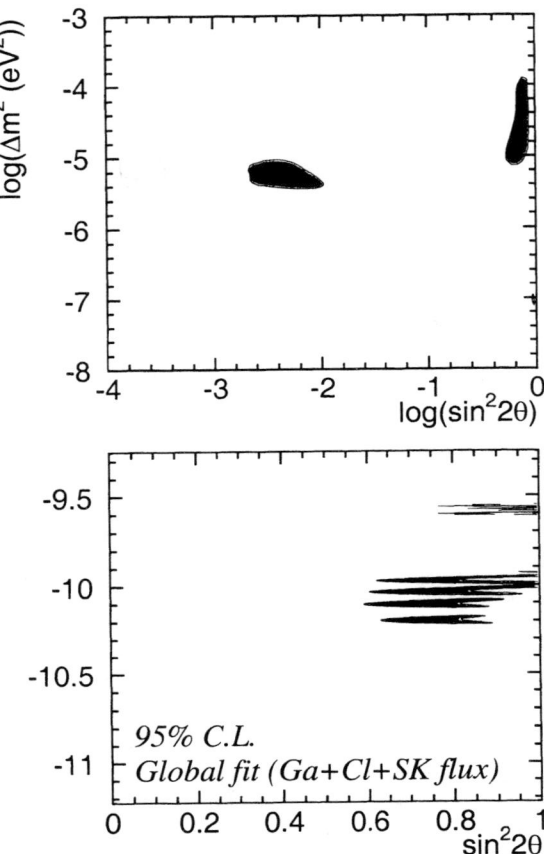

FIGURE 27. The allowed regions in $(\sin^2 2\Theta, \Delta m^2)$ phase space remaining after the solar neutrino fluxes measured in three types of detectors are combined in a global fit. The upper plot shows the large and small angle MSW solutions, as well as the so-called "low" region. The lower plot, at much lower Δm^2, shows the vacuum (or "just-so") oscillation solutions. Shaded areas are allowed at the 95% confidence level for oscillations into active neutrino species.

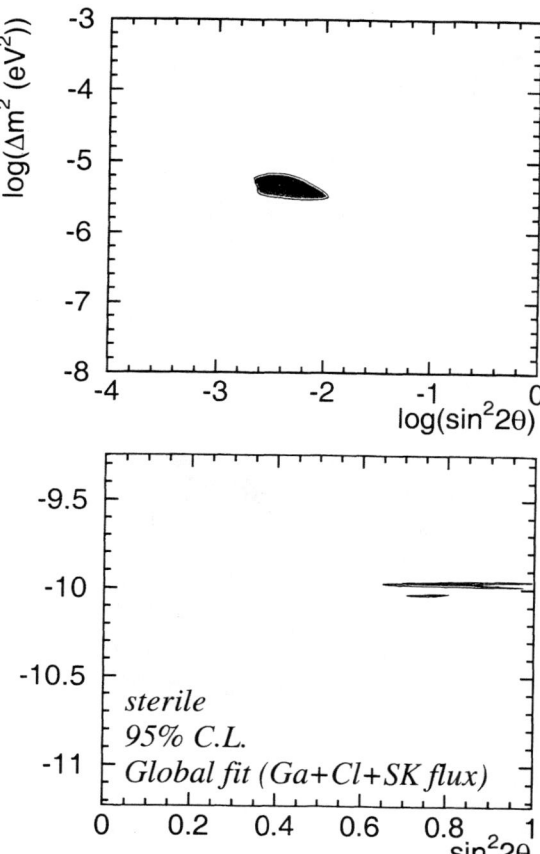

FIGURE 28. The allowed regions in ($\sin^2 2\Theta, \Delta m^2$) phase space remaining after the solar neutrino fluxes measured in three types of detectors are combined in a global fit. The upper plot shows the small angle MSW solution, while the lower plot, at much lower Δm^2, shows the vacuum (or "just-so") oscillation solutions. Shaded areas are allowed at the 95% confidence level for oscillations into sterile neutrinos.

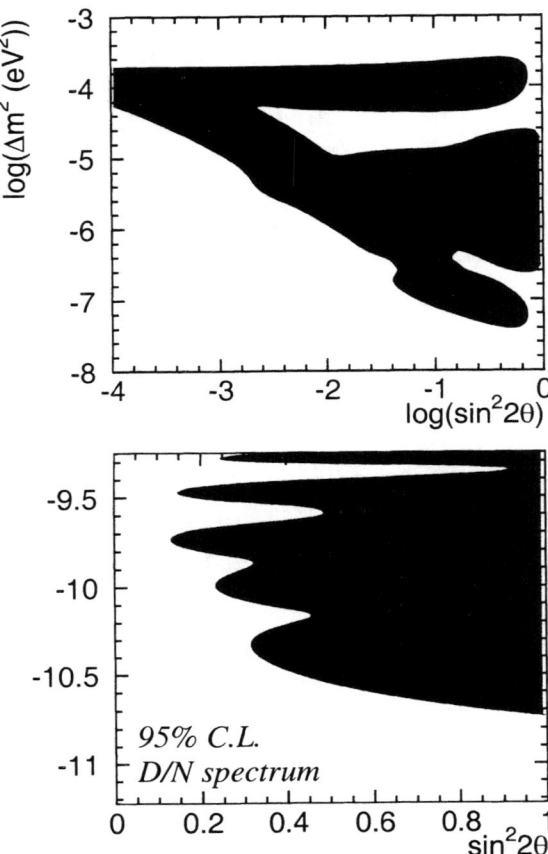

FIGURE 29. The excluded regions in $(\sin^2 2\Theta, \Delta m^2)$ phase space resulting from a combination of day and night spectral information from Super–K. Shaded areas are excluded at the 95% confidence level for oscillations into active neutrino species.

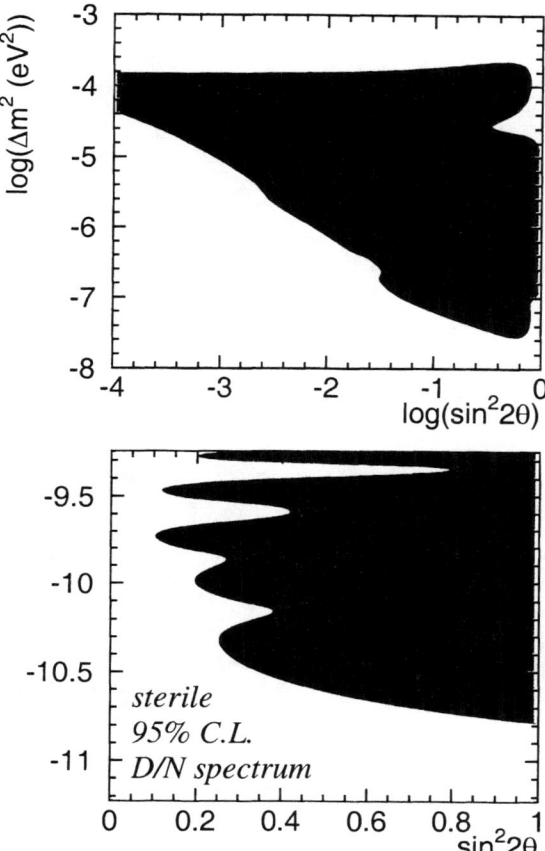

FIGURE 30. The excluded regions in $(\sin^2 2\Theta, \Delta m^2)$ phase space resulting from a combination of day and night spectral information from Super–K. Shaded areas are excluded at the 95% confidence level for oscillations into sterile neutrinos.

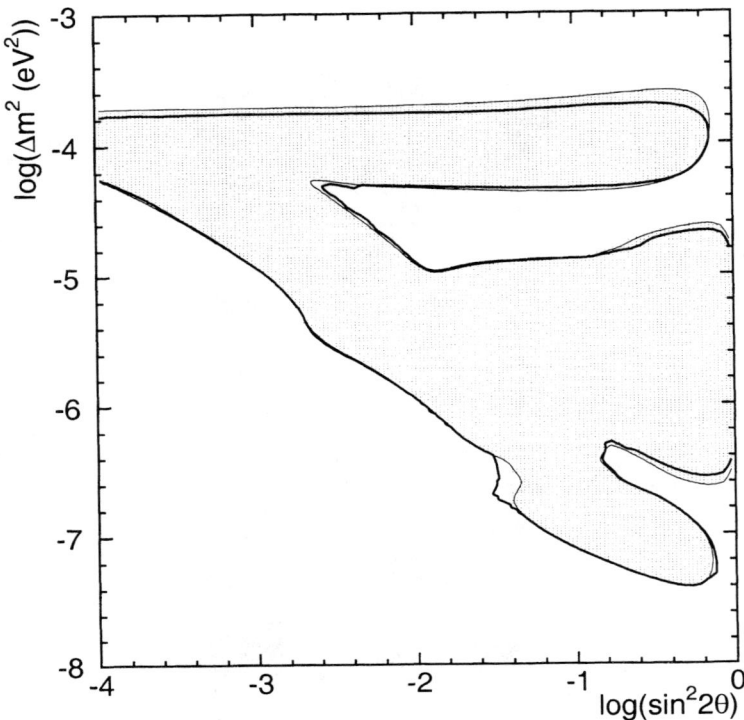

FIGURE 31. The excluded area in the MSW region of $(\sin^2 2\Theta, \Delta m^2)$ phase space resulting from a combination of day and night spectral information from Super–K. Shaded areas are excluded at the 95% confidence level for oscillations into active neutrino species with the hep flux fixed to the SSM prediction. The dark line shows the edge of the excluded region if the hep flux contribution is allowed to float to its best-fit value. The regions excluded in the two cases are obviously quite similar.

4 Solar Neutrino Oscillations — What's Left?

By comparing Figures 27 and 29 it can be seen that, in the active case, Super–K can exclude both the small mixing angle MSW solution and the vacuum solution from the global flux analysis completely at the 95% confidence level, while simultaneously cutting the allowed large mixing angle MSW region in half.

Similarly, superimposing Figures 28 and 30 shows that, in the sterile case, *all* global-fit allowed regions are excluded at 95%!

These powerful results depend, to varying degrees, on the combined flux results from several experiments. Suppose we want to see Super–K's reach in phase space all by itself. Figures 32 and 33 show what we can do all by ourselves if we add our flux measurement to our day and night spectra. Once again, the small mixing angle MSW solution is eliminated at better than 95% in both the active and sterile cases, while the vacuum solution is *nearly* ruled out, though it continues to hang on in a sliver of phase space. The main difference for the Super–K only analysis is that the large mixing angle and low solutions remain for oscillation into sterile neutrinos, since these regions are strongly suppressed by the low Homestake (Cl) flux.

F Low Energy Summary and Conclusions

The past three years have seen increasingly beautiful results come out of the Super–Kamiokande solar neutrino analysis. Measurements have been made which were simply impossible before Super–K came on-line four and a half years ago. The number of events we have collected in those 4.5 years has far surpassed all similar, previous experiments (Kamiokande's total solar neutrino sample, collected over ten years of running, had been equaled by Super–Kamiokande after only about two months of operations).

We have seen the first hints of hep neutrinos and perhaps even a small non-zero day/night effect. We have views of the Sun and a map of the sky made in low-energy neutrino "light." Most importantly, we are seriously constraining the phase space remaining for solar neutrino oscillations. The small mixing angle MSW solution is now definitely ruled out at the 95% confidence level, while the vacuum region is similarly disfavored. The remaining large mixing angle region has been cut in half, and oscillations into sterile neutrinos are looking very unlikely indeed.

The lower end of the ^{8}B solar neutrino energy spectrum is yielding to our efforts, and in many areas we are nearing the point where theory predicts we will either see evidence of solar neutrino oscillations (if indeed they are the origin of the long-standing solar neutrino problem) or be able to convincingly rule them out (if they are not). Hopefully the solar neutrino problem will not be a problem for very much longer.

This has been an exciting and highly productive time for us. The near future promises more stimulating developments, as various projects, including an even

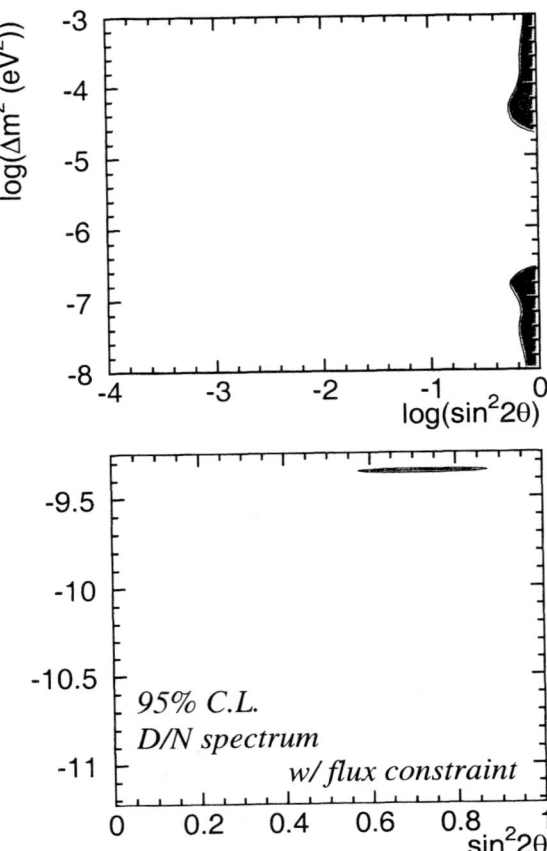

FIGURE 32. The allowed regions in $(\sin^2 2\Theta, \Delta m^2)$ phase space remaining after the solar neutrino flux measured by Super–K is combined with our day and night spectral information. Shaded areas are allowed at the 95% confidence level for oscillations into active neutrino species.

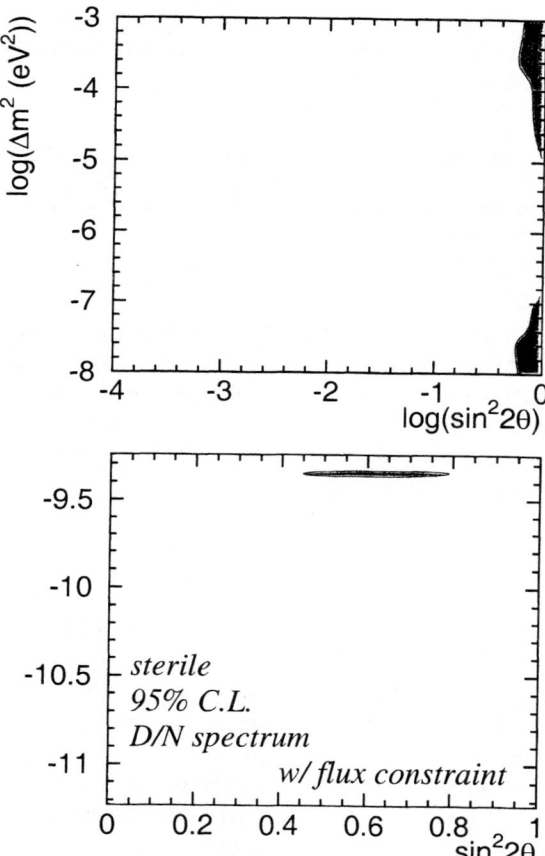

FIGURE 33. The allowed regions in $(\sin^2 2\Theta, \Delta m^2)$ phase space remaining after the solar neutrino flux measured by Super–K is combined with our day and night spectral information. Shaded areas are allowed at the 95% confidence level for oscillations into sterile neutrinos.

more powerful Intelligent Trigger, an enhanced low-energy Monte Carlo, and an improved understanding (and reduction) of our backgrounds all continue to deliver the weakly-interacting goods.

III HIGH ENERGY ANALYSIS — ATMOSPHERIC NEUTRINOS AT SUPER–KAMIOKANDE

A Atmospheric Intro

When we talk about Super–Kamiokande's "high-energy" analysis, we generally mean events with greater than 100 MeV of visible energy deposited in the detector. Our primary source of these events (since proton decay has proven quite elusive!) are the result of interactions of cosmic ray particles with the upper atmosphere.

Neutrinos of all energies (above about 1 GeV, the flux is well-described by an $E^{2.7}$ power law) are produced in the atmosphere by cosmic ray showers and the cascades

$$\pi/K \to \mu \to e. \qquad (8)$$

Each muon produced is associated with a ν_μ and each muon which decays produces a ν_e in addition to a second ν_μ.

Neutrinos arrive at the detector from all directions, with the upward and downward fluxes equal to within about $\pm 10\%$. Neutrinos arriving from above have traveled only about 15 km from their point of production in the atmosphere, while those arriving from below have traversed a distance comparable to the earth's diameter (13,000 km). Thus the atmospheric neutrino flux, spanning many decades of both E and L, provides an ideal beam for studying L/E-dependent oscillation effects.

We have obtained the world's largest sample of atmospheric neutrino events. Our most recent results, representing 1144 live days of analyzed high-energy data, will be presented below.

B Recent Results of the Atmospheric Neutrino Analysis

1 Data Reduction

Although the high-energy data is, by its nature, much cleaner than the low-energy data due to a lack of physics backgrounds (the most pernicious are downward-going, cosmic ray muons), it is still necessary to reduce the data to produce a final sample. Figure 34 shows an outline of the process. Note that looking for hits in Super-K's anticounter is a key part of this reduction. Even after this reduction there is a small contamination of the data with "stealthy" muons. Figure 35 shows the need

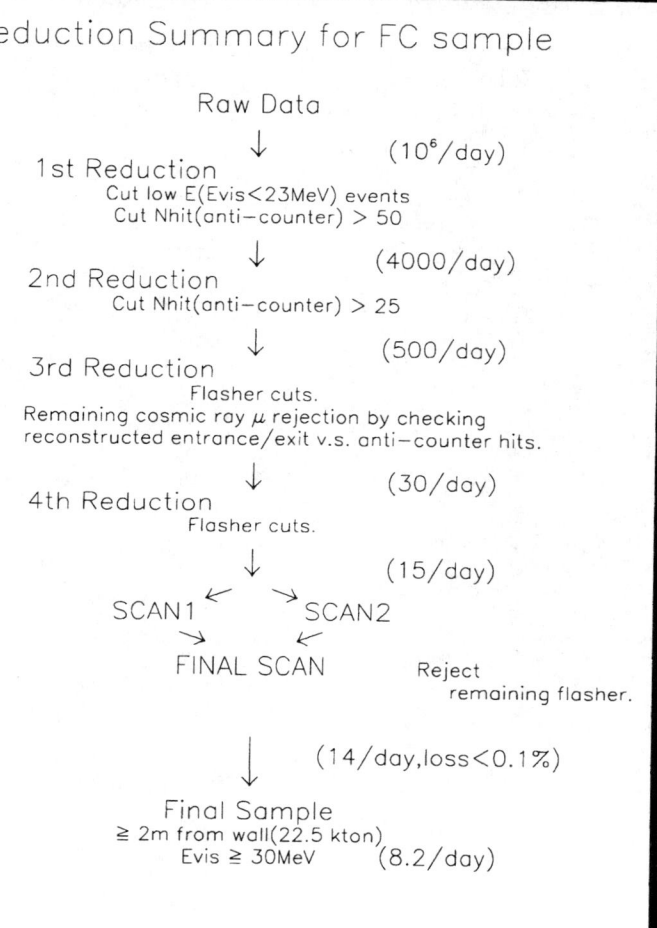

FIGURE 34. Steps and key cuts in the offline reduction of high-energy events.

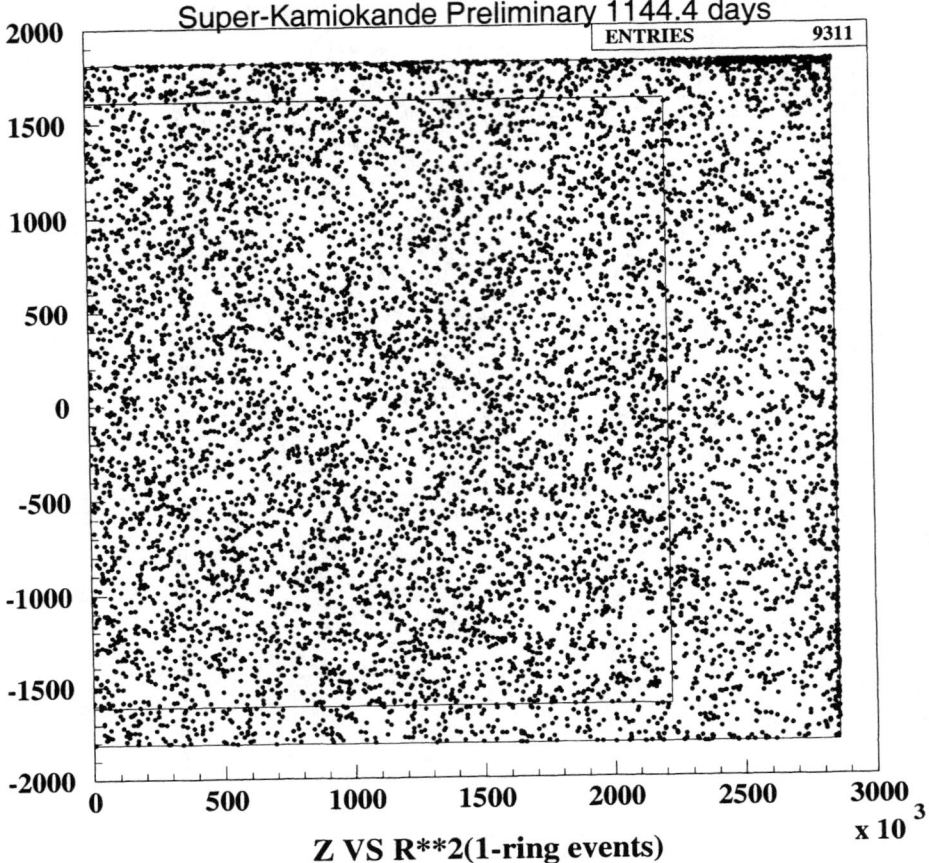

FIGURE 35. Distribution of atmospheric neutrino vertices in the Super–Kamiokande detector in z (the vertical direction) and the radius squared. Units are cm and cm-squared. Note the contamination of events due to downward-going cosmic ray muons in the upper right of the plot. This is why a 2 meter fiducial volume cut, indicated by the thin line, is needed.

for a 2 meter fiducial volume cut, resulting in our familiar 22.5 kton volume and our final high-energy sample.

Figure 36 depicts our particle identification for e-like and μ-like events; it is assumed that the progenitor neutrino is of the same leptonic flavor. A clear supression of μ-like data verses the Monte Carlo prediction is seen, although from this plot alone it is not clear if the muons are supressed or the electrons are enhanced with respect to expectations. In order to get away as much as possible from theoretical uncertainties, we generally form the quantity known as "R", the "ratio of ratios":

$$R = \frac{(\mu/e)_{\text{Data}}}{(\mu/e)_{\text{Monte Carlo}}} \quad (9)$$

For largely historical reasons, our atmospheric data are split into a "sub-GeV" sample and a "multi-GeV" sample. The data are further broken up by the number of Cherenkov rings visible in Super–K, and into fully contained [FC] events (no light in the anticounter) and higher-energy, partially contained [PC] events (some light in the anticounter, but not through-going). Figures 37 and 38 summarize the number of events in our final sample. Note that the values for "R" are remarkably consistent in every one of our data sets.

2   The Evidence for Oscillations

A low ratio of ratios is suggestive that something odd is going on with the atmospheric neutrinos but, just like measuring a reduced flux of solar neutrinos relative to SSM predictions, it does not by itself constitute an inescapable "smoking gun" signal of neutrino oscillations. Such a smoking gun does exist in the high-energy realm: the zenith-angle distribution of our amtospheric events.

In Figure 39 we see plots of the zenith-angle dependence of various high-energy data sets. Here, $\cos\theta = 1.0$ indicates the events which are coming from directly overhead, $\cos\theta = 0.0$ are those events coming from the horizon, and those with $\cos\theta = -1.0$ come from directly underfoot. The top two plots contain sub-GeV events, and the lower two are filled with multi-GeV data. Those two on the left are due to ν_e interactions in Super–K, and the two on the right come from ν_μ interactions. The upper line in the ν_μ plots are the theoretical predictions if there are no oscillations, while the lower line, which passes through almost every data point, is the shape one would expect for $(\sin^2 2\Theta, \Delta m^2) = (1.00, 3.2 \times 10^{-3} \text{ eV}^2)$. Note that the ν_e-generated events show no angular deviations from the expected (no oscillation) case, while the ν_μ-generated events start to drop off the further from straight down they become. The muon neutrinos seem to know how long they've traveled before being caught by Super–K! This is our smoking gun. What's more, because the e-like events show no excess, we can tell that the oscillations we are seeing are *not* $\nu_\mu \to \nu_e$, but rather must be either $\nu_\mu \to \nu_\tau$ or $\nu_\mu \to \nu_{\text{sterile}}$.

Figure 39 contains events generated inside the fiducial volume of Super–K by electron and muon neutrinos interacting with the water in the detector. Of course,

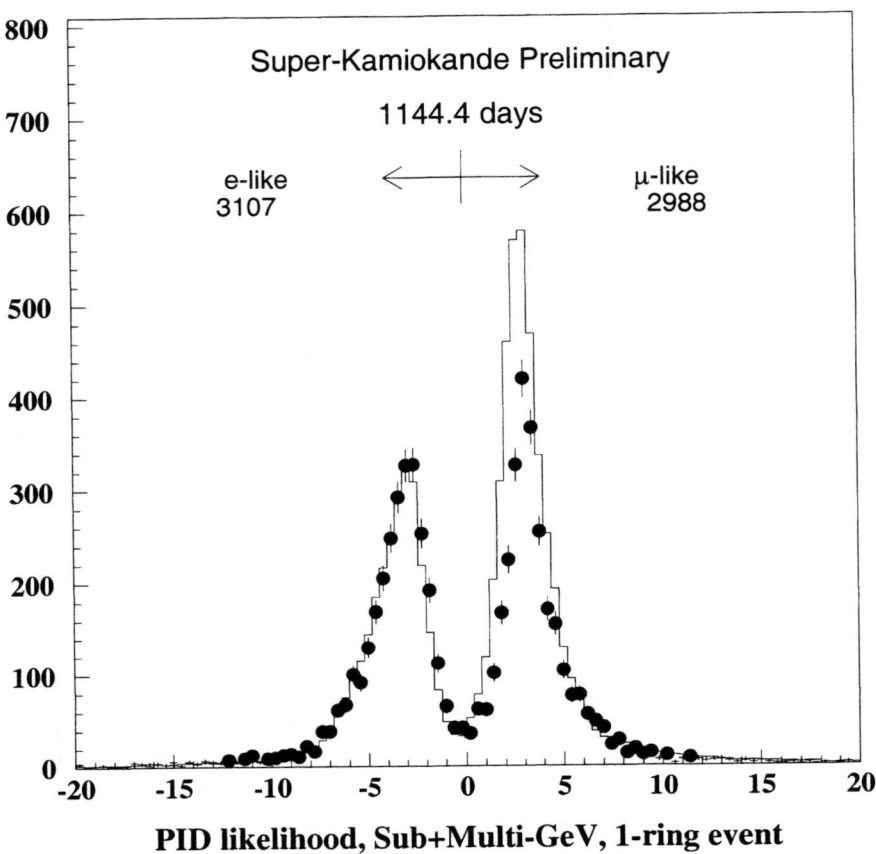

FIGURE 36. Particle identification, where data (points) are compared with Monte Carlo predictions (solid line), for events with one Cherenkov ring.

Sub-GeV event Summary

Evis < 1.33GeV
$P_e > 100$MeV/c
$P_\mu > 200$MeV/c

	DATA	MC(Honda)	MC(Bartol)
1R	5017	6023.5	5880.1
e-like	2531	2402.6	2364.9
µ-like	2486	3620.9	3515.2
2R	1311	1568.5	1543.3
≥3R	574	753.0	751.3
TOTAL	6902	8345.0	8174.7

$$\frac{(\mu/e)_{DATA}}{(\mu/e)_{MC}} = 0.652 \pm {}^{0.019}_{0.018} \pm 0.051 \text{ (Honda)}$$

stat.   sys.

$$= 0.661 \pm {}^{0.019}_{0.018} \pm 0.052 \text{ (Bartol)}$$

stat.   sys.

FIGURE 37. Sub-GeV Final Sample.

Multi-GeV event Summary

(1) FC (Evis > 1.33GeV)

	DATA	MC(Honda)	MC(Bartol)
1R	1078	1294.1	1314.8
e-like	576	555.4	576.1
μ-like	502	738.7	738.7
2R	454	566.5	579.9
≥3R	744	903.6	943.5
TOTAL	2276	2764.2	2838.2

(2) PC

	DATA	MC(Honda)	MC(Bartol)
TOTAL	665	945.1	997.4

*All events are assumed to be μ-like.
*Fraction of CC ν_μ, $\bar\nu_\mu$ events in the PC sample is estimated to be (97-98)%.

$$\frac{(\mu/e)_{DATA}}{(\mu/e)_{MC}}$$

FC + PC:
$= 0.668 \pm ^{0.035}_{0.033}\text{ stat.} \pm 0.079\text{ sys.}$ (Honda)

$= 0.672 \pm ^{0.035}_{0.033}\text{ stat.} \pm 0.080\text{ sys.}$ (Bartol)

FC only:
$= 0.655 \pm ^{0.041}_{0.039}\text{ stat.} \pm 0.096\text{ sys.}$ (Honda)

$= 0.680 \pm ^{0.043}_{0.040}\text{ stat.} \pm 0.099\text{ sys.}$ (Bartol)

FIGURE 38. Multi-GeV Final Sample.

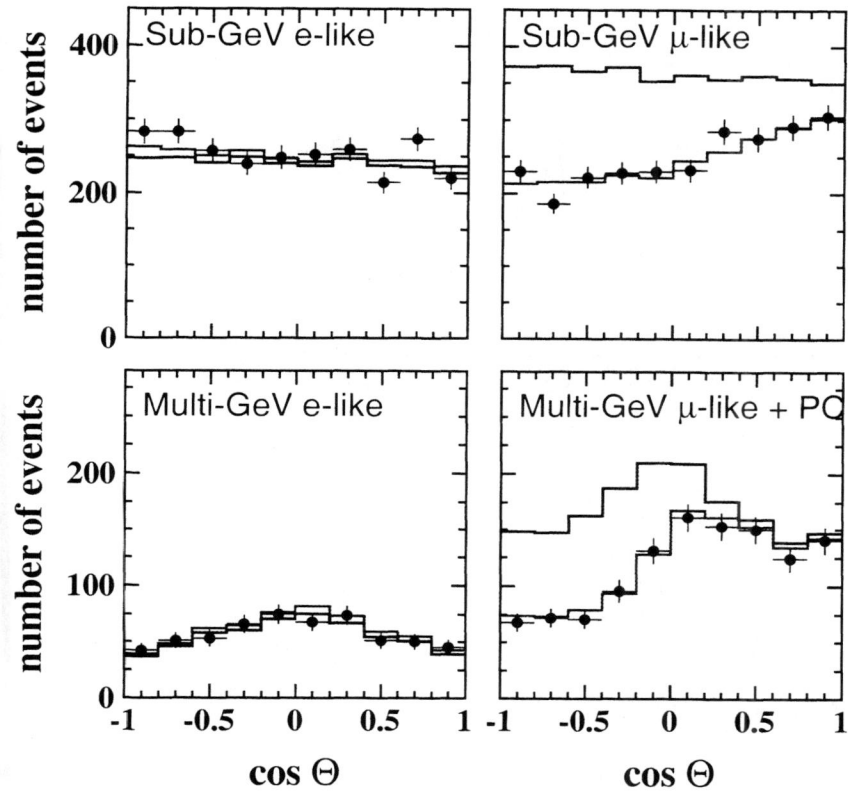

FIGURE 39. Zenith-angle distributions for 1144 days of atmospheric neutrino data. The upper lines in the μ-like plots are what is expected for no oscillations, and the fit to the data is for the oscillation solution $(\sin^2 2\Theta, \Delta m^2) = (1.00, 3.2 \times 10^{-3}\text{ eV}^2)$. These four plots are filled with events caused by interactions between atmospheric electron and muon neutrinos with the water in Super–K.

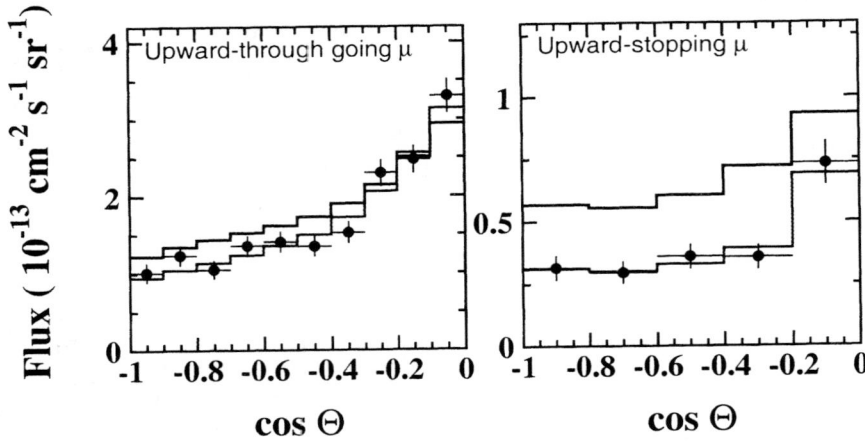

FIGURE 40. Zenith-angle distributions for 1144 days of atmospheric neutrino data. The upper lines in the right plots are what is expected for no oscillations, and the fit to the data is for the oscillation solution $(\sin^2 2\Theta, \Delta m^2) = (1.00, 3.2 \times 10^{-3} \text{ eV}^2)$. These plots are filled with very high-energy muons which were produced by interactions between atmospheric muon neutrinos and the rock beneath the detector.

neutrinos can also interact in the surrounding rock before they reach the detector. Electrons so produced will be immediatedly absorbed, but muons, if their initial energies are high enough, can travel though the surrounding rock and enter the detector. Figure 40 is made using these kind of events — only events originating below the horizon are considered due to the large number of downward-going muons from cosmic ray showers. These upward-going muons probe higher energy neutrinos, but once again the upper line in the plots are the theoretical predictions if there are no oscillations, while the lower line, which passes through almost every data point, is the shape one would expect for $(\sin^2 2\Theta, \Delta m^2) = (1.00, 3.2 \times 10^{-3} \text{ eV}^2)$. Figure 41 shows the allowed $(\sin^2 2\Theta, \Delta m^2)$ phase space when the data from Figures 39 and 40 are combined. The χ^2_{min} for oscillations is 135.3/152 degrees of freedom, while χ^2_{min} for *no* oscillations is 316.2/154 degrees of freedom.

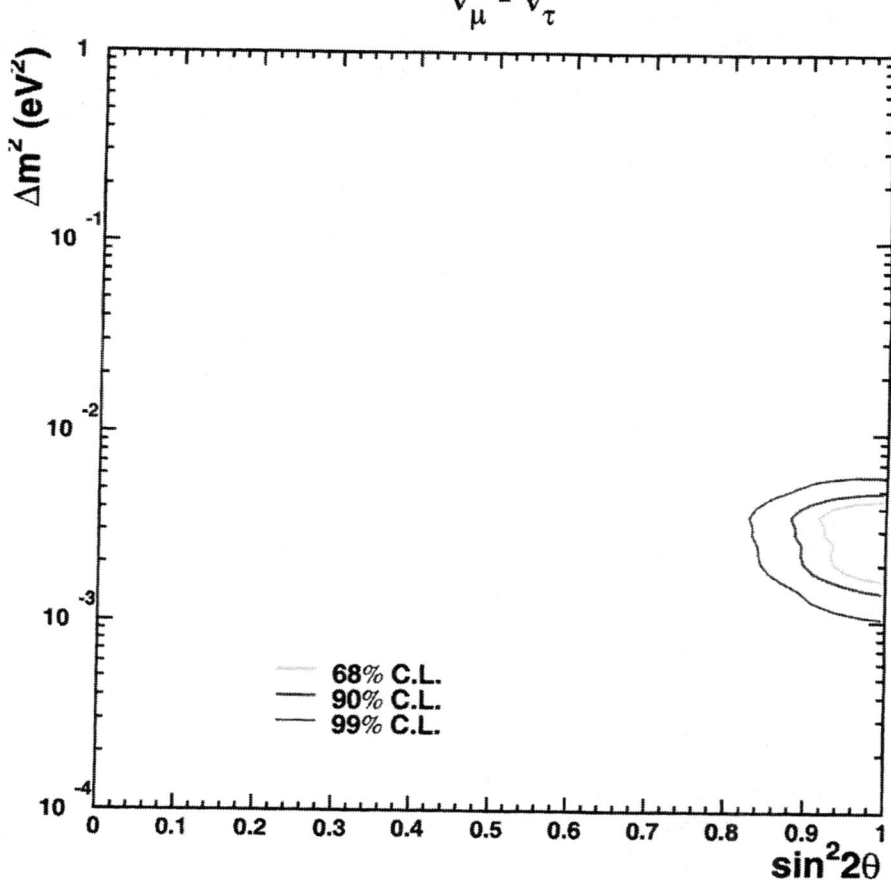

FIGURE 41. Allowed region for atmospheric neutrino oscillations for 1144 days of data.

3 $\nu_\mu \to \nu_\tau$ or $\nu_\mu \to \nu_s$?

Given the evidence presented above, it seems quite certain that Super-Kamiokande has observed the disappearance of atmospheric ν_μ's via oscillations. Furthermore, since no excess of ν_e's is seen, the ν_μ's must be changing into an unseen species, either ν_τ's or ν_{sterile}'s. Can we tell which one? As it turns out, there are two ways in which conversion into sterile neutrinos is distinctive:

1. Charged current [CC] interactions are suppressed by the τ mass, while neutral current [NC] interaction are not. Sterile neutrinos, on the other hand, have *neither* CC or NC interactions.

2. Due to the lack of interactions with electrons in the Earth, ν_{sterile} oscillations are perturbed by matter in the same way, but with opposite sign, as ν_e.

In the first case, we can examine a NC-enhanced sample of atmospheric neutrino events, and look for differences between upward-going and downward-going event rates. This is shown in Figure 42, while the expected up vs. down ratio for oscillations into ν_τ's and into ν_s's as a function of Δm^2 is compared with data in Figure 43.

In the second case, the factors $\sin^2 2\Theta$ and L which appear in

$$P(\nu_a \to \nu_b) = \sin^2 2\Theta \sin^2(1.27 \Delta m^2 [\text{eV}^2] L[\text{km}]/E[\text{GeV}])$$

must be replaced with the effective mixing angle and oscillation length in matter:

$$L_m = \frac{L}{\sqrt{(\zeta - \cos 2\theta)^2 + \sin^2 2\theta}}, \tag{10}$$

and

$$\sin^2 2\theta_m = \frac{\sin^2 2\theta}{(\zeta - \cos 2\theta)^2 + \sin^2 2\theta}, \tag{11}$$

where

$$\zeta = \frac{2 V_{ab} E_\nu}{\Delta m^2} \tag{12}$$

and the density-dependent factor V_{ab} reflects the difference in interaction potential between ν_μ and ν_s. V_{ab} reflects the fact that ν_μ undergo neutral current interactions with matter while ν_s do not. When E becomes large compared to $\Delta m^2/V_{ab}$, the ζ term in the denominator of the effective mixing angle also becomes large, and drives the effective mixing angle to zero. Thus, in the $\nu_\mu \to \nu_s$ case, matter effects works to prevent oscillation at high energy. So, if we look at especially high-energy events and *see* a suppression in the upward-going events this argues against $\nu_\mu \to \nu_s$ and for $\nu_\mu \to \nu_\tau$.

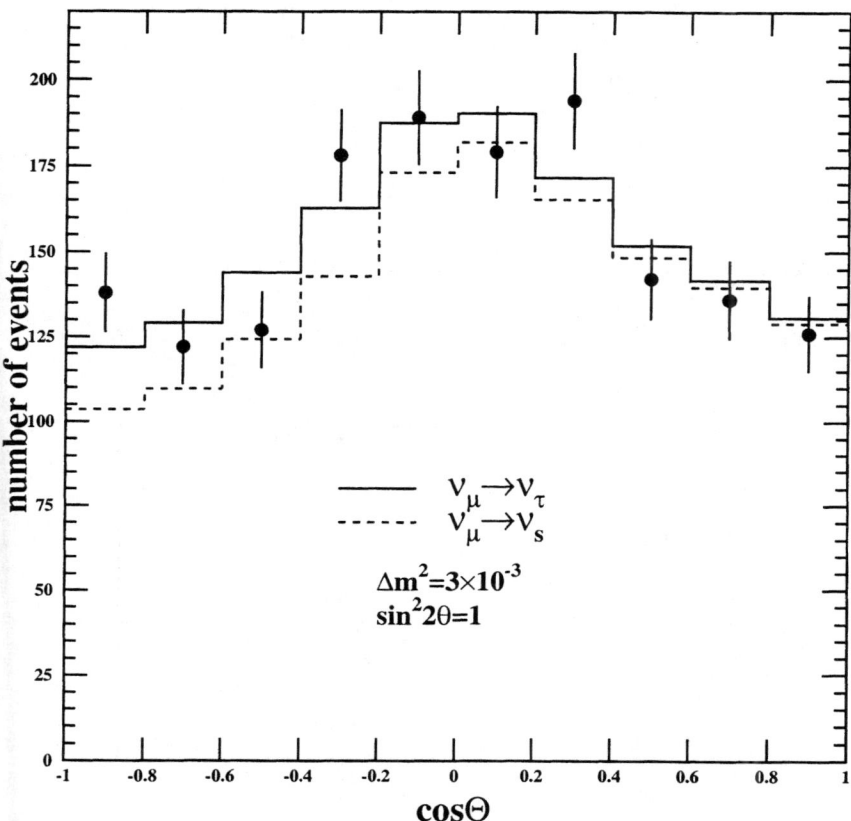

FIGURE 42. Neutral current-enhanced sample of atmospheric neutrinos for 1144 days of data. Oscillations of ν_μ into ν_τ would lead to more upward-going events than would oscillations into sterile neutrinos due to ν_τ's undergoing neutral current interactions (which $\nu_{sterile}$'s do not).

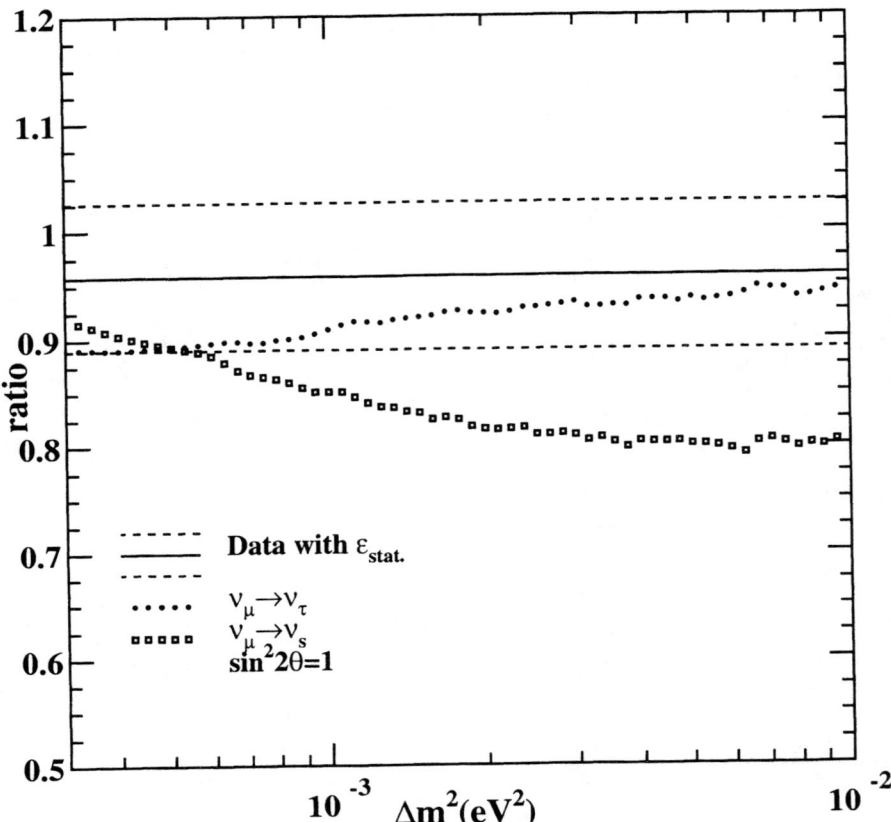

FIGURE 43. Up/Down ratio in the neutral current-enhanced sample of atmospheric neutrinos for 1144 days of data. The solid line is our best-fit value (with dashed lines representing one-sigma errors), while the curves represent theortical predictions of $\nu_\mu \to \nu_\tau$ and $\nu_\mu \to \nu_s$ as a function of Δm^2. The $\nu_\mu \to \nu_\tau$ solution is clearly favored by the data.

In Figure 44 we look at the zenith-angle distribution of high-energy partially contained events, while in Figure 46 we look at upward-going muons which pass entirely though the detector; these sample the very highest-energy neutrinos which Super–K is capable of seeing. Figures 45 and 47 depict the up/down ratio of the PC events and the vertical/horizontal ratio of the through-going muons. When our data is compared to the theoretical expectations for $\nu_\mu \to \nu_s$ and $\nu_\mu \to \nu_\tau$ it is clear that $\nu_\mu \to \nu_\tau$ is strongly favored.

Combining the information from these three studies leads to a very powerful result. Figure 48 shows the allowed regions for oscillations, a la Figure 41, for three possible oscillation scenarios. Overlaid on them are the regions excluded by the aforementioned $\nu_\mu \to \nu_\tau$ vs. $\nu_\mu \to \nu_s$ analyses — areas below these lines in each plot are excluded. The upshot is that while the entire allowed region for $\nu_\mu \to \nu_\tau$ survives, the entire allowed region for $\nu_\mu \to \nu_s$ is completely excluded at better than 99% regardless of whether or not ν_μ is more massive than ν_s. We therefore conclude that we are indeed seeing the result of $\nu_\mu \to \nu_\tau$ oscillations in Super–Kamiokande and believe $\nu_\mu \to \nu_s$ oscillations to be conclusively ruled out by our data.

4 Exotic Solutions

There have been a number of theoretical discussions proposing that perhaps the zenith-angle-dependent reduction in our ν_μ flux is not due to oscillations at all, but rather is the result of some more exotic physics, like the violation of Lorentz invariance. These models generally require that the probablilty of ν_μ survival does not go as L/E, but rather as E to some power other than the -1 called for by

$$P(\nu_a \to \nu_b) = \sin^2 2\Theta \sin^2(1.27 \Delta m^2 [\text{eV}^2] L[\text{km}]/E[\text{GeV}]).$$

To test this, we have performed an oscillation analysis with

$$P(\nu_\mu \to \nu_\tau) = \sin^2 2\Theta \sin^2(\beta L \bullet E^n) \qquad (13)$$

where n has been allowed to vary between -2.0 and 1.0 and $\sin^2 2\Theta$ has been varied between 0.7 and 1.3. For each set of parameters the minumum χ^2 fit to our atmospheric data was computed. Figure 49 shows the results: the global minimum occurs at $\chi^2 = 74.7/81$ degrees of freedom, when $n = -1.06 \pm 0.14$. This strongly rules out some of the more esoteric hypotheses which attempt to explain our results without invoking oscillations.

C Nucleon Decay

In many ways, this whole large water Cherenkov business got its start in the early 1980's as a means to search for proton decay. In those days, the leading unified theory, SU(5), predicted that protons would decay at a rate which would be

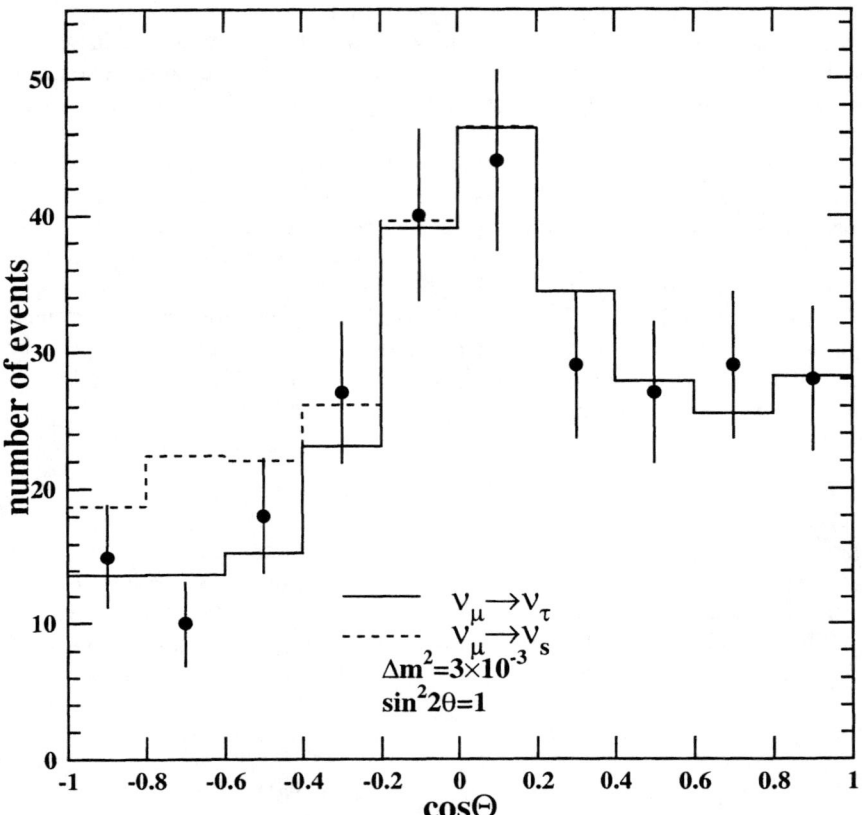

FIGURE 44. High-energy partially contained sample of atmospheric neutrinos for 1144 days of data. Oscillations of ν_μ into ν_τ would lead to less upward-going events than would oscillations into sterile neutrinos due to matter effects suppressing oscillation into ν_s for these high-energy ν_μ's.

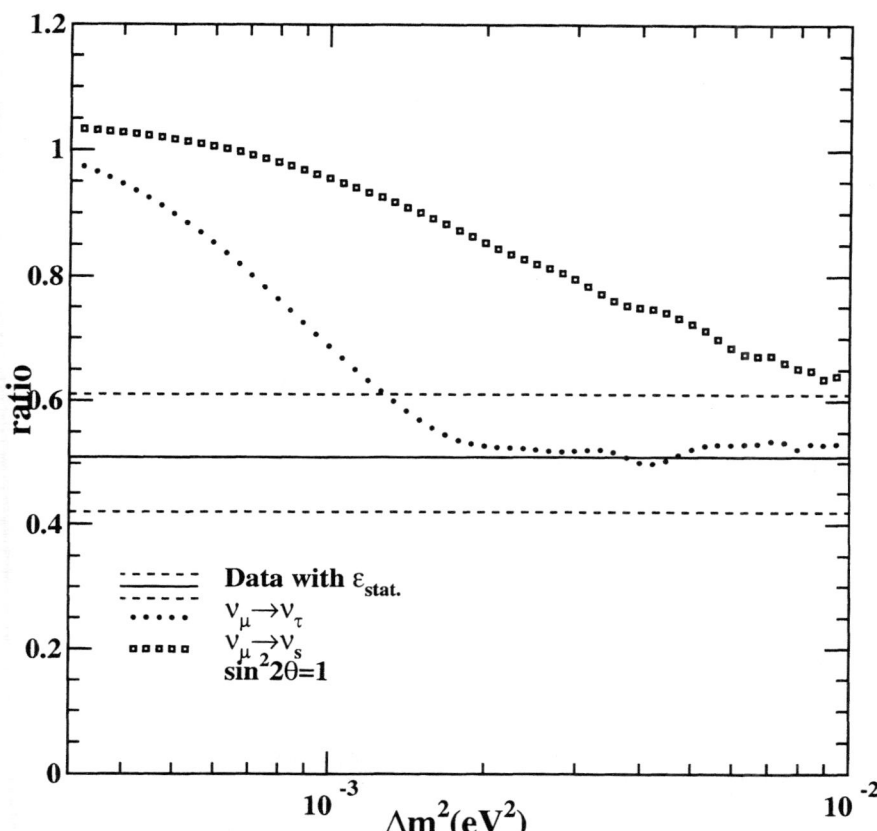

FIGURE 45. Up/Down ratio in the high-energy partially contained sample of atmospheric neutrinos for 1144 days of data. The solid line is our best-fit value (with dashed lines representing one-sigma errors), while the curves represent theortical predictions of $\nu_\mu \to \nu_\tau$ and $\nu_\mu \to \nu_s$ as a function of Δm^2. The $\nu_\mu \to \nu_\tau$ solution is strongly favored by the data.

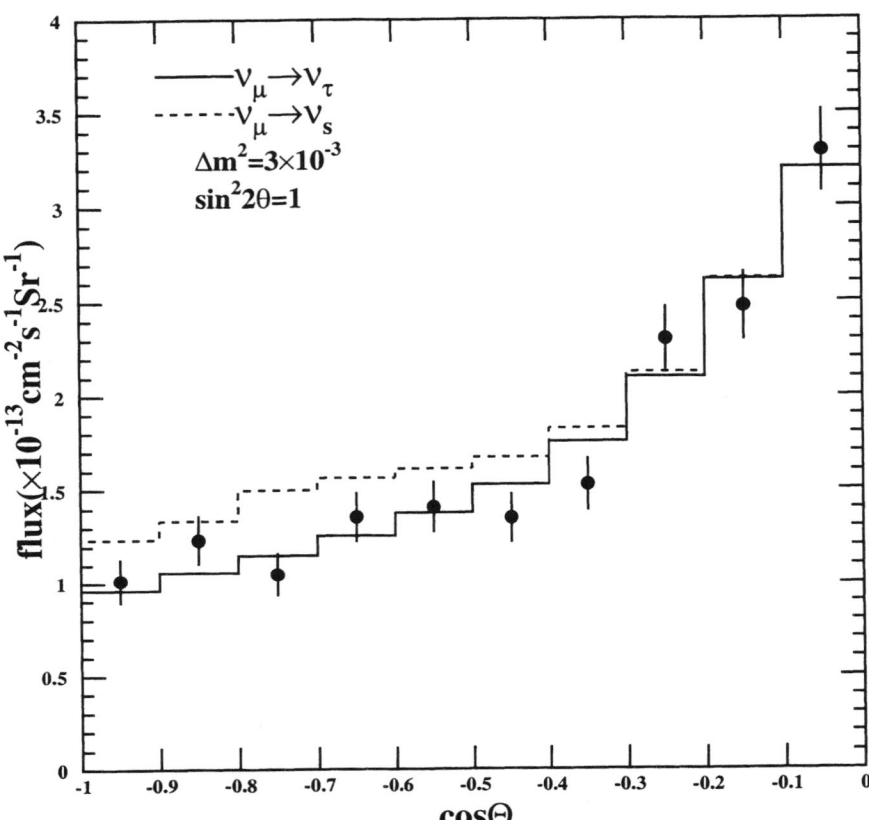

FIGURE 46. Very high-energy upward through-going muon sample for 1138 days of data. Oscillations of ν_μ into ν_τ would lead to less upward-going events than would oscillations into sterile neutrinos due to matter effects suppressing oscillation into ν_s for these high-energy ν_μ's. This plot only extends to the horizontal due to the large background flux of downward through-going muons from cosmic ray showers.

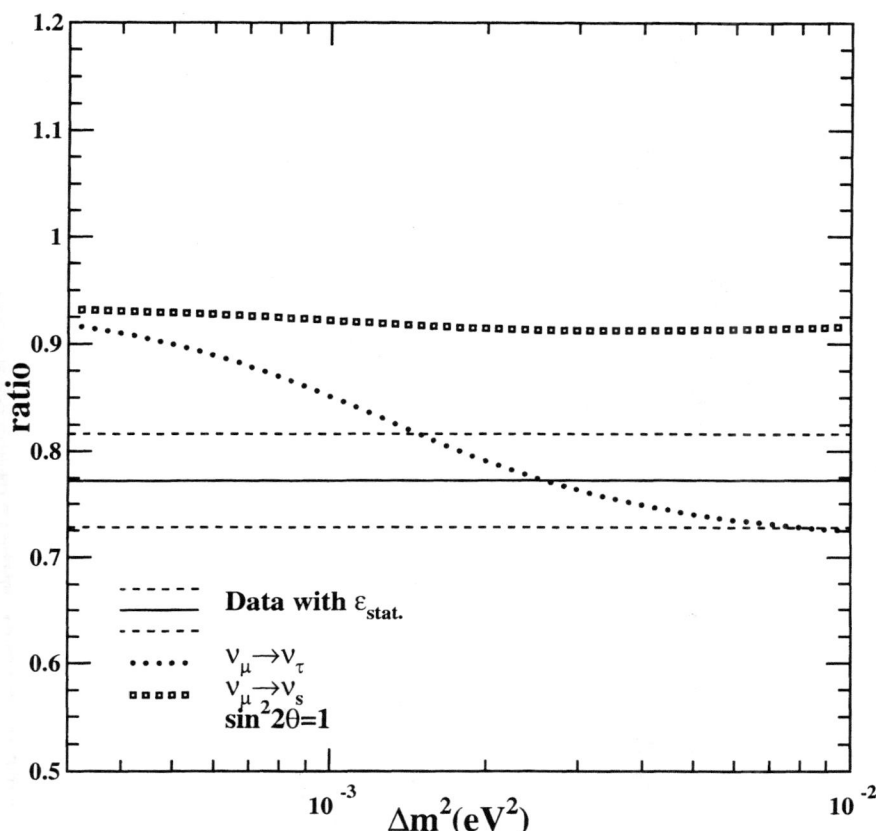

FIGURE 47. Vertical/Horizontal ratio in the very high-energy upward through-going muon sample for 1138 days of data. The solid line is our best-fit value (with dashed lines representing one-sigma errors), while the curves represent theortical predictions of $\nu_\mu \to \nu_\tau$ and $\nu_\mu \to \nu_s$ as a function of Δm^2. The $\nu_\mu \to \nu_\tau$ solution is very strongly favored by the data, especially considering that our best fit for Δm^2 is 3.2×10^{-3} eV^2.

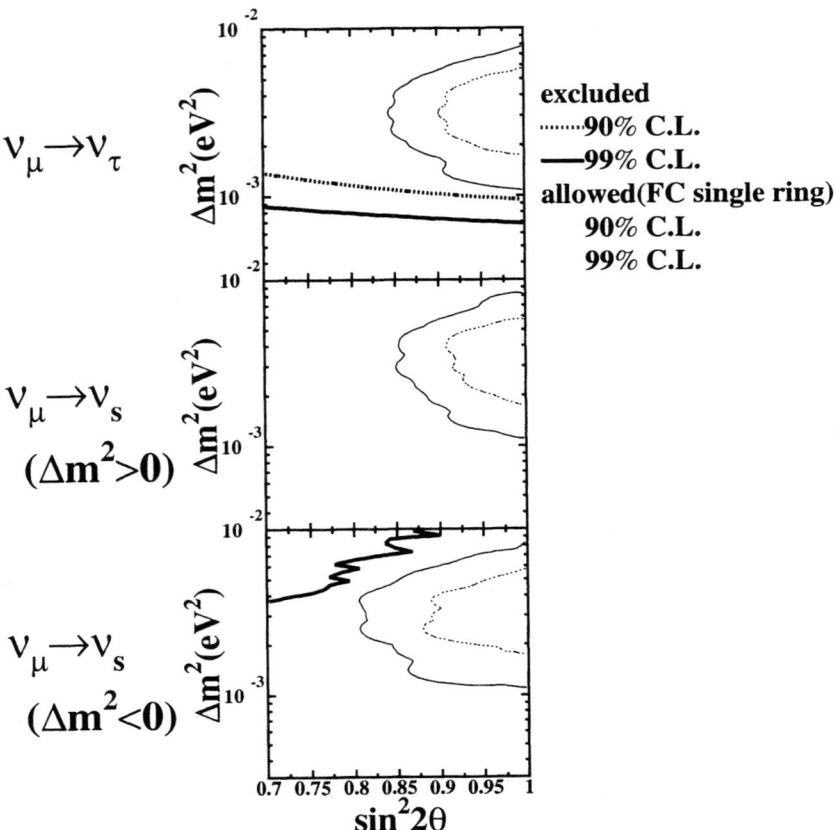

FIGURE 48. The allowed regions for oscillations in three possible oscillation hypotheses overlaid with the excluded regions (areas below the lines are excluded) from our $\nu_\mu \to \nu_\tau$ vs. $\nu_\mu \to \nu_s$ analyses. The $\nu_\mu \to \nu_\tau$ allowed region is completely unconstrained, while the entire phase space for $\nu_\mu \to \nu_s$ ($\Delta m^2 > 0$) and all of the allowed region for $\nu_\mu \to \nu_s$ ($\Delta m^2 < 0$) are excluded at better than the 99% confidence level. We therefore conclude that Super–K is seeing the results of $\nu_\mu \to \nu_\tau$ oscillations.

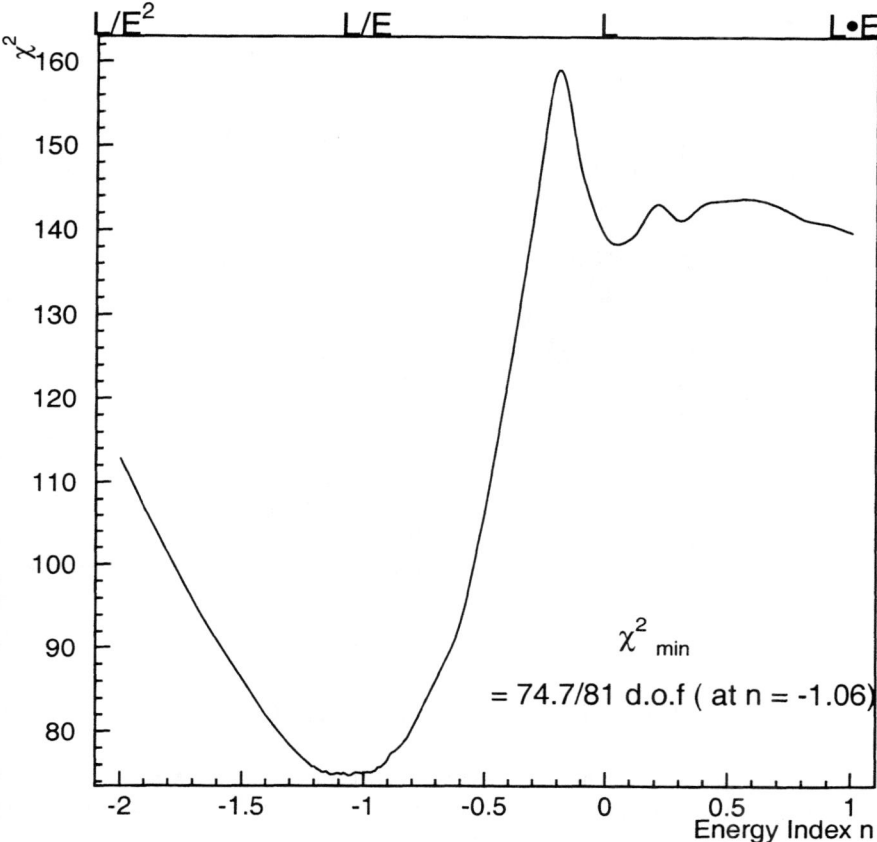

FIGURE 49. The minimum χ^2 for 1140 days of FC and PC data and 1117 days of upward-going muon data versus different powers of E. The global minimum is found at $n = -1.06 \pm 0.14$, strongly disfavoring more exotic models explaining our observed suppression of ν_μ's.

easily observable in such detectors. Of course, it has turned out that protons have a lifetime at least 10,000 times greater than that predicted by minimal SU(5), and some twenty years later we are still looking for our first gold-plated event. Large water detectors are now seen primarily as neutrino observatories, not proton decay experiments — the meaning of the trailing "nde" was quietly changed from "nucleon decay experiment" in Kamiokande to "neutrino detection experiment" in Super-Kamiokande. Nevertheless, because nucleon decays would produce background events in our atmospheric neutrino sample (a somewhat ironic inversion of what was expected in the past), we can get nucleon decay lifetime limits from our high-energy data at very little additional cost.

Figures 50 and 51 show our current lifetime limits for various decay modes. Although there seems to be an excess of events beyond what is predicted for background in two of the modes involving $K^0 \to \pi^+\pi^-$ decays in the final state, the lack of a corresponding excess in the twice-as-common $K^0 \to \pi^0\pi^0$ final state modes make us think that these are not real nucleon decays but rather the result of some novel background.

Proton decay remains one of the Holy Grails of both theoretical and experimental particle physics. There are currently discussions under way, both in the U.S. and Japan, concerning the construction of a next-generation large water Cherenkov detector at least twenty times the volume of Super-K which would be primarily focused on this continuing quest (see Figure 52).

D The K2K Experiment

Convincing as they are on their own, it is nevertheless highly desirable to confirm Super-K's dramatic oscillation results under conditions in which the incoming neutrinos can be measured just after production as well as after they have traveled a significant distance. To accomplish this, long-baseline neutrino oscillation programs are approved and under way at national accelerator facilities in the U.S., Europe, and Japan. While those in the U.S. and Europe are expected to become operational sometime in the middle of this decade, the Japanese project is already up and running. Beginning in late spring of 1999, a new experiment, KEK to Kamioka [K2K], started taking data. As seen in Figure 53, this project directs a beam of artificially produced muon neutrinos through Japan from the KEK accelerator laboratory outside of Tokyo all the way through central Japan to Super-K, some 250 kilometers distant. Figure 54 shows the near detector complex, located 300 meters from the end of the neutrino production region. The one kiloton water Cherenkov detector, essentially a 2% scale model of Super-K, is being used to normalize the neutrino flux. Deviations in the number of events seen in Super-K from the expected number based on simple geometic considerations would serve as additional evidence of oscillations.

With about 20% of its scheduled data collected, it is still a bit too early to say what the outcome of the K2K experiment will be, although a suppression in

Summary of Nucleon Decay Searches

mode	exposure (kt·yr)	εB_m (%)	observed event	B.G.	τ/B limit (10^{32} yrs)
$p \to e^+ + \pi^0$	70	43	0	0.1	44
$p \to \mu^+ + \pi^0$	70	32	0	0.4	34
$p \to e^+ + \eta$	45	17	0	0.3	11
$p \to \mu^+ + \eta$	45	12	0	0	7.8
$n \to \bar{\nu} + \eta$	45	21	5	9	5.6
$p \to e^+ + \rho$	61	6.8	0	0.6	6.1
$p \to e^+ + \omega$	61	3.3	0	0.3	2.9
$p \to e^+ + \gamma$	70	71	0	0.1	73
$p \to \mu^+ + \gamma$	70	60	0	0.2	61
$p \to \bar{\nu} + K^+$	70				19
$K^+ \to \nu\mu^+$ (spectrum)		34	--	--	4.3
prompt $\gamma + \mu^+$		9.3	0	1.1	9.5
$K^+ \to \pi^+\pi^0$		6.8	0	1.9	6.9
$n \to \bar{\nu} + K^0$	70				1.8
$K^0 \to \pi^0\pi^0$		9.6	27	30.5	2.2
$K^0 \to \pi^+\pi^-$		4.6	11	5.9	0.83
$p \to e^+ + K^0$	70				5.4
$K^0 \to \pi^0\pi^0$		11.8	1	1.4	8.8
$K^0 \to \pi^+\pi^-$					
2-ring		6.2	6	1.0	1.5
3-ring		1.4	0	0.2	1.4
$p \to \mu^+ + K^0$	70				10
$K^0 \to \pi^0\pi^0$		6.1	0	1.1	6.2
$K^0 \to \pi^+\pi^-$					
2-ring		5.3	0	1.5	5.4
3-ring		2.8	1	0.2	1.8

FIGURE 50. Nucleon decay limits for 1144 live days (70.4 kiloton years) of data.

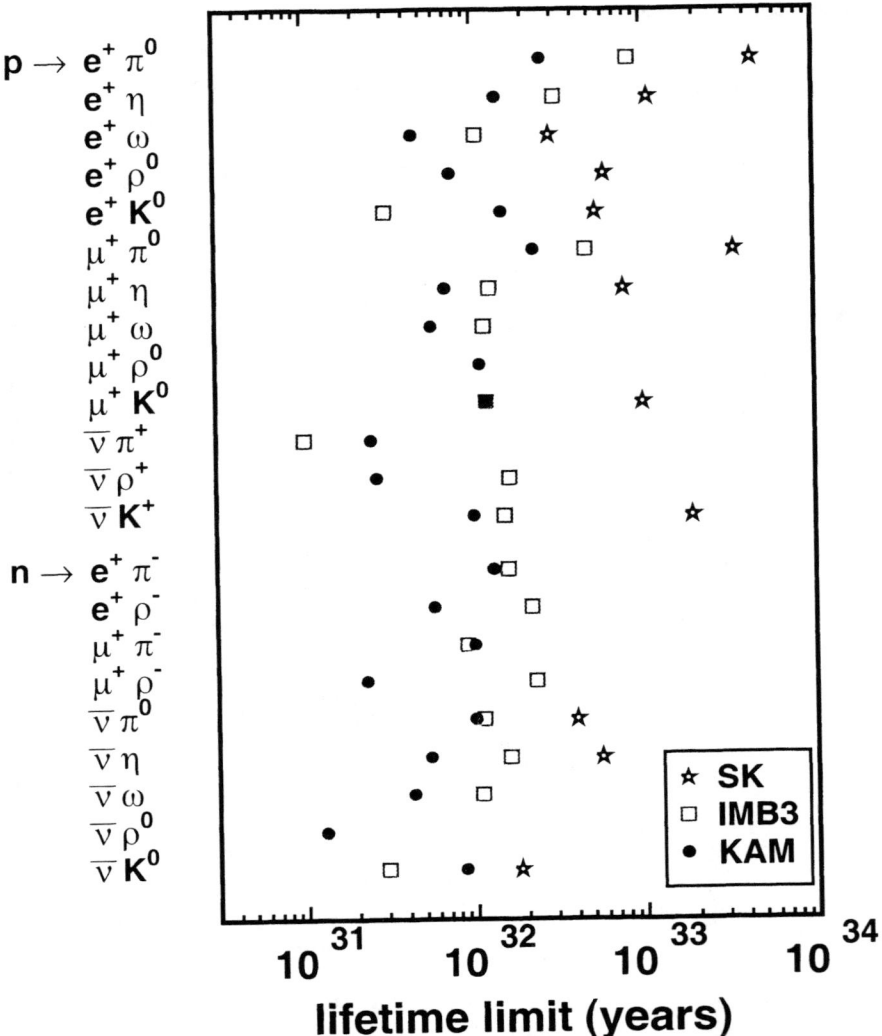

FIGURE 51. Comparitive nucleon decay limits for 1144 live days (70.4 kiloton years) of Super-K data and about ten years of both IMB and Kamiokande data. It is expected that Super-K will eventually reach well past 10^{34} years in some modes.

The search for proton decay is a search for processes which will ultimately dissolve the very universe as we know it — hence my suggested name, mascot, and slogan for the project:

A Really Massive And Gigantic Experiment for Detecting Decay Of Nucleons

"Have a nice decay!"

FIGURE 52. The next nucleon decay experiment?

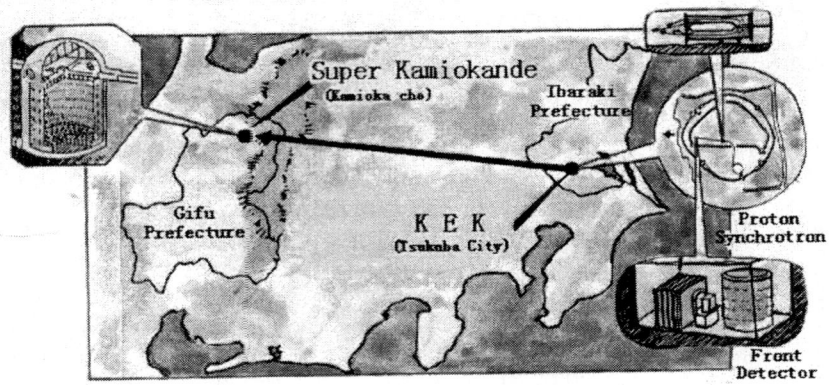

FIGURE 53. The KEK to Kamioka [K2K] experiment.

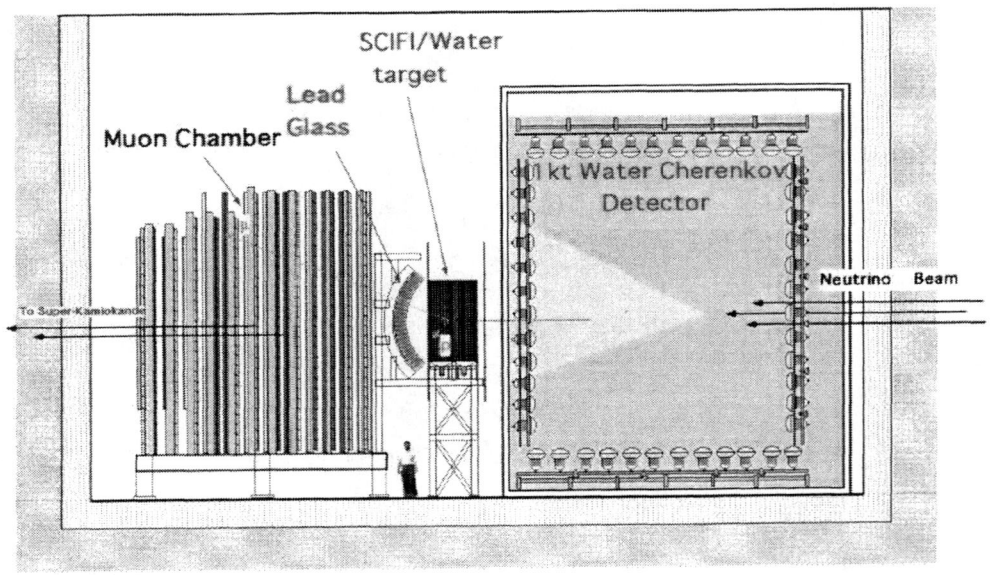

FIGURE 54. The K2K experiment's near detector complex. Note the water Cherenkov detector, which is basically a 2% scale model of Super–Kamiokande.

the event rate at Super–K vs. that seen in the near detectors is currently being observed.

E High Energy Summary and Conclusions

Super–Kamiokande's high-energy analysis of atmospheric neutrinos has produced the world's first conclusive evidence of physics beyond the Standard Model: our zenith-angle dependence of the ν_μ flux provides powerful proof of neutrino oscillations and hence proof of the existence of massive neutrinos.

We have pretty much completely excluded $\nu_\mu \to \nu_e$ and $\nu_\mu \to \nu_s$ oscillations, as well as more exotic explanations of the disappearance of the ν_μ's. It now looks quite certain that $\nu_\mu \to \nu_\tau$ oscillations are being observed, though final confirmation will likely come from long-baseline accelerator experiments currently under way or under construction.

No nucleon decay has yet been observed, but we have extended the limits on the lifetime of the proton and bound neutron by an order of magnitude over the former best values, perhaps providing motivation for yet another generation of water Cherenkov detectors in the process. And so it goes...

Naturally, we will continue to collect and refine high-energy data in the coming years, most likely focusing on the exciting possiblity of seeing atmospheric tau *appearance* in our detector, as well as searching more of the potential nucleon decay modes for that long-sought-after golden event. Super–Kamiokande has a committment for at least thirty (!) more years of operations, so who knows what the future may bring? Barring an extension of the data-taking run, see you in 2030 for a final wrap-up of our results.

ACKNOWLEDGEMENTS

We gratefully acknowledge the cooperation of the Kamioka Mining and Smelting Company. This work was partly supported by the Japanese Ministry of Education, Science and Culture and the U.S. Department of Energy.

REFERENCES

Interested readers are directed to one or more of our fine publications:

(1) "Measurement of a small atmospheric ν_μ/ν_e ratio," The Super–Kamiokande Collaboration, Physics Letters **B433**, 9 (1998)
(2) "Measurements of the Solar Neutrino Flux from Super–Kamiokande's First 300 Days," The S–K Collab., Physical Review Letters **81**, 1158 (1998)
(3) "Evidence for oscillation of atmospheric neutrinos," The S–K Collab., Physical Review Letters **81**, 1562 (1998)
(4) "Study of the atmospheric neutrino flux in the multi–GeV energy range,"

The S–K Collab., Physics Letters **B436**, 33 (1998)
(5) "Search for Proton Decay via $p \to e^+\pi^0$ in a Large Water Cherenkov Detector,"
The S–K Collab., Physical Review Letters **81**, 3319 (1998)
(6) "Calibration of Super–Kamiokande using an electron LINAC,"
The S–K Collab., Nuclear Instruments and Methods **A421**, 113 (1999)
(7) "Measurement of Radon Concentrations at Super–Kamiokande,"
The S–K Collab., Physics Letters **B452**, 418 (1999)
(8) "Neutrino Induced Upward Stopping Muons in Super–Kamiokande,"
The S–K Collab., Physics Letters **B467**, 185 (1999)
(9) "Constraints on Neutrino Oscillation Parameters from the Measurement of Day Night Solar Fluxes at Super–Kamiokande," The S–K Collab., Physical Review Letters **82**, 1810 (1999)
(10) "Measurement of the Solar Neutrino Energy Spectrum Using Neutrino Electron Scattering," The S–K Collab., Physical Review Letters **82**, 2430 (1999)
(11) "Measurement of the Flux and Zenith Angle Distribution of Upward Through Going Muons," The S–K Collab., Physical Review Letters **82**, 2644 (1999)
(12) "Observation of an East–West Anisotropy of the Atmospheric Neutrino Flux," The S–K Collab., Physical Review Letters **82**, 5194 (1999)
(13) "Search for Proton Decay via $p \to \bar{\nu}K^+$ in a Large Water Cherenkov Detector," The S–K Collab., Physical Review Letters **83**, 1529 (1999)

SNO Detector Status

by R.G. Van de Water,
representing the SNO collaboration

University of Pennsylvania
Dept. of Physics and Astronomy
Philadelphia, PA, 19104-6396
U.S.A.

Abstract. The Sudbury Neutrino Observatory (SNO) is a 1 kiloton heavy water, 10,000 phototube, real-time, state of the art second generation water Cerenkov detector designed primarily to study solar neutrinos. The SNO detector has been taking high quality production neutrino data for the last seven months. Water assays and many electronic and source calibration runs have been performed to study the detector cleanliness, response and systematics. It has met or exceeded all design goals and expectations.

The Sudbury Neutrino Observatory (SNO) is a 1 kt water Cerenkov detector situated 2km deep underground in an active mine at the INCO Creighton #9 shaft, near Sudbury, Canada. The use of heavy water permits detection of neutrinos though the three main reactions,

$$(ES) \quad \nu_e + e^- \to \nu_e + e^- \tag{1}$$

$$(CC) \quad \nu_e + d \to e^- + p + p \tag{2}$$

$$(NC) \quad \nu_x + d \to \nu_x + n + p \tag{3}$$

where ν_x refers to any active flavor of neutrino. Both the elastic scattering (ES) and charge current (CC) reactions are detected via the recoil/emitted electron that subsequently produces Cerenkov light. In the case of the neutral current (NC) reaction, the liberated neutron captures on deuterium, producing a 6.25 MeV photon that subsequently Compton scatters and produces Cerenkov photons.

For more detector details, technical specifications, and description of calibration sources mentioned in this paper, the reader is referred to [1].

I COMMISSIONING PHASE

The water fill was completed in April, 1999. From early May thru to October, 1999, the SNO detector went through its final commissioning phase. In this period, the individual detector components and triggers were turned on and studied in detail. The detector behavior was monitored carefully, and operating parameters were tuned for optimal performance and stability.

Many sources of instrumental backgrounds were identified and characterized, e.g. flashing phototubes, electronic noise, light from static discharges in the acrylic vessel (AV) neck area, seismic activity, etc. They were either reduced to tolerable levels, or tagged by the addition of new hardware (e.g. neck phototubes, hydrophone), or analysis cuts developed to efficiently identify these events. During this period initial phototube and optical calibrations were performed using a diffusion laser source.

Initial assays of the water to determine the radioactivity background levels was also begun, as well as analysis of the PMT data. Initial results of these activities were encouraging. Some improvements in the radon seal above the light and heavy water was warranted, and successfully completed during this stage.

After a brief shutdown in October 1999 for final detector tuning and upgrades, the detector state was essentially frozen and full production neutrino data taking begun.

II   PRODUCTION DATA PHASE: PURE D_2O

Since November 2, 1999, the detector has been running with pure D_2O in a stable and well characterized state. The overall detector livetime up to May, 2000, is over 95%, while the fraction of the total time that is good for neutrino analysis is about 80%. The missing neutrino livetime is due to detector calibration and maintenance periods, however, the detector is still supernova live during these activities. The 5% detector deadtime is due mostly to power outages from mining activities. Figure 1 shows the total livetime since the beginning of production running.

A   Detector Performance

Since the start of the pure D_2O phase, high quality production data has been taken and detector performance has met or exceeded design requirements on most subsystems. The detector is extremely stable, running at a fairly steady rate around 20 Hz. Low hardware trigger thresholds have been achieved, with few electronics or DAQ run problems. Table 1 shows the various simultaneous triggers enabled and the corresponding thresholds and detector rates.

The main NHIT (100 nsec) trigger threshold is 18 hits, when energy calibrated using the ^{16}N gamma source, this corresponds to ~ 2 MeV, while the 90% trigger turn-on level is about 1/3 MeV.

The electronics and DAQ readout chain is very robust and stable, especially at neutrino run rates, but can also handle much higher rates. For example, during calibration runs with the 6.1 MeV gamma source the detector can handle rates up

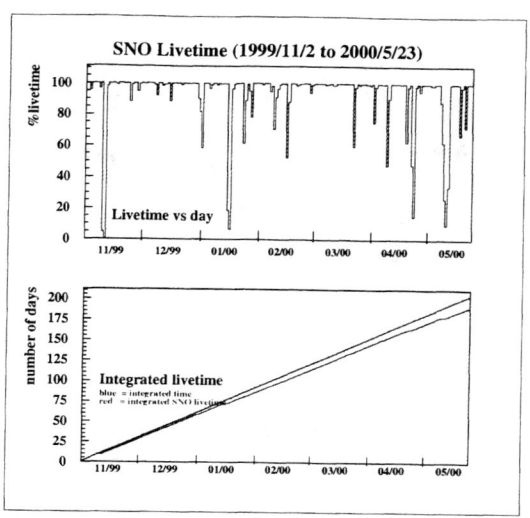

FIGURE 1. Detector livetime during the pure D_2O production run phase.

to 300 Hz sustained. As well, the detector has burst capabilities up to 2 MHz, to handle supernova and other short time scale physics events.

Figure 2 shows the detector rates since the start of production running. It demonstrates the overall detector stability, with only a few episodes of rate excursions. The third plot in Figure 2 shows the unbiased 5 Hz pulsed trigger mean NHIT. It averages about two background hits over a 400 nsec window. This corresponds to less than half a hit of random background in a 100 nsec in-time analysis window. Also shown is an overall downward trend in mean NHIT which indicates a gradual quieting of the detector, i.e. decreasing radon rates, instrumental backgrounds, etc.

The inner detector consists of 9366 phototubes (PMT's), 94 outer looking tubes for muon identification, 4 tubes for AV neck event identification, and a further 22

Trigger Type	Threshold	Rate
100 ns NHIT coincidence	18 PMTs	5 – 10 Hz
20 ns NHIT coincidence	18 PMTs	~1 Hz
Energy Sum	~ 150 pe	~5 Hz
Pulsed Trigger	Zero Bias	5 Hz
Prescaled (1:1000)	11 PMTs	< 1 Hz
OWL nhit	12 PMTs	< 0.1 Hz

TABLE 1. Detector trigger thresholds and rates during a typical neutrino run.

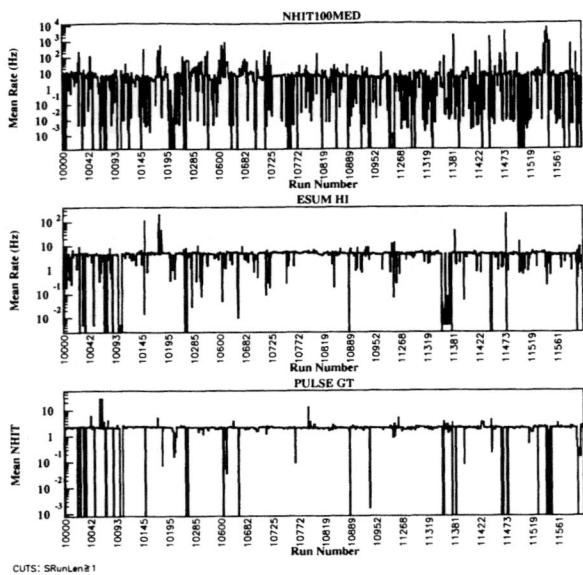

FIGURE 2. Main detector trigger (NHIT100, ESUM, PULSE-GT) trends over a six month period starting Nov 2, 1999.

connector testing tubes. Of the total, to date, only 186 PMT's are not functional. The phototubes and high voltage water connectors have been operating stably since the re-gasification of the light water in April 1999. Since the start of commissioning, the tube death rate from water ingress, failed PMT components, etc, is around 0.6% per year (see Figure 3). This is consistent with the long term experience of other water based Cerenkov detectors.

B Detector Calibrations

The electronic channel setup constants, charge pedestals, time slopes, high voltage gains, and thresholds, have been shown to be extremely stable. Electronics calibrations, i.e. charge pedestals and TAC slopes, are done twice a week, with average variations typically less than 0.05 photoelectrons for charge, and 0.1 nsec for time measurements.

Phototube calibrations are done once a month. This calibration checks the PMT gain stability, timing characteristics/delays, and discriminator walk corrections. Figure 4 shows the typical single photo-electron (PE) response of all the tubes summed together. The average channel charge threshold is found to be 0.25 PE, and is well matched around the detector. This corresponds to a electronic channel efficiency of over 90% for PMT's operating at 10^7 gain. At these thresholds and

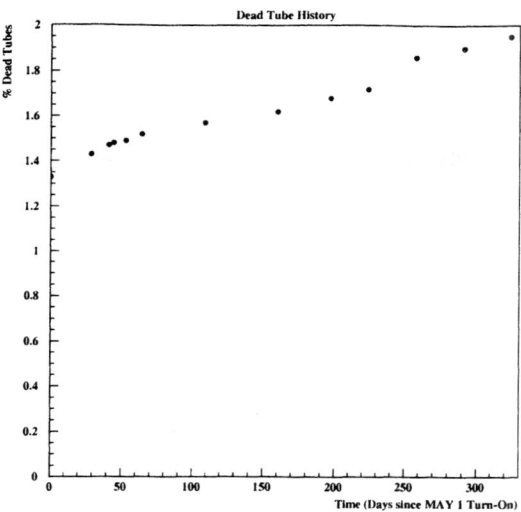

FIGURE 3. Phototube death rate during that past year of operation. The jump at day 250 is a result of turning off a number of long term low occupancy channels.

gains, the typical phototube rate in the light water, at 10°C, is 250 to 500 Hz.

Another calibration done with the diffusion laser source is a check of the PMT timing response. Figure 5 shows a typical distance and walk corrected time response for all channels summed together. The timing spread corresponds to only 1.7 nsec, which is a convolution of the phototube response to a single photon, and a much smaller component due to the laser's intrinsic width.

A single point detector energy calibration has been done using the ^{16}N, primarily monoenergetic (6.1 MeV), gamma ray source. With magnetic field compensation coils on, and all PMT's at nominal high voltage and charge thresholds, the detector response is approximately $\sim$9 hits/MeV. Figure 6 shows the detector NHIT response for various high voltage and discriminator settings. This source has only become fully operational in the last few months, and will be deployed on a monthly basis to check for detector response stability and energy calibration.

Other sources over the next six months will be deployed in the pure D_2O phase to expand on the energy calibration, study event fitter systematics, check neutron response, and calibrate low energy backgrounds. The full list of all sources to be deployed in the first year of operation is shown below.

- Electronic pulser - charge pedestals and time slope constants.
- Diffusion laser - phototube calibrations, detector optics.
- ^{16}N - detector energy and angular response, 6.1 MeV gamma ray.
- pT - detector energy and angular response, 20 MeV gamma ray.

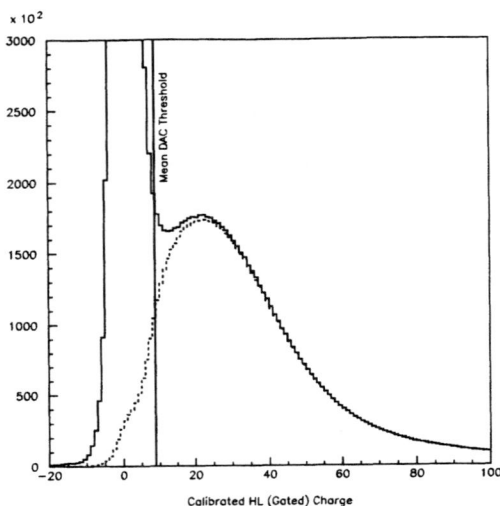

FIGURE 4. The single photon charge response distribution (using the diffusion laser) of all inner detector PMT's summed together. The solid line represents gated charge, while the dashed is threshold discriminated data. The solid vertical line is the mean discriminator threshold setting. The horizontal axis is raw channel charge in units of pedestal subtracted ADC counts.

- ^{252}Cf - detector response to neutrons.
- ^{8}Li - β spectrum, 15 MeV endpoint - fitting reconstruction systematics.
- Low energy backgrounds - ^{238}U and ^{232}Th response and rate calibration.

C Water Status

Since the summer of 1999, both the light and heavy water have been circulated and filtered on a regular basis. Initial optical calibrations with the diffusion laser ball shows that the optical properties of the water are excellent, with an attenuation length much greater than 75 m.

As well, a program of water and cover gas assays are done on a weekly basis. The levels of Radon, Uranium, and Thorium, have been measured and found to be at or below target levels. These radioactive elements can contribute to either neutron backgrounds, or produce low energy β particles. The target level for neutron contaminations is about 10% the SSM prediction for the neutral current flux. The target for the β emitter backgrounds is an analysis threshold in the D_2O volume of 5 MeV. Given the low assay levels, and initial detector data analysis, we will easily reach this target, and probably do better.

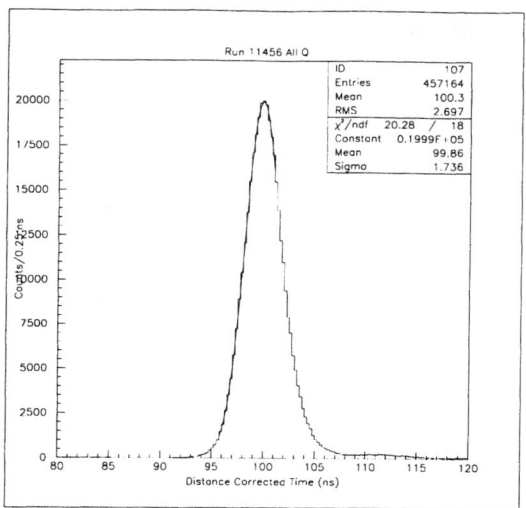

FIGURE 5. The single photon time response distribution (using the diffusion laser) of all inner detector PMT's summed together. The horizontal axis is channel time, corrected for tube distance and walk, in units of nanoseconds.

III OUTLOOK AND CONCLUSIONS

There are three planned detector run phases for SNO; pure D_2O, D_2O + salt, and D_2O + NCD. The physics goals of SNO, i.e. the solution to the solar neutrino deficit, will be realized via the successful completion of these three distinct detector configurations.

The first, which will last a total of one year, is the pure D_2O phase, and will allow a clean determination of the CC flux, energy spectrum, and initial estimates of backgrounds. The second phase will add two tones of salt (NaCl) to the heavy water to significantly improve the neutral current detection efficiency. In this phase, which will also last one year, a more accurate estimate of the neutral current rate will be made, and comparison with the CC rate will provide conclusive evidence for, or against, neutrino oscillations. The final, and longest phase, will see the addition of an array of Helium filled neutral current detectors (NCD) to allow an independent measurement of the NC rate.

To conclude, the extreme attention paid to cleanliness and design/construction details has paid off handsomely. The SNO detector is stable and taking quality solar neutrino data. The water is clean, trigger thresholds low, and the detector is being throughly calibrated and understood. Quantitative results on the pure D_2O phase will be forthcoming soon.

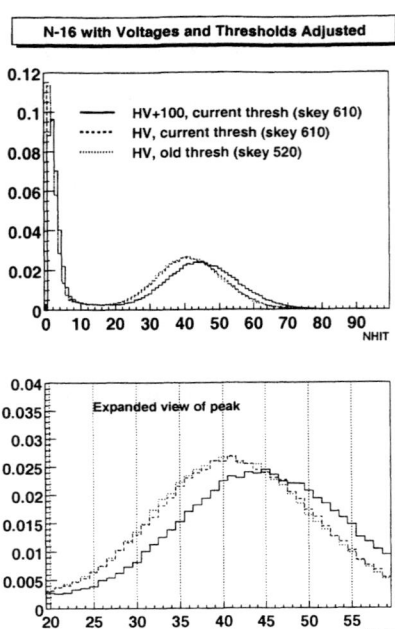

FIGURE 6. The detector NHIT (400 nsec window) response for the ^{16}N source. The solid line represents the production detector configuration for the pure D_2O phase.

REFERENCES

1. *The Sudbury Neutrino Observatory*, J. Boger, et al. (SNO), to be published in Nucl. Inst. Meth. A, July 2000, LANL preprint NUCL-EX/9910016.

NEUTRINOS IN COSMOLOGY

Neutrinos in Supernovae and Extra Dimensions

David O. Caldwell

Department of Physics, University of California, Santa Barbara, CA 93106-9530, USA

Abstract. A neutrino mass-mixing scheme which explains qualitatively all present evidence for neutrino mass (the solar and atmospheric neutrino deficits, LSND, and significant hot dark matter) also successfully avoids the "alpha effect", allowing r-process nucleosynthesis in the neutrino-heated ejecta of supernovae. The properties of neutrinos required to ensure production of heavy nuclei provides independent evidence for (1) at least one light sterile neutrino, ν_s; (2) a near maximally-mixed ν_μ-ν_τ doublet (which also explains the atmospheric anomaly and provides hot dark matter) split from a lower mass ν_e-ν_s doublet (needed also for the solar ν_e deficit); (3) ν_μ-ν_e mixing $\gtrsim 10^{-4}$; and (4) a splitting between the doublets (measured by the ν_μ-ν_e mass difference) $\gtrsim 1$ eV^2, favoring the upper part of the LSND range. There is a quantitative problem with the solar observations, which do not in detail fit this or any other model. If, however, the ν_s is a bulk neutrino in extra dimensions the nearness in mass of the zero mode to the ν_e provides vacuum oscillations, while the Kaluza-Klein states give MSW oscillations, and all the solar data can be fit.

INTRODUCTION

The deficit of ν_e from the sun, the ν_μ/ν_e ratio coming from the atmosphere, and either the existence of hot dark matter or the results of the LSND experiment require four types of neutrinos. The measured width of the Z^0 boson allows only three light neutrinos, so the fourth must not have the usual weak interaction and hence is called a sterile neutrino, ν_s. This four-neutrino scheme at least qualitatively explains all the observations: light $\nu_e \to \nu_s$ for solar, heavier $\nu_\mu \to \nu_\tau$ for atmospheric and to share the hot dark matter role, and $\nu_\mu \to \nu_e$ for the LSND oscillation data. In the next section these particle physics motivations for the neutrino scheme are discussed, with special emphasis on recent results from the LSND experiment.

Surprisingly there is quite independent and detailed evidence for exactly this same neutrino scheme from a very different source, heavy-element nucleosynthesis in supernovae. While there is strong evidence [1] that the heaviest elements are produced in the neutrino-heated material ejected relatively long (~ 10 s) after the explosion of a Type II or Type I b/c supernova, present calculations [2,3] show conditions which would prevent this rapid-neutron-capture (or r) process from occurring. Though general relativistic effects [4] and multi-dimensional hydrodynamic outflow [5] have been invoked to solve these problems, these solutions are at best exceedingly finely tuned. In contrast, the solution [6] presented here is extremely robust, and the neutrino mass-mixing scheme it requires is exactly that needed if one is to explain all present evidence for neutrino mass.

All this evidence for this four-neutrino mass scheme makes it important to test the scheme quantitatively. The evidence from four solar experiments of three types provides such a check, especially as these extensive data apparently are not explained in detail by any neutrino scheme. The usual non-interacting sterile neutrino is usually said to provide a small-angle MSW [7] oscillation solution, or a vacuum-oscillation ("just so") solution, since the two possible large-angle MSW solutions are both very bad fits and are ruled out by big-bang (light-element) nucleosynthesis considerations. In fact, neither allowed solution really works, but the data give hints of both. If large extra dimensions exist, the sterile neutrino can reside in the bulk, whereas active neutrinos are confined to the brane. If the zero mode of the ν_s is close in mass to ν_e, then vacuum oscillations can occur. Such a ν_s will have higher mass Kaluza-Klein modes, and the spacing of these makes their mass difference from the ν_e such that MSW oscillations can occur. The combination of MSW and vacuum oscillations provides a quantitative fit to the solar data.

PARTICLE PHYSICS EVIDENCE FOR A FOUR-NEUTRINO SCHEME

The need for at least one light sterile (i.e., not having the usual weak interaction) neutrino in addition to the known three active neutrinos was proposed [8] as a way to explain the solar ν_e deficit, the anomalous ν_μ/ν_e ratio from atmospheric neutrinos, and the apparent need for appreciable hot dark matter. The atmospheric anomaly, due to $\nu_\mu \to \nu_\tau$, requires a mass-squared difference between the ν_μ and ν_τ of $\Delta m^2_{\mu\tau} = 0.003$ eV^2 and maximal mixing ($\sin^2 2\theta_{\mu\tau} = 1.0$), with the latter property being quite important to solving the r-process problem, as shown below. The solar ν_e deficit is explained by $\nu_e \to \nu_s$ with $\Delta m^2_{es} \approx 10^{-5}$ eV^2 and a small mixing angle ($\sin^2 2\theta_{es} \approx 0.01$) or $\Delta m^2_{es} \sim 10^{-10}$ eV^2 and $\sin^2 2\theta_{es} \sim 1$. The ν_e-ν_s pair is of lower mass than the ν_μ-ν_τ pair, which provide the hot dark matter. For the originally favored [9] critical-mass-density universe the ν_μ and ν_τ needed masses of around 2 eV each. Given not only the failure of low-mass density models to fit universe structure on all scales, but also the small size of the second Doppler peak in the cosmic microwave background radiation, the critical-mass-density cold plus hot dark matter model, which fits universe structure remarkably well, could be revived if it becomes necessary to add new physics to the standard cosmology.

This phenomenology was subsequently given theoretical bases in two 1993 papers [10], and has been utilized since in a large number of publications. Since that time the scheme has received support from the LSND experiment, the results of which provide some measure of the mass difference between the ν_e-ν_s and ν_μ-ν_τ pairs and hence of the neutrino contribution to dark matter.

In its 1996 publication [11], LSND claimed a signal in $\bar\nu_\mu \to \bar\nu_e$ on the basis of seeing events of the type $\bar\nu_e p \to e^+ n$, with the e^+ identified by both scintillation and Cherenkov light and the n by the 2.2-MeV γ from $np \to d\gamma$. Unfortunately, the likelihood analysis used to show favored regions of Δm^2 vs. $\sin^2 2\theta$ did not have a Gaussian likelihood distribution (the integral being infinite), but the likelihood contour labeled "90%" was obtained by going down a factor of 10 from the maximum, as in the Gaussian case. The contours in the LSND plot have been widely misinterpreted as confidence levels—which they certainly are not—because they were plotted along with confidence-level limits from other experiments. Recently the difficult, computer-intensive analysis in terms of real confidence levels has been done [12]. The likelihood for a grid in ($\sin^2 2\theta$, Δm^2) space, including backgrounds, has been computed and compared with numerous Monte Carlo experiments to obtain a 90% confidence region. While the equivalency varies from point to point in the Δm^2-$\sin^2 2\theta$ plane, a typical value for the 90% confidence level is down a factor of 20 from the likelihood maximum. Thus the LSND allowed regions are considerably broader in $\sin^2 2\theta$ than in the plots published so far, and other experiments constrain allowed Δm^2 regions less.

The confusion of comparing likelihood levels for LSND with confidence levels from other experiments may be exacerbated by using the 20–36 MeV region for the LSND data. While this higher background energy range due to $\nu_e{}^{12}C \to e^- X$ makes some difference for the 1993-5 data, it could have had an appreciable effect for the parasitic 1996-8 runs, which were at a low event rate, increasing the effect of cosmic ray background. This could raise the low end of the supposed signal energy spectrum, especially as the one LSND distribution which was statistically worrisome was the ratio (R) of real to accidental events. Some accidental events in this 20–36 MeV region would favor low values of $\Delta m^2_{e\mu}$, making the higher Δm^2 values desirable for dark matter appear less likely.

Nevertheless, when a joint analysis [12] was made of the LSND and KARMEN [13] experiments even using the 20–36 MeV range for LSND, the region around 5.5 eV^2 is as probable as the banana-shaped region at lower Δm^2, as shown in Fig. 1. Frequently ignored by theorists, this higher mass region is favored by the $\nu_\mu \to \nu_e$ LSND data. Of course in the $\nu_\mu \to \nu_e$ case [11], using 60 to 200 MeV ν_μ from π^+ decay in flight and detecting ν_e by $\nu_e{}^{12}C \to e^- X$, the backgrounds are higher and hence yield much poorer statistics than for $\bar\nu_\mu \to \bar\nu_e$ with $\bar\nu_\mu$ from μ^+ at rest. In addition to the Δm^2 issue, the important point of Fig. 1 is that although the KARMEN data are consistent with background and apparently partly conflicts with LSND, the joint analysis of the $\bar\nu_\mu \to \bar\nu_e$ data from the two experiments shows an appreciable region for a signal. KARMEN is continuing to take data, and LSND now has an improved analysis. This new analysis has produced an excellent R distribution for all the data (even down to 20 MeV) and an energy distribution using 20–60 MeV data with reduced contributions at the low end,

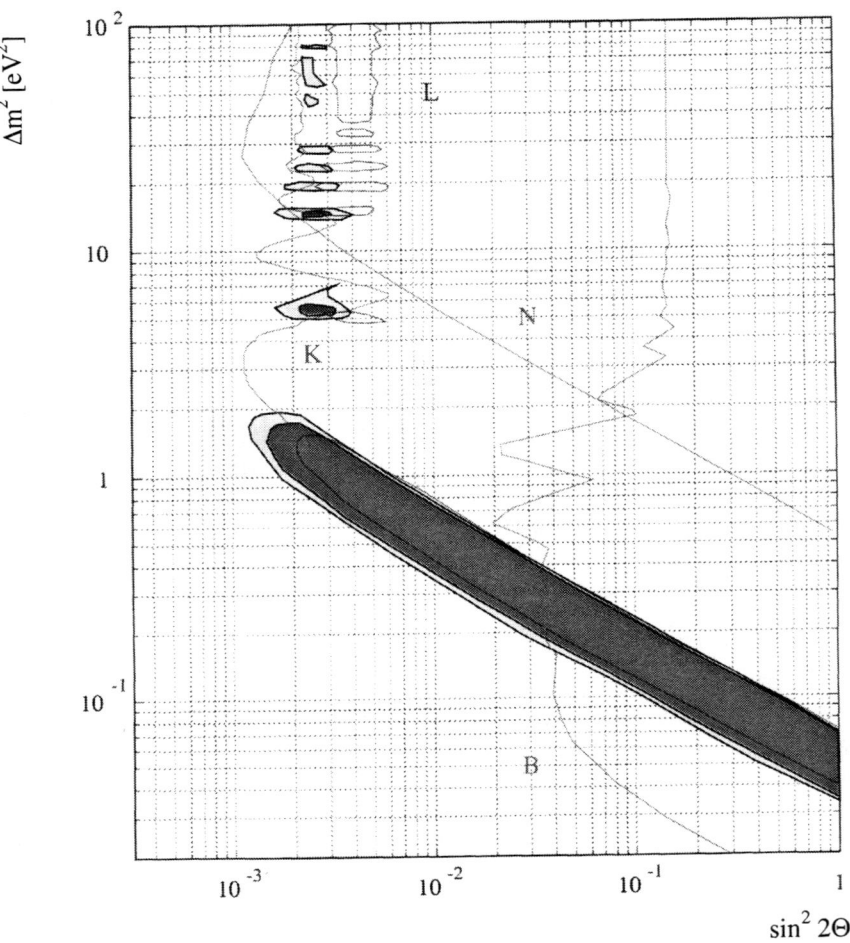

FIGURE 1. Filled in areas are 90% and 95% confidence regions based on the product of the KARMEN and LSND Feldman-Cousins likelihood ratios. Also shown is the Feldman-Cousins 90% confidence region for LSND alone ("L"). Left of the "K", "N", and "B" curves are exclusion regions of KARMEN, NOMAD, and Bugey.

favoring higher Δm^2_μ values than does Fig. 1. However, it has been decided to present a global analysis, using all the data from 20 to 200 MeV. Too small a $\nu_\mu \to \nu_e$ signal in the 60–200 MeV region is observed in the global analysis, unlike the more thorough analysis [11] specifically looking for $\nu_\mu \to \nu_e$. That reduced result is more compatible with smaller Δm^2 values at smaller mixing angles than is the case for larger Δm^2, and hence including it makes especially the ~ 6 eV^2 region seem less probable. In the author's opinion, the 60–200 MeV tail of small, systematics prone signal is wagging the dog of the 20–60 MeV region where the much larger and better determined signal is found. Generally, comparing LSND and KARMEN now does not result in much change from Fig. 1, since the former's oscillation probability is reduced to 0.25% from 0.3%, while the latter's limits have also shifted to the left because of longer running time.

PROBLEMS WITH SYNTHESIS OF THE HEAVIEST ELEMENTS

While in the next section we will find that the r process of rapid neutron capture in supernovae provides strong support for the double doublet of neutrinos, initially the reverse appeared to be true, with the r process apparently placing stringent limits on ν_μ-ν_e mixing. The origin of these limits is that energetic ν_μ ($\langle E \rangle \approx 25$ MeV) coming from deep in the supernova core could convert via an MSW transition to ν_e inside the region of the r-process, producing ν_e of much higher energy than the thermal ν_e ($\langle E \rangle \approx 11$ MeV). The latter, because of their charged-current interactions, emerge from farther out in the supernova where it is cooler. Since the cross section for $\nu_e n \to e^- p$ rises as the square of the energy, these converted energetic ν_e would deplete neutrons, stopping the r-process. Calculations [14] of this effect limit $\sin^2 2\theta$ for $\nu_\mu \to \nu_e$ to $\lesssim 10^{-4}$ for $\Delta m^2_{e\mu} \gtrsim 2$ eV^2, in conflict with at least the higher mass region of the LSND results, which will be of particular interest here.

More recently, serious problems have been found with the r process itself. First, simulations [2] have revealed the r-process region to be insufficiently neutron-rich, since about 10^2 neutrons are required for each seed nucleus, such as iron. This was bad enough, but the recent realization of the full effect of α-particle formation has created a disaster for the r process [3]. At a radial region inside where the r process should occur, all available protons swallow up neutrons to form the very stable α particles, following which $\nu_e n \to e^- p$ reactions reduce the neutrons further and create more protons which make more α particles, and so on. The depletion of neutrons rapidly shuts off the r process, and essentially no nuclei above $A = 95$ are produced.

To solve this problem the ν_e flux has to be removed before the r process site, while leaving a very large ν_e flux at a smaller radius for material heating and ejection. The obvious difficulty of accomplishing this has led to searches for other possible sites for the r process, such as neutron star mergers.

NEUTRINO SOLUTION FOR A SUCCESSFUL R PROCESS

The apparent miracle of having a huge ν_e flux disappear before it reaches the radius of the supernova where α particles form can be accomplished [6] if there is (1) a sterile neutrino, (2) approximately maximal $\nu_\mu \to \nu_\tau$ mixing, (3) $\nu_\mu \to \nu_e$ mixing $\gtrsim 10^{-4}$, and (4) an appreciable ($\gtrsim 1$ eV^2) mass-squared difference between ν_s and the ν_μ-ν_τ. This is precisely the neutrino mass pattern required to explain the solar and atmospheric anomalies and the LSND result, plus providing some hot dark matter!

Such a mass-mixing pattern creates two level crossings. The inner one, which is outside the neutrinosphere (beyond which neutrinos can readily escape) is near where the $\nu_{\mu,\tau}$ potential $\propto (n_{\nu_e} - n_n/2)$ goes to zero. Here n_{ν_e} and n_n are the numbers of ν_e and neutrons, respectively. The $\nu_{\mu,\tau} \to \nu_s$ transition which occurs depletes the dangerous high-energy $\nu_{\mu,\tau}$ population. Outside of this level crossing, another occurs where the density is appropriate for a matter-enhanced MSW transition corresponding to whatever $\Delta m^2_{e\mu}$ LSND is observing. Because of the $\nu_{\mu,\tau}$ reduction at the first level crossing, the dominant process in the MSW region reverses from the deleterious $\nu_{\mu,\tau} \to \nu_e$, becoming $\nu_e \to \nu_{\mu,\tau}$ and dropping the ν_e flux. For an appropriate value of $\Delta m^2_{e\mu}$, the two level crossings are separate but sufficiently close so that the transitions are coherent. Then with adiabatic transitions (as calculations show) and maximal ν_μ-ν_τ mixing, the neutrino flux emerging from the second level crossing is 1/4 ν_μ, 1/4 ν_τ, and 1/2 ν_s, and no ν_e.

A more exact way to explain this in the four-neutrino formalism is to transform the four mass eigenstates into the flavor states via the mixing angles ϕ for solar $\nu_e \to \nu_s$, ω for LSND $\nu_\mu \to \nu_e$, and $\pi/4$ (maximal mixing) for atmospheric $\nu_\mu \to \nu_\tau$. Symbolically,

$$\begin{pmatrix} |\nu_s\rangle \\ |\nu_e\rangle \\ |\nu_\mu\rangle \\ |\nu_\tau\rangle \end{pmatrix} = \begin{pmatrix} \text{Atmospheric} \\ \nu_\mu \to \nu_\tau \end{pmatrix} \begin{pmatrix} \text{Solar} \\ \nu_e \to \nu_s \end{pmatrix} \begin{pmatrix} \text{LSND} \\ \nu_\mu \to \nu_e \end{pmatrix} \begin{pmatrix} |\nu_1\rangle \\ |\nu_2\rangle \\ |\nu_3\rangle \\ |\nu_4\rangle \end{pmatrix},$$

giving

$$\begin{pmatrix} |\nu_e\rangle \\ |\nu_s\rangle \\ |\nu_\mu^*\rangle \\ |\nu_\tau^*\rangle \end{pmatrix} = \begin{pmatrix} \cos\phi & \sin\phi\cos\omega & \sin\phi\sin\omega & 0 \\ -\sin\phi & \cos\phi\cos\omega & \cos\phi\sin\omega & 0 \\ 0 & -\sin\omega & \cos\omega & 0 \\ 0 & 0 & 0 & 1 \end{pmatrix} \begin{pmatrix} |\nu_1\rangle \\ |\nu_2\rangle \\ |\nu_3\rangle \\ |\nu_4\rangle \end{pmatrix},$$

where

$$|\nu_\mu^*\rangle = 1/\sqrt{2}(|\nu_\mu\rangle - |\nu_\tau\rangle) = -\sin\omega|\nu_2\rangle + \cos\omega|\nu_3\rangle$$

$$|\nu_\tau^*\rangle = 1/\sqrt{2}(|\nu_\mu\rangle + |\nu_\tau\rangle) = |\nu_4\rangle, \text{ a mass eigenstate.}$$

In this formalism what occurs at the first level crossing is $\nu_\mu^* \to \nu_s$, and at the second, $\nu_e \to \nu_\mu^*$, while ν_τ^* being a mass eigenstate goes through both regions unaffected. Again this gives 1/4 ν_μ, 1/4 ν_τ, 1/2 ν_s, and no ν_e at all.

Note that the $\bar{\nu}_e$ flux is also unaffected at the level crossings, so $\bar{\nu}_e p \to e^+ n$ enhances the neutron number in the r process region, since the protons have not been depleted by α particle formation. It should be emphasized that this mechanism is quite robust, not depending on details of the supernova dynamics, especially as it occurs quite late in the explosive expansion.

It is essential that the two level crossings be in the correct order, and this provides a requirement on $\Delta m^2_{e\mu}$, since the MSW transition depends on density and hence on radial distance from the protoneutron star. Detailed calculations have been made for $\Delta m^2_{e\mu} \sim 6$ eV^2, which works very well. Possibly $\Delta m^2_{e\mu}$ as low as 2 eV^2 or maybe even 1 eV^2 would work, but that is speculative. At any rate, the mass difference needed in this scheme, which is the only one surely consistent with all manifestations of neutrino mass and which rescues the r process [15], implies appreciable hot dark matter.

POSSIBLE EFFECT OF EXTRA LARGE DIMENSIONS

Since this four-neutrino model provides an explanation for all observations of neutrino mass and is strongly supported by apparently being required to make heavy-element nucleosynthesis work, it ought to be in quantitative agreement with all the evidence. At this time that appears possible in all areas except one, the solar neutrino observations [16]. Although there are four "solutions" for active neutrinos and two for sterile neutrinos for the solar ν_e deficit, in fact none of these really fits all the data in detail. For sterile neutrinos these are the small-angle MSW [7] and vacuum oscillation solutions. The Super-Kamiokande energy distribution for ^{8}B neutrinos is fit very poorly by the former and quite well by the latter, but over a wider range of energies the former gives a fair fit to the Ga, Cl, and water Cherenkov experiments average absolute rates, while the latter fit is 30% (3.4 σ) too high for Cl relative to the other two. There is a small (1.3 σ) indication for a day-night rate difference in Super-Kamiokande, which would be characteristic of MSW solutions, but the Ga (especially GALLEX), Cl, and (to a slight extent) Super-Kamiokande experiments provide some evidence for the seasonal variation characteristic of vacuum oscillations [17].

Thus at present there is no good solution for $\nu_e \to \nu_s$, and $\nu_e \to \nu_\mu$ or ν_τ is little better. If, however, large extra dimensions exist, then ν_s becomes a particle which can exist in the extra dimension(s), or in other words it inhabits the bulk, while active neutrinos are confined to the brane. There could be more than one extra dimension and more than one brane, but for simplicity the discussion here is limited to one of each. A characteristic of a bulk particle (such as the graviton) is that it is really a series of states; this Kaluza-Klein tower has mass values $m_n \approx n/R$, where R is the size of the extra dimension.

While many papers have been written about bulk sterile neutrinos, since this provides a means of getting sterile neutrinos of small mass which have some mixing with active neutrinos, generally these theories do not produce three Δm^2 or attempt to explain all the evidence for neutrino mass. Solar vacuum oscillation is usually ignored, but this can be accommodated if $m_{\nu_e} \approx m_o$, the ground state or zero mode of the Kaluza-Klein tower, with $\Delta m^2 \sim 10^{-10}$ eV^2 being required. In the present model the mixing angle is given by $\cos\theta = 1/N$, where

$$N^2 = 1 + (\pi m_0 R)^2 + (m_n/m_o)^2 \tag{1}$$

for the nth Kaluza/Klein state. As we shall see, $m_o R \ll 1$, so the second term can be neglected. Thus for the zero mode, $N^2 \approx 1 + (m_o/m_o)^2 = 2$, or $\cot\theta = 1/\sqrt{2}$, or $\theta = \pi/4$, which is the maximal mixing required by experiment. The magnitude of R is limited by tests of gravity down to small distances, and that puts the spacing ($\approx \Delta n/R$) into the region which would give MSW solar oscillations for states higher than the zero mode. For these oscillations to be effective, however, there has to be enough coupling to ν_e. For $n \geq 1$ using $m_n \approx n/R$ in Eq. 1 gives $N^2 \approx 1 + (n/m_o R)^2 \approx (n/m_o R)^2$. While one determines $m_o R$ by fitting to the solar data, the approximate size of the various parameters can be illustrated by choosing values from the standard small-angle MSW solution, which Super-Kamiokande now claims is ruled out for $\nu_e \to \nu_\mu$. Taking $\Delta m^2 \approx 4 \times 10^{-6}$ eV$^2 \approx (\Delta n/R)^2$ for $\Delta n = 1$ gives $1/R = 2 \times 10^{-3}$ eV, and $\sin^2 2\theta \approx 8 \times 10^{-3} \approx 4\cos^2\theta \approx 4(m_o R)^2$ gives $m_o R \approx 0.05$, justifying the above approximations. This gives $m_o \approx 10^{-4}$ eV, whereas in most models of bulk ν_s the zero mode has zero mass. Asked to produce a model which would work, R.N. Mohapatra added two singlets (η^+, Δ^{++}) to the Higgs sector on the brane and generated the needed neutrino masses at the two-loop level. The resulting parameters in this model for all the neutrinos appear to be quite reasonable.

Since R can be chosen to suppress by MSW oscillation the rate in the Cl experiment relative to those in the Ga experiments and Super-Kamiokande, the larger effect of a vacuum oscillation can be made to work. The latter is often called the "just-so" solution, since it depends on a coincidence between Δm^2 and the earth-sun distance, but in this case it is much less of a coincidence, since a range of values will work with the two means of suppressing ν_e flux at the earth.

CONCLUSIONS

It is quite remarkable that the profound problems of producing the heaviest elements by supernovae can be solved in a manner which requires no adjustment of parameters if the arrangement of masses and mixings of neutrinos is exactly that required to explain the solar ν_e deficit, the atmospheric neutrino anomaly, and the observations of the LSND experiment (or alternatively the need for hot dark matter). This apparently successful four-neutrino scheme fails quantitatively (as do other models) to explain all the solar ν_e data, unless the essential sterile neutrino is a bulk neutrino of extra large dimensions. The resulting Kaluza-Klein tower of states provides both MSW and vacuum oscillations, explaining the otherwise confusing solar data. This gives evidence for extra large dimensions and, if these indeed exist, the totally successful neutrino scheme shows a need for appreciable hot dark matter.

ACKNOWLEDGMENTS

The author thanks R.N. Mohapatra for joint early work on the neutrino scheme, as well as the recent collaboration on the issue of extra dimensions, for which S.J. Yellin is also thanked. I am grateful also to G.M. Fuller and Y.-Z. Qian, with whom the heavy-element nucleosynthesis work was done. My work was partially supported by the U.S. Department of Energy under contract DE-FG03-91ER40618.

REFERENCES

1. Haxton W.C., Langanke K., Qian Y.-Z., and Vogel P., *Phys. Rev. Lett.* **78**, 2694 (1997); Qian Y.-Z., Haxton W.C., Langanke K., and Vogel P., *Phys. Rev.* C **55**, 1532 (1997); Sneden C., Burles S., Fuller G.M., Cowan J.J., Beers T., and Burris D.L., in preparation (1999).
2. Hoffman R.D., Woosley S.E., and Qian Y.-Z., *Astrophys. J.* **482**, 951 (1996); Meyer B.S., and Brown J.S., *Astrophys. J. Suppl.* **112**, 199 (1997).
3. Fuller G.M., and Meyer B.S., *Astrophys. J.* **453**, 792 (1995); Meyer B.S., McLaughlin G.C., and Fuller G.M., *Phys. Rev.* C **58**, 3696 (1998).
4. Qian Y.-Z., and Woosley S.E., *Astrophys. J.* **471**, 331 (1996); Fuller G.M., and Qian Y.-Z., *Nucl. Phys.* A **606**, 167 (1996); Cardall C.Y., and Fuller G.M., *Astrophys. J.* **486**, L111 (1997); Samuelson J., and Wilson J.R., Astrophys. J., in press (1999); Otsuki K., Tagoshi H., Kajino T., and Wanajo S., astro-ph/9911164 (to be published in Astrophys. J.).
5. McLaughlin G.C., and Fuller G.M., *Astrophys. J.* **472**, 440 (1996).
6. Caldwell D.O., Fuller G.M., and Qian Y.-Z., astro-ph/9910175 (to be published in Phys. Rev. D).
7. Mikheyev S.P., and Smirnov A. Yu., *Sov. J. Nucl. Phys.* **24**, 913 (1985); Wolfenstein L., *Phys. Rev.* D **17**, 2369 (1978).
8. Caldwell D.O., *Perspectives in Neutrinos, Atomic Physics and Gravitation* (Editions Frontières, Gif-sur-Yvette, France, 1993), p. 187.
9. Primack J.R., Holtzman J., Klypin A., and Caldwell D.O., *Phys. Rev. Lett.* **74**, 2160 (1995); Gawiser E., and Silk J., *Science* **280**, 1405 (1998).
10. Caldwell D.O., and Mohapatra R.N., *Phys. Rev.* D **48**, 3259 (1993); Peltoniemi J.T., and Valle J.W.F., *Nucl. Phys.* B **406**, 409 (1993).
11. Athanassopoulos C., et al., *Phys. Rev. Lett.* **75**, 2650 (1995); *Phys. Rev.* C **54**, 2685 (1996); *Phys. Rev. Lett.* **77**, 3082 (1996); *Phys. Rev. Lett.* **81**, 1774 (1998).
12. Eitel K., hep-ex/9909036; New Jour. Phys. 2:1.1 (2000).
13. Zeitnitz B., et al., *Prog. Part. Nucl. Phys.* **40**, 169 (1998), but see ref. 12 for more recent information.
14. Qian Y.-Z., et al., *Phys. Rev. Lett.* **71**, 1965 (1993); Qian Y.-Z., and Fuller G.M., *Phys. Rev.* D **51**, 1479 (1995); Sigl G., *Phys. Rev.* D **51**, 4035 (1995).

15. McLaughlin G.C., Fetter J.M., Balantekin A.B., and Fuller G.M., *Phys. Rev. C* **59**, 2873 (1999) employs a neutrino scheme with a heavy sterile neutrino and also rescues the r process, but it is likely to be in trouble with big-bang nucleosynthesis, as well as with the Super-Kamiokande atmospheric up-down asymmetry [see Bilenky S.M., Giunti C., Grimus W., and Schwatz T., *Phys. Rev. D* **60**, 073007 (1999)].
16. Reviews include Haxton W.C., *Ann. Rev. Astron. Astrophys.* **33**, 459 (1995); Bahcall J.N., *Astrophys. J.* **467**, 475 (1996); Castellani V., et al., *Phys. Rep.* **281**, 309 (1997).
17. Berezinsky V., Fiorentini G., and Lissia M., *Astropart. Phys.* **12**, 299 (2000).

Relic Neutrino Asymmetries[1]

Raymond R Volkas

School of Physics
Research Centre for High Energy Physics
The University of Melbourne
Victoria 3010 Australia

Abstract. I review the topic of relic neutrino asymmetry generation through active-sterile neutrino and antineutrino oscillations in the early universe. Applications to (i) the suppression of sterile neutrino production, and (ii) the primordial Helium abundance are briefly presented.

Reasonably large relic neutrino asymmetries will be generated by active-sterile neutrino and antineutrino oscillations in the early universe provided that certain mild conditions are met [1]. "Reasonably large" in this context means that the neutrino asymmetry, defined by

$$L_{\nu_\alpha} \equiv \frac{n_{\nu_\alpha} - n_{\bar{\nu}_\alpha}}{n_\gamma}, \qquad (1)$$

where $\alpha = e, \mu, \tau$ and the n's are number densities, can have a final value as large as about 3/8 [2,3]. The mild conditions are that the vacuum mixing angle θ_0 should be small and that the Δm^2 should be negative. The flavour and mass eigenstates are related by

$$|\nu_\alpha\rangle = \cos\theta_0 |\nu_a\rangle + \sin\theta_0 |\nu_b\rangle \qquad (2)$$
$$|\nu_s\rangle = -\sin\theta_0 |\nu_a\rangle + \cos\theta_0 |\nu_b\rangle,$$

so $\Delta m^2 < 0$ means that the predominantly sterile mass eigenstate is lighter than the predominantly active mass eigenstate. We focus on that epoch of the early universe subsequent to the disappearance of a significant muon/antimuon component to the plasma, and ending with the period of Big Bang Nucleosynthesis (BBN).

One reason the generation of neutrino or lepton asymmetries is interesting is evident from the effective matter potential,

$$V_{eff} = \frac{\Delta m^2}{2p}(-a + b) \qquad (3)$$

---

[1] This work was supported by the Australian Research Council

where p is the neutrino momentum, and

$$a \equiv -\frac{4\sqrt{2}\zeta(3)}{\pi^2}\frac{G_F T^3 p}{\Delta m^2}L^{(\alpha)}, \qquad (4)$$

$$b \equiv -\frac{4\sqrt{2}\zeta(3)}{\pi^2}A_\alpha\frac{G_F T^4 p^2}{m_W^2 \Delta m^2}. \qquad (5)$$

The function a is just the generalisation of the usual Wolfenstein effective potential [4] pertinent to the considered epoch of the early universe. G_F is the Fermi constant, T is the temperature and the "effective asymmetry" $L^{(\alpha)}$ is given by

$$L^{(\alpha)} \equiv L_{\nu_\alpha} + L_{\nu_e} + L_{\nu_\mu} + L_{\nu_\tau} + \eta, \qquad (6)$$

where $\eta \sim 10^{-10}-10^{-9}$ is related to the baryon and electron asymmetries required to produce the universe we observe. When lepton asymmetries are large, the effective potential is also large, leading to small matter-affected mixing angles and thus also to the suppression of sterile neutrino production. The asymmetry generation effect must be taken into account when exploring the implications for BBN of neutrino scenarios involving light sterile degrees of freedom. In particular, $\nu_\mu \leftrightarrow \nu_s$ oscillation parameters motivated by the atmospheric neutrino deficit are far from being necessarily ruled out by BBN [1,2,5,6]. The function b is the leading finite-T correction to the Wolfenstein term [7]. The constant m_W is the W-boson mass, while A_α is a numerical factor given by $A_e \simeq 17$ and $A_{\mu,\tau} \simeq 4.9$. At high temperatures, b is large enough to suppress sterile neutrino production by itself. However, it decreases with temperature as T^6, so a large asymmetry is eventually required to keep sterile neutrino production suppressed at lower temperatures.

Another reason lepton asymmetries are interesting is conveyed by the reactions,

$$\nu_e n \leftrightarrow e^- p, \qquad \bar{\nu}_e p \leftrightarrow e^+ n, \qquad (7)$$

which play an important role in setting the n/p ratio at weak freeze-out just prior to BBN. A sufficiently large electron neutrino asymmetry (a few percent) will alter this ratio and hence change the predicted Helium abundance [3,8,9]. In particular, successful BBN can be achieved with a higher than usual baryon density, provided that L_{ν_e} is positive and of the correct magnitude. The threatened overproduction of Helium is compensated by the preferential conversion of neutrons into protons due to the positive L_{ν_e}.

Light sterile neutrinos are themselves of considerable interest independent of their possible role in asymmetry generation. For instance, two-fold maximal mixing is easily achieved through a pseudo-Dirac structure [10] or through ordinary-mirror neutrino mixing [11]. (Mirror neutrinos are strictly speaking not sterile because they feel mirror weak interactions. However, they are effectively sterile from the point of view of ordinary weak interactions.) Two-fold maximal mixing is of great interest in light of the solar and atmospheric neutrino deficits. In addition, it is well known that least one light sterile flavour is required if oscillations are to simultaneously explain the solar, atmospheric and LSND neutrino data.

As a final comment about motivation, let me also add that collision and matter affected neutrino oscillation dynamics is a fascinating subject in its own right. As we will see, neutrino asymmetry generation is driven by a subtle interplay between quantal coherence and decoherence.

Consider for simplicity a two-flavour active-sterile system $\nu_\alpha + \nu_s$ together with its antiparticle counterpart. The appropriate dynamical variables are the 1-body reduced density matrices for the neutrinos and antineutrinos,

$$\rho = \frac{1}{2}(P_0 + \vec{P} \cdot \sigma), \quad \bar{\rho} = \frac{1}{2}(\bar{P}_0 + \vec{\bar{P}} \cdot \sigma), \tag{8}$$

respectively. The decomposition with respect to the 2 × 2 identity matrix and the Pauli matrices proves to be convenient. All of these quantities are functions of neutrino momentum p and time t (or equivalently temperature T).

The diagonal entries ρ are appropriately normalised distribution functions for ν_α and ν_s:

$$\rho_{\alpha\alpha} = \frac{1}{2}(P_0 + P_z) = \frac{N_\alpha}{N^{eq}(0)},$$

$$\rho_{ss} = \frac{1}{2}(P_0 - P_z) = \frac{N_s}{N^{eq}(0)}, \tag{9}$$

where $n_f = \int_0^\infty N_f dp$ and $N^{eq}(\xi)$ is the Fermi-Dirac (FD) distribution with dimensionless chemical potential $\xi \equiv \mu_{\nu_\alpha}/T$,

$$N^{eq}(\xi) = \frac{1}{2\pi^2} \frac{p^2}{\exp(\frac{p}{T} - \xi) + 1}. \tag{10}$$

We have chosen to normalise the neutrino distribution to the FD distribution with zero chemical potential.

The off-diagonal entries,

$$\rho_{\alpha s} = \rho_{s\alpha}^* = \frac{1}{2}(P_x - iP_y), \tag{11}$$

are *coherences*. They quantify the amount of quantal or phase coherence enjoyed by the neutrinos. These quantities are needed because non-forward neutrino scattering off the background plasma decreases quantal coherence. The entries of $\bar{\rho}$ have a similar interpretation.

The evolution of ρ is given by the Quantum Kinetic (or Boltzmann) Equations (QKEs) [12],

$$\frac{\partial \vec{P}}{\partial t} = \vec{V} \times \vec{P} - D\vec{P}_T + \frac{\partial P_0}{\partial t}\hat{z},$$

$$\frac{\partial P_0}{\partial t} \simeq \Gamma \left[\frac{N^{eq}(\xi)}{N^{eq}(0)} - \frac{1}{2}(P_0 + P_z) \right], \tag{12}$$

where $D = \Gamma/2$ and $\vec{P}_T \equiv P_x\vec{x} + P_y\vec{y}$ with $\vec{x}, \vec{y}, \vec{z}$ being unit vectors in the stated directions. The second equation above has an approximate equality sign because the righthand side assumes that all background fermions have thermal FD distributions, and that ν_α is distributed in an approximately thermal manner.

The $\vec{V} \times \vec{P}$ term is just a re-expression of coherent matter-affected oscillatory neutrino evolution. It is equivalent to the usual Schrödinger Equation governing matter-affected neutrino oscillations. Note, however, that in the early universe this evolution is non-linear because of neutrino scattering off the background neutrinos of the same flavour. The vector $\vec{V}$ is given by

$$\vec{V} = \beta\vec{x} + \lambda\vec{z}, \tag{13}$$

where

$$\beta = \frac{\Delta m^2}{2p}\sin 2\theta_0, \quad \lambda = -\frac{\Delta m^2}{2p}\cos 2\theta_0 + V_{eff}. \tag{14}$$

The non-linear effects enter through V_{eff}.

The $-D\vec{P}_T$ term drives *collisional quantal decoherence*. The decoherence rate is equal to half of the collision rate Γ for a neutrino of momentum p, where

$$\Gamma \simeq y_\alpha G_F^2 T^5 \frac{p}{\langle p \rangle}. \tag{15}$$

The quantity $\langle p \rangle \simeq 3.15T$ is the average momentum for a FD distribution with zero chemical potential, while $y_e \simeq 4$ and $y_{\mu,\tau} \simeq 2.9$. This expression for Γ is correct provided that thermal equilibrium holds and asymmetries are not very large. The $-D\vec{P}_T$ term tries to exponentially damp the coherences $P_{x,y}$ to zero. Since the collision rate goes as T^5, quantal coherence is expunged at high temperatures. In typical applications, the $\vec{V} \times \vec{P}$ term begins to dominate over the $-D\vec{P}_T$ term as T approaches a few MeV.

The $\partial P_0/\partial t$ equation describes the repopulation of the ν_α distribution from the background plasma heat bath. It is proportional to the total weak collision rate Γ, and the term in square brackets on the righthand side acts to drive the actual ν_α distribution function $(P_0 + P_z)N^{eq}(0)/2$ to FD form $N^{eq}(\xi)$.

The neutrino asymmetry L_{ν_α} enters these equations through the function a in V_{eff}, and through the chemical potential ξ in $\partial P_0/\partial t$. For small L_{ν_α},

$$L_{\nu_\alpha} \simeq \frac{T^3}{6n_\gamma}\xi. \tag{16}$$

In thermal equilibrium the existence of a neutrino asymmetry is equivalent to the existence of a nonzero ξ (where the chemical potential for antineutrinos is equal and opposite that of the neutrinos above the chemical decoupling temperature).

The antineutrino QKEs are of the same form as the above equations with the substitutions $L_{\nu_\alpha} \to -L_{\nu_\alpha}$ and $L^{(\alpha)} \to -L^{(\alpha)}$.

We are primarily interested in the evolution of the asymmetry L_{ν_α}. This is indirectly given by the neutrino and antineutrino QKEs. It is easy enough, however, to derive a redundant but useful direct evolution equation for the asymmetry. Employing the QKEs and $\alpha + s$ lepton number conservation one obtains

$$\frac{dL_{\nu_\alpha}}{dt} = \frac{1}{2n_\gamma} \int_0^\infty \beta(P_y - \overline{P}_y) N^{eq}(0) dp. \qquad (17)$$

This equation is numerically useful because, during most of its evolution, L_{ν_α} is the difference of two large numbers (the neutrino and antineutrino number densities per photon). The use of Eq.(17) circumvents this numerical difficulty. This equation is also a very useful starting point for developing approximate evolution equations [2,13].

We now discuss the behaviour of the QKEs. We will suppose that the initial asymmetries are very small or zero, and that a negligible fraction of the primordial plasma is in the form of sterile states. Also, we consider $|\Delta m^2|$ values higher than about 10^{-4} eV^2, because for lower values asymmetry growth begins when decoherence is negligible, and it tends to be oscillatory [14].

For sufficiently high temperatures T, neutrino and antineutrino oscillations are severely damped or frozen. There are two reasons for this. First, the decoherence function $D \sim T^5$ is large and drives $P_{x,y}$ and $\overline{P}_{x,y}$ to zero, and hence also the righthand side of Eq.(17) (Quantum Zeno Effect). Second, the b-term in V_{eff} also rises as T^5, so the effective matter mixing angle is very small anyway.

As T decreases with the expansion of the universe, the oscillations begin to unfreeze. Their initial non-trivial evolution is, however, still dominated by non-forward scattering. During this phase, a very interesting phenomenon will occur provided that the mild conditions discussed at the beginning of this article are met: small θ_0 and $\Delta m^2 < 0$. There is a *critical temperature* T_c, roughly given by

$$T_c \simeq (15 \to 18 \text{ MeV}) \left[\cos 2\theta_0 \frac{|\Delta m^2|}{\text{eV}^2} \right]^{\frac{1}{6}}. \qquad (18)$$

Above T_c, the evolution equations drive L_{ν_α} such as to impel the effective asymmetry $L^{(\alpha)}$ towards zero (in other words, L_{ν_α} evolves from zero such as to cancel the η term in the effective asymmetry). We can call this the *asymmetry destruction phase*, and $L^{(\alpha)} = 0$ is a stable fixed point of the evolution equations. At T_c, this stable fixed point becomes unstable and runaway positive feedback leads to an exponential increase in L_{ν_α} during the *explosive growth phase*. The approximately exponential growth is cut off by the non-linear terms in the QKEs after L_{ν_α} reaches a value which is several orders of magnitude higher than the baryon and electron asymmetries. A *power law growth phase*, as per $L_{\nu_\alpha} \sim T^{-4}$, then sets in. Typically, collision dominated evolution gives way to oscillation dominated evolution during the power law phase. This is important, because physically there is a "hand over" from collisional asymmetry growth to non-linear MSW driven growth during this phase. Finally, asymmetry growth stops when L_{ν_α} reaches about $0.2 - 0.3$.

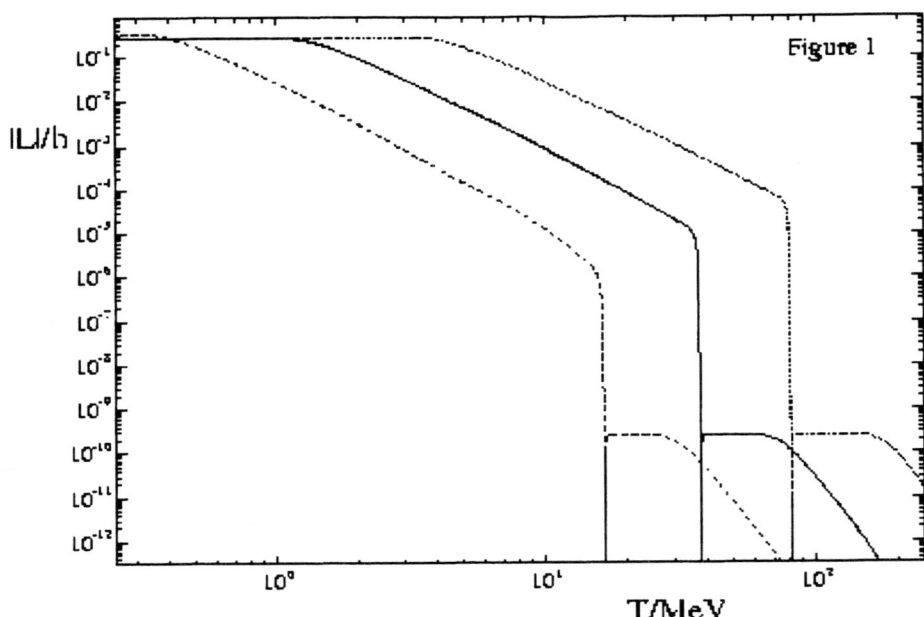

FIGURE 1. Examples of lepton asymmetry growth curves driven by $\nu_\tau \leftrightarrow \nu_s$ and the corresponding antineutrino oscillations. The mixing angle is selected to be $\sin^2 2\theta = 10^{-8}$. The three curves correspond to $\Delta m^2 = -0.5, -50, -5000$ eV^2, reading left to right.
Reprinted from *Astroparticle Physics* **10**, R. Foot, p. 253, Copyright 1999, with permission from Elsevier Science.

Some examples of asymmetry evolution are given in Fig.1, which is taken from Ref. [5]. These curves were produced by numerically solving the QKEs together with Eq.(17). An analytical understanding of why these curves take the displayed form has been given in the literature [1–3,13]. The key idea is to take the adiabatic limit of the QKEs. At high T, it is then possible to define precisely what is meant by collision dominated evolution, and to derive an approximate dL_{ν_α}/dt equation which makes the critical behaviour at T_c manifest. At low T, when collisions can be neglected, the adiabatic limit of the QKEs is exactly the same as the adiabatic limit for non-linear MSW evolution. See Refs. [2,3,13] for a complete discussion.

I will now review two applications, beginning with the suppression of sterile neutrino production. Suppose that we wish to solve the atmospheric neutrino problem by maximal $\nu_\mu \to \nu_s$ oscillation with $|\Delta m^2_{\mu s}|$ in the range 10^{-3} eV^2 to 10^{-2} eV^2.[2] If the $\nu_\mu + \nu_s$ subsystem does not mix with any other neutrino, then the sterile

---

[2] SuperKamiokande claim that present data disfavour the $\nu_\mu \to \nu_s$ possibility relative to the $\nu_\mu \to \nu_\tau$ mode [15]. However, the dust has yet to settle on this issue.

neutrinos (and antineutrinos) will certainly be brought into thermal equilibrium with the rest of the plasma prior to BBN (unless there are large pre-existing lepton asymmetries [16]). The resulting increase in the expansion rate of the universe will lead to a higher than standard Helium-4 abundance. I will assume for the sake of the example that primordial abundance observations cannot tolerate such an increase in the expansion rate. (A complete account of the somewhat complicated status of primordial abundance observations vis-à-vis BBN is beyond the scope of this talk.)

Small mixing between the ν_s and, say, a more massive ν_τ can completely change the conclusion that ν_s is brought into thermal equilibrium. The ν_τ/ν_s oscillation parameters, as chosen in the previous sentence, satisfy the requirements for large L_{ν_τ} generation. If the large L_{ν_τ} does not have its effects cancelled (see below), and if it is generated early enough, then the large matter potential consequently generated for the $\nu_\mu \leftrightarrow \nu_s$ mode will suppress these oscillations and hence also ν_s production. A numerical calculation is required to properly analyse the outcome, because the maximally mixed $\nu_\mu \leftrightarrow \nu_s$ mode always tends to induce $L^{(\mu)} \to 0$. In other words, the large L_{ν_τ} that is being created by the $\nu_\tau \leftrightarrow \nu_s$ mode could be compensated by L_{ν_μ} creation driven by $\nu_\mu \leftrightarrow \nu_s$, with the overall effect being that the μ-like effective asymmetry $L^{(\mu)}$ is driven to zero, and consequently that ν_s production through the ν_μ channel is unsuppressed after all.

Figure 2 shows the outcome of such a numerical calculation [2,5,6]. It is a plot in ν_τ/ν_s oscillation parameter space. The solid lines correspond to $|\Delta m^2_{\mu s}| = 10^{-3}, 10^{-2.5}, 10^{-2}$ eV^2 in ascending order up the page. Choose your favourite atmospheric $\Delta m^2_{\mu s}$. In the parameter region above the relevant solid line, the L_{ν_τ} asymmetry is *not* cancelled such that $L^{(\mu)} \to 0$. Below the solid line, it is cancelled. The transition is very sharp. So, above the solid line, ν_s production via the $\nu_\mu \to \nu_s$ mode is very suppressed. The dot-dashed line refers to ν_s production via the other available mode, $\nu_\tau \to \nu_s$. This small angle mode receives very little "self-suppression" from its own L_{ν_τ} creation activity, so its oscillation parameters must satisfy a different sort of bound. This bound is actually very similar to the old constraints calculated in Ref. [17] with asymmetry generation artificially switched off. For illustrative purposes, the dot-dashed line assumes that no more than 0.6 extra effective neutrino flavours are allowed.

The second application concerns the direct effect of a ν_e asymmetry on the neutron to proton ratio at weak freeze out. The implications of neutrino asymmetry generation in multi-flavour scenarios must be studied on a case-by-case basis. Here I will consider the scenario of Ref. [18]: there is one sterile flavour, which is used to solve the solar neutrino problem via an MSW style solution. The more massive ν_μ and ν_τ states are maximally mixed, with the mass gap between (ν_e, ν_s) and (ν_μ, ν_τ) set in the LSND range. In this scenario, both the $\nu_\mu \to \nu_s$ and $\nu_\tau \to \nu_s$ modes satisfy the conditions for asymmetry generation. Once L_{ν_μ} and L_{ν_τ} have been generated, these asymmetries can be reprocessed into L_{ν_e} via $\nu_{\mu,\tau} \leftrightarrow \nu_e$ oscillations [8]. The L_{ν_e} outcome is insensitive to the $\nu_e/\nu_{\mu,\tau}$ mixing angles for a large range

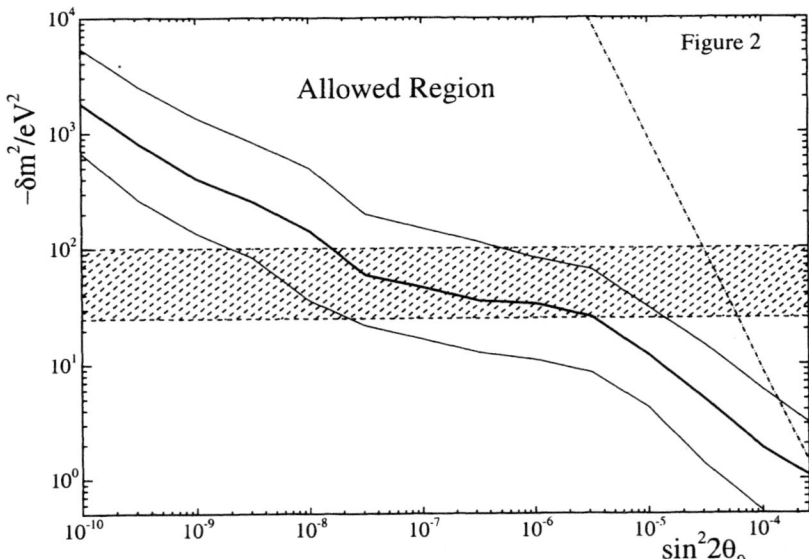

FIGURE 2. Allowed region according to BBN for $\nu_\tau \leftrightarrow \nu_s$ oscillation parameters in order to suppress maximal $\nu_\mu \leftrightarrow \nu_s$ oscillations via a large L_{ν_τ}. See text for a complete discussion. Reprinted from *Astroparticle Physics* **10**, R. Foot, p. 253, Copyright 1999, with permission from Elsevier Science.

of these parameters because the MSW transitions are adiabatic. It is, however, somewhat sensitive to Δm^2 between (ν_e, ν_s) and (ν_μ, ν_τ). Figure 3 shows a plot of the change in the effective number of neutrino flavours during BBN as a function of this Δm^2 parameter [8]. The "effective number of neutrino flavours" is just a convenient measure of the change in the Helium mass fraction Y_p relative to standard BBN, roughly obeying the relation $\delta Y_p \simeq 0.012 \delta N_\nu^{eff}$. It is affected by both expansion rate alterations and L_{ν_e}. Figure 3 incorporates both effects.

There are a few other interesting neutrino models that have been analysed in the literature. A particularly interesting case is the Exact Parity or Mirror Matter Model [11]. The full analysis is very complicated; please see Ref. [19] for all of the details.

To conclude: If light sterile or mirror neutrinos exist, then they should play a major role in cosmology. The oscillation generated neutrino asymmetry phenomenon would be a central feature. Sufficiently large neutrino asymmetries would suppress sterile or mirror neutrino production, and an electron-neutrino asymmetry of the right magnitude and sign would affect primordial Helium abundance. In particular, a positive L_{ν_e} would allow a larger baryon density to exist without the overproduction of Helium. We note with interest that the recent Boomerang/Maxima cosmic microwave background anistropy measurements favour a larger than standard baryon density [20]. While the combined solar, atmospheric and LSND data require at least one light sterile neutrino if oscillations are to simultaneously re-

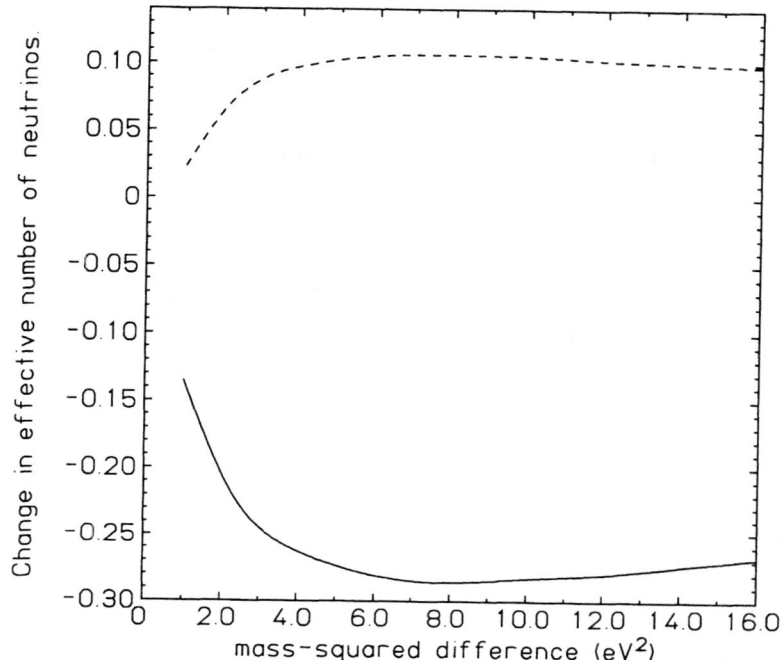

FIGURE 3. The change in the effective number of neutrino flavours during BBN for a certain neutrino scenario, plotted as a function of a Δm^2 parameter that was taken to be in the LSND range. See the text for a more complete discussion. This figure is taken from Ref. [8]

solve all of the anomalies, we especially look forward to future experiments that could provide further evidence for light sterile neutrinos: SNO, MiniBooNe, the longbaseline experiments, as well as further SuperKamiokande data.

ACKNOWLEDGMENTS

I would like to thank Jose Nieves, Arthur Halprin, Terry Leung and Qaisar Shafi and all of the very helpful staff for organising an extremely enjoyable workshop. Thanks to Yvonne Wong for technical assistance.

REFERENCES

1. R. Foot, M. J. Thomson and R. R. Volkas, Phys. Rev. D53, 5349 (1996).
2. R. Foot and R. R. Volkas, Phys. Rev. D55, 5147 (1997).
3. R. Foot and R. R. Volkas, Phys. Rev. D56, 6653 (1997); Erratum-ibid, D59, 029901 (1999).
4. L. Wolfenstein, Phys. Rev. D17, 2369 (1978); S. P. Mikheyev and A. Yu. Smirnov, Nuovo Cim. C9, 17 (1986); see also V. Barger et al., Phys. Rev. D22, 2718 (1980).
5. R. Foot, Astropart. Phys. 10, 253 (1999).
6. P. Di Bari, P. Lipari and M. Lusignoli, Int. J. Mod. Phys. (in press), hep-ph/9907548.
7. D. Notzold and G. Raffelt, Nucl. Phys. B307, 924 (1988).
8. N. F. Bell, R. Foot and R. R. Volkas, Phys. Rev. D58, 105010 (1998).
9. R. Foot, Phys. Rev. D61, 023516 (2000).
10. J. Bowes and R. R. Volkas, J. Phys. G24, 1249 (1998); A. Geiser, Phys. Lett. B444, 358 (1999); P. Langacker, Phys. Rev. D58, 093017 (1998); Y. Koide and H. Fusaoka, Phys. Rev. D59, 053004 (1999); W. Krolikowski, hep-ph/9808307; see also, C. Giunti, C. W. Kim and U. W. Kim, Phys. Rev. D46, 3034 (1992); M. Kobayashi, C. S. Lim and M. M. Nojiri, Phys. Rev. Lett. 67, 1685 (1991).
11. R. Foot, H. Lew and R. R. Volkas, Phys. Lett. B272, 67 (1991). R. Foot, H. Lew and R. R. Volkas, Mod. Phys. Lett. A7, 2567 (1992); R. Foot, Mod. Phys. Lett. A9, 169 (1994); R. Foot and R. R. Volkas, Phys. Rev. D52, 6595 (1995); see also Z. Silagadze, Phys. At. Nucl. 60, 272 (1997).
12. R. A. Harris and L. Stodolsky, Phys. Lett. 116B, 464 (1982); Phys. Lett. B78, 313 (1978); A. Dolgov, Sov. J. Nucl. Phys. 33, 700 (1981); L. Stodolsky, Phys. Rev. D36, 2273 (1987); M. Thomson, Phys. Rev. A45, 2243 (1991); K. Enqvist, K. Kainulainen and J. Maalampi, Nucl. Phys. B349, 754 (1991). B. H. J. McKellar and M. J. Thomson, Phys. Rev. D49, 2710 (1994).
13. N. F. Bell, R. R. Volkas and Y. Y. Y. Wong, Phys. Rev. D59, 113001 (1999); R. R. Volkas and Y. Y. Y. Wong (in preparation).
14. D. P. Kirilova and M. V. Chizhov, Phys. Lett. B393, 375 (1997); hep-ph/9806441.
15. M. Vagins, talk at this Workshop.
16. R. Foot and R. R. Volkas, Phys. Rev. Lett. 75, 4350 (1995).
17. P. Langacker, University of Pennsylvania Preprint, UPR 0401T, September (1989); R. Barbieri and A. Dolgov, Phys. Lett. B237, 440 (1990); Nucl. Phys. B349, 743

(1991); K. Kainulainen, Phys. Lett. B244, 191 (1990); K. Enqvist, K. Kainulainen and M. Thomson, Nucl. Phys. B 373, 498 (1992); J. Cline, Phys. Rev. Lett. 68, 3137 (1992); X. Shi, D. N. Schramm and B. D. Fields, Phys. Rev. D48, 2568 (1993); G. Raffelt, G. Sigl and L. Stodolsky, Phys. Rev. Lett. 70, 2363 (1993); C. Y. Cardall and G. M. Fuller, Phys. Rev. D54, 1260 (1996).

18. D. O. Caldwell and R. N. Mohapatra, Phys. Rev. D48, 3259 (1993); D50, 3477 (1994); J. T. Peltoniemi and J. W. F. Valle, Nucl. Phys. B406, 409 (1993).

19. R. Foot and R. R. Volkas, Astropart. Phys. 7, 283 (1997); Phys. Rev. D61, 043507 (2000).

20. P. de Bernardis et al., Nature 404, 955 (2000); A. E. Lange et al., astro-ph/0005004; S. Hanany et al., astro-ph/0005123; A. Balbi et al., astro-ph/0005124.

SUPERNOVA COSMOLOGY
AND ASTROPHYSICS

Evidence from Type Ia Supernovae for an Accelerating Universe

Alexei V. Filippenko*

Department of Astronomy, University of California, Berkeley, CA 94720-3411
(alex@astro.berkeley.edu)

Adam G. Riess*

Space Telescope Science Institute, 3700 San Martin Dr., Baltimore, MD 21218
(ariess@stsci.edu)

(* On behalf of the High-z Supernova Search Team)

Abstract. We review the use of Type Ia supernovae for cosmological distance determinations. Low-redshift SNe Ia ($z \lesssim 0.1$) demonstrate that the Hubble expansion is linear, that $H_0 = 65 \pm 2$ (statistical) km s^{-1} Mpc^{-1}, and that the properties of dust in other galaxies are similar to those of dust in the Milky Way. We find that the light curves of high-redshift ($z = 0.3$–1) SNe Ia are stretched in a manner consistent with the expansion of space; similarly, their spectra exhibit slower temporal evolution (by a factor of $1 + z$) than those of nearby SNe Ia. The luminosity distances of our first set of 16 high-redshift SNe Ia are, on average, 10–15% farther than expected in a low mass-density ($\Omega_M = 0.2$) universe without a cosmological constant. Preliminary analysis of our second set of 9 SNe Ia is consistent with this. Our work supports models with positive cosmological constant and a current acceleration of the expansion. We address the main potential sources of systematic error; at present, none of them appears to reconcile the data with $\Omega_\Lambda = 0$ and $q_0 \geq 0$. The dynamical age of the Universe is estimated to be 14.2 ± 1.7 Gyr, consistent with the ages of globular star clusters.

INTRODUCTION

Supernovae (SNe) come in two main varieties (see reference [1] for a review). Those whose optical spectra exhibit hydrogen are classified as Type II, while hydrogen-deficient SNe are designated Type I. SNe I are further subdivided according to the appearance of the early-time spectrum: SNe Ia are characterized by strong absorption near 6150 Å (now attributed to Si II), SNe Ib lack this feature but instead show prominent He I lines, and SNe Ic have neither the Si II nor the He I lines. SNe Ia are believed to result from the thermonuclear disruption of carbon-oxygen white dwarfs, while SNe II come from core collapse in massive supergiant stars. The latter mechanism probably produces most SNe Ib/Ic as well, but the progenitor stars previously lost their outer layers of hydrogen or even helium.

It has long been recognized that SNe Ia may be very useful distance indicators for a number of reasons; see [2,3], and references therein). (1) They are exceedingly luminous, with peak absolute blue magnitudes averaging -19.2 if the Hubble constant, H_0, is 65 km s^{-1} Mpc^{-1}. (2) "Normal" SNe Ia have small dispersion among their peak absolute magnitudes ($\sigma \lesssim 0.3$ mag). (3) Our understanding of the progenitors and explosion mechanism of SNe Ia is on a reasonably firm physical basis. (4) Little cosmic evolution is expected in the peak luminosities of SNe Ia, and it can be modeled. This makes SNe Ia superior to galaxies as distance indicators. (5) One can perform *local* tests of various possible complications and evolutionary effects by comparing nearby SNe Ia in different environments.

Research on SNe Ia in the 1990s has demonstrated their enormous potential as cosmological distance indicators. Although there are subtle effects that must indeed be taken into account, it appears that SNe Ia provide among the most accurate values of H_0, q_0 (the deceleration parameter), Ω_M (the matter density), and Ω_Λ (the cosmological constant, $\Lambda c^2/3H_0^2$).

There are now two major teams involved in the systematic investigation of high-redshift SNe Ia for cosmological purposes. The "Supernova Cosmology Project" (SCP) is led by Saul Perlmutter of the Lawrence Berkeley Laboratory, while the "High-Z Supernova Search Team" (HZT) is led by Brian Schmidt of the Mt. Stromlo and Siding Springs Observatories. One of us (A.V.F.) has worked with both teams, but his primary allegiance is now with the HZT. In this paper we present results from the HZT.

HOMOGENEITY AND HETEROGENEITY

The traditional way in which SNe Ia have been used for cosmological distance determinations has been to assume that they are perfect "standard candles" and to compare their observed peak brightness with those of SNe Ia in galaxies whose distances have been independently determined (e.g., Cepheids). The rationale is that SNe Ia exhibit relatively little scatter in their peak blue luminosity ($\sigma_B \approx 0.4$–0.5 mag; [4]), and even less if "peculiar" or highly reddened objects are eliminated from consideration by using a color cut. Moreover, the optical spectra of SNe Ia are usually rather homogeneous, if care is taken to compare objects at similar times relative to maximum brightness ([5] and references therein). Over 80% of all SNe Ia discovered through the early 1990s were "normal" [6].

From a Hubble diagram constructed with unreddened, moderately distant SNe Ia ($z \lesssim 0.1$) for which peculiar motions should be small and relative distances (as given by ratios of redshifts) are accurate, Vaughan *et al.* [7] find that

$$< M_B(\text{max}) > = (-19.74 \pm 0.06) + 5\log(H_0/50) \text{ mag}. \qquad (1)$$

In a series of papers, Sandage et al. [8] and Saha et al. [9] combine similar relations with *Hubble Space Telescope (HST)* Cepheid distances to the host galaxies of seven SNe Ia to derive $H_0 = 57 \pm 4$ km s^{-1} Mpc^{-1}.

Over the past decade it has become clear, however, that SNe Ia do *not* constitute a perfectly homogeneous subclass (e.g., [1,10]). In retrospect this should have been obvious: the Hubble diagram for SNe Ia exhibits scatter larger than the photometric errors, the dispersion actually *rises* when reddening corrections are applied (under the assumption that all SNe Ia have uniform, very blue intrinsic colors at maximum; [11,12]), and there are some significant outliers whose anomalous magnitudes cannot possibly be explained by extinction alone.

Spectroscopic and photometric peculiarities have been noted with increasing frequency in well-observed SNe Ia. A striking case is SN 1991T; its pre-maximum spectrum did not exhibit Si II or Ca II absorption lines, yet two months past maximum brightness the spectrum was nearly indistinguishable from that of a classical SN Ia [13,14]. The light curves of SN 1991T were slightly broader than the SN Ia template curves, and the object was probably somewhat more luminous than average at maximum. The reigning champion of well observed, peculiar SNe Ia is SN 1991bg [15-17]. At maximum brightness it was subluminous by 1.6 mag in V and 2.5 mag in B, its colors were intrinsically red, and its spectrum was peculiar (with a deep absorption trough due to Ti II). Moreover, the decline from maximum brightness was very steep, the I-band light curve did not exhibit a secondary maximum like normal SNe Ia, and the velocity of the ejecta was unusually low. The photometric heterogeneity among SNe Ia is well demonstrated by Suntzeff [18] with five objects having excellent $BVRI$ light curves.

COSMOLOGICAL USES: LOW REDSHIFTS

Although SNe Ia can no longer be considered perfect "standard candles," they are still exceptionally useful for cosmological distance determinations. Excluding those of low luminosity (which are hard to find, especially at large distances), most of the nearby SNe Ia that had been discovered through the early 1990s were *nearly* standard ([6], but see Li et al. [19] for recent evidence of a higher intrinsic peculiarity rate). Also, after many tenuous suggestions (e.g., [20-22]), convincing evidence has finally been found for a *correlation* between light-curve shape and luminosity. Phillips [23] achieved this by quantifying the photometric differences among a set of nine well-observed SNe Ia using a parameter, $\Delta m_{15}(B)$, which measures the total drop (in B magnitudes) from maximum to $t = 15$ days after B maximum. In all cases the host galaxies of his SNe Ia have accurate relative distances from surface brightness fluctuations or from the Tully-Fisher relation. In B, the SNe Ia exhibit a total spread of ~ 2 mag in maximum luminosity, and the intrinsically bright SNe Ia clearly decline more slowly than dim ones. The range in absolute magnitude is smaller in V and I, making the correlation with $\Delta m_{15}(B)$ less steep than in B, but it is present nonetheless.

Using SNe Ia discovered during the Calán/Tololo survey ($z \lesssim 0.1$), Hamuy et al. [24,25] confirm and refine the Phillips [23] correlation between $\Delta m_{15}(B)$ and $M_{max}(B,V)$: it is not as steep as had been claimed. Apparently the slope is steep only at low luminosities; thus, objects such as SN 1991bg skew the slope of the best-fitting single straight line. Hamuy et al. reduce the scatter in the Hubble diagram of normal, unreddened SNe Ia to only 0.17 mag in B and 0.14 mag in V; see also [26].

In a similar effort, Riess, Press, & Kirshner [27] show that the luminosity of SNe Ia correlates with the detailed shape of the overall light curve. They form a "training set" of light-curve shapes from 9 well-observed SNe Ia having known relative distances, including very peculiar objects (e.g., SN 1991bg). When the light curves of an independent sample of 13 SNe Ia (the Calán/Tololo survey) are analyzed with this set of basis vectors, the dispersion in the V-band Hubble diagram drops from 0.50 to 0.21 mag, and the Hubble constant rises from 53 ± 11 to 67 ± 7 km s^{-1} Mpc^{-1}, comparable to the conclusions of Hamuy et al. [24,25]. About half of the rise in H_0 results from a change in the position of the "ridge line" defining the linear Hubble relation, and half is from a correction to the luminosity of some of the local calibrators which appear to be unusually luminous (e.g., SN 1972E).

By using light-curve shapes measured through several different filters, Riess, Press, & Kirshner [28] extend their analysis and objectively eliminate the effects of interstellar extinction: a SN Ia that has an unusually red $B - V$ color at maximum brightness is assumed to be *intrinsically* subluminous if its light curves rise and decline quickly, or of normal luminosity but significantly *reddened* if its light curves rise and decline slowly. With a set of 20 SNe Ia consisting of the Calán/Tololo sample and their own objects, they show that the dispersion decreases from 0.52 mag to 0.12 mag after application of this "multi-color light curve shape" (MLCS) method. The results from a recent, expanded set of nearly 50 SNe Ia indicate that the dispersion decreases from 0.44 mag to 0.15 mag (Figure 1). The resulting Hubble constant is 65 ± 2 (statistical) km s^{-1} Mpc^{-1}, with an additional systematic and zero-point uncertainty of ± 5 km s^{-1} Mpc^{-1}. Riess et al. [28] also show that the Hubble flow is remarkably linear; indeed, SNe Ia now constitute the best evidence for linearity. Finally, they argue that the dust affecting SNe Ia is *not* of circumstellar origin, and show quantitatively that the extinction curve in external galaxies typically does not differ from that in the Milky Way (cf. [2], but see [29]).

The advantage of systematically correcting the luminosities of SNe Ia at high redshifts rather than trying to isolate "normal" ones seems clear in view of evidence that the luminosity of SNe Ia may be a function of stellar population. If the most luminous SNe Ia occur in young stellar populations [24,31,32], then we might expect the mean peak luminosity of high-redshift SNe Ia to differ from that of a local sample. Alternatively, the use of Cepheids (Population I objects) to calibrate local SNe Ia can lead to a zero point that is too luminous. On the other hand, as long as the physics of SNe Ia is essentially the same in young stellar populations locally and at high redshift, we should be able to adopt the luminosity correction methods (photometric and spectroscopic) found from detailed studies of low-redshift SNe Ia.

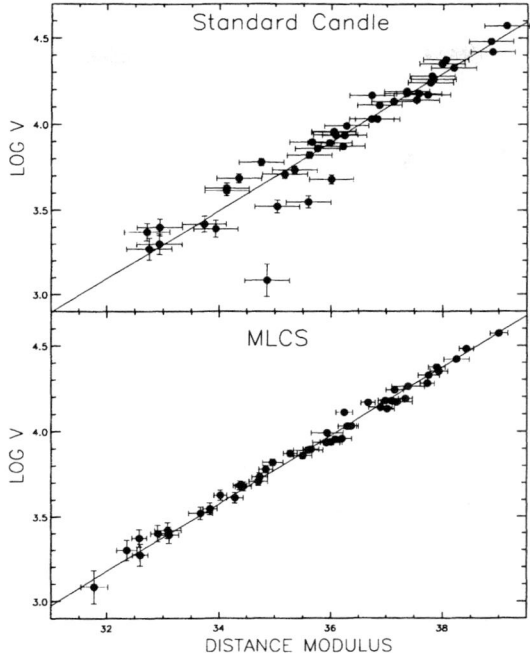

Figure 1: Hubble diagrams for SNe Ia [30] with velocities (km s^{-1}) in the COBE rest frame on the Cepheid distance scale. *Top:* The objects are assumed to be *standard candles* and there is no correction for extinction; the result is $\sigma = 0.42$ mag and $H_0 = 58 \pm 8$ km s^{-1} Mpc^{-1}. *Bottom:* The same objects, after correction for reddening and intrinsic differences in luminosity. The result is $\sigma = 0.15$ mag and $H_0 = 65 \pm 2$ (statistical) km s^{-1} Mpc^{-1}.

Large numbers of nearby SNe Ia are now being found by the Lick Observatory Supernova Search (LOSS) conducted with the 0.76-m Katzman Automatic Imaging Telescope (KAIT; [19,34]). CCD images are taken of about 1000 galaxies per night and compared with KAIT "template images" obtained earlier; the templates are automatically subtracted from the new images and analyzed with computer software. The system reobserves the best candidates the same night, to eliminate star-like cosmic rays, asteroids, and other sources of false alarms. The next day, undergraduate students at UC Berkeley examine all candidates, including weak ones, and they glance at all subtracted images to locate SNe that might be near bright, poorly subtracted stars or galactic nuclei. LOSS discovered 20 SNe (of all types) in 1998 and 40 SNe in 1999, making it the world's most successful search for nearby SNe. The most important objects were photometrically monitored through $BVRI$ (and sometimes U) filters (e.g., [33]), and unfiltered follow-up observations were made of all of them during the course of the SN search. This growing sample of well-observed SNe Ia should allow us to more precisely calibrate the MLCS method, as well as to look for correlations between the observed properties of the SNe and their environment (Hubble type of host galaxy, metallicity, stellar population, etc.).

COSMOLOGICAL USES: HIGH REDSHIFTS

These same techniques can be applied to construct a Hubble diagram with high-redshift SNe Ia, from which the value of q_0 can be determined. With enough objects spanning a range of redshifts, we can determine Ω_M and Ω_Λ independently (e.g., [35]). Contours of peak apparent R-band magnitude for SNe Ia at two redshifts have different slopes in the Ω_M–Ω_Λ plane, and the regions of intersection provide the answers we seek.

The Search

Based on the pioneering work of Norgaard-Nielsen et al. [36], whose goal was to find SNe in moderate-redshift clusters of galaxies, the SCP [37] and our HZT [38] devised a strategy that almost guarantees the discovery of many faint, distant SNe Ia on demand, during a predetermined set of nights. This "batch" approach to studying distant SNe allows follow-up spectroscopy and photometry to be *scheduled* in advance, resulting in a systematic study not possible with random discoveries. Most of the searched fields are equatorial, permitting follow-up from both hemispheres.

Our approach is simple in principle; see [38] for details, and for a description of our first high-redshift SN Ia (SN 1995K). Pairs of first-epoch images are obtained with the CTIO or CFHT 4-m telescopes and wide-angle imaging cameras during the nights just after new moon, followed by second-epoch images 3–4 weeks later. (Pairs of images permit removal of cosmic rays, asteroids, and distant Kuiper-belt objects.) These are compared immediately using well-tested software, and new SN candidates are identified in the second-epoch images (Figure 2). Spectra are obtained as soon as possible after discovery to verify that the objects are SNe Ia and determine their redshifts. Each team has already found over 80 SNe in concentrated batches, as reported in numerous *IAU Circulars* (e.g., [39], 11 SNe with $0.16 \lesssim z \lesssim 0.65$; [40], 17 SNe with $0.09 \lesssim z \lesssim 0.84$).

Intensive photometry of the SNe Ia commences within a few days after procurement of the second-epoch images; it is continued throughout the ensuing and subsequent dark runs. In a few cases *HST* images are obtained. As expected, most of the discoveries are *on the rise or near maximum brightness*. When possible, the SNe are observed in filters which closely match the redshifted B and V bands; this way, the K-corrections become only a second-order effect [41]. Custom-designed filters for redshifts centered on 0.35 and 0.45 are used by our HZT [38], when appropriate. We try to obtain excellent *multi-color* light curves, so that reddening and luminosity corrections can be applied [28,31,25].

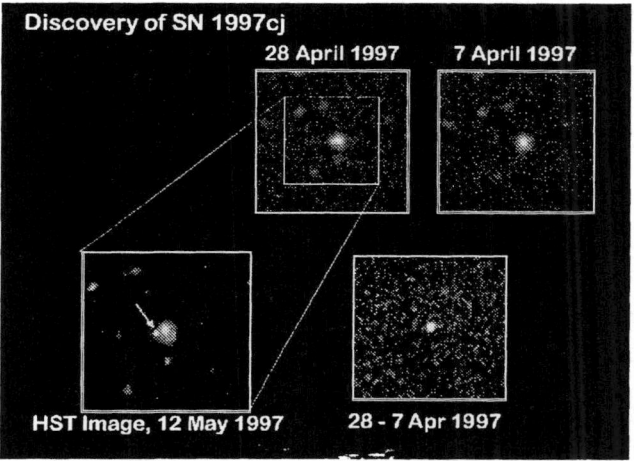

Figure 2: Discovery image of SN 1997cj (28 April 1997), along with the template image (7 April 1997) and an *HST* image obtained subsequently. The net (subtracted) image is also shown.

Although SNe in the magnitude range 22–22.5 can sometimes be spectroscopically confirmed with 4-m class telescopes, the signal-to-noise ratios are low, even after several hours of integration. Certainly Keck is required for the fainter objects (mag 22.5–24.5). With Keck, not only can we rapidly confirm a large number of candidate SNe, but we can search for peculiarities in the spectra that might indicate evolution of SNe Ia with redshift. Moreover, high-quality spectra allow us to measure the age of a supernova: we have developed a method for automatically comparing the spectrum of a SN Ia with a library of spectra corresponding to many different epochs in the development of SNe Ia [5]. Our technique also has great practical utility at the telescope: we can determine the age of a SN "on the fly," within half an hour after obtaining its spectrum. This allows us to rapidly decide which SNe are best for subsequent photometric follow-up, and we immediately alert our collaborators on other telescopes.

Results

First, we note that the light curves of high-redshift SNe Ia are broader than those of nearby SNe Ia; the initial indications [42,43] are amply confirmed with our larger samples. Quantitatively, the amount by which the light curves are "stretched" is consistent with a factor of $1 + z$, as expected if redshifts are produced by the expansion of space rather than by "tired light." We were also able to demonstrate this *spectroscopically* at the 2σ confidence level for a single object: the spectrum of SN 1996bj ($z = 0.57$) evolved more slowly than those of nearby SNe Ia, by a factor consistent with $1 + z$ [5]. More recently, we have used observations of SN 1997ex ($z = 0.36$) at three epochs to conclusively verify the effects of time dilation: temporal changes in the spectra are slower than those of nearby SNe Ia by roughly the expected factor of 1.36 [44].

Following our Spring 1997 campaign, in which we found a SN with $z = 0.97$ (SN 1997ck), and for which we obtained *HST* follow-up images of three SNe, we published our first substantial results concerning the density of the Universe [45]: $\Omega_M = 0.35 \pm 0.3$ under the *assumption* that $\Omega_{\text{total}} = 1$, or $\Omega_M = -0.1 \pm 0.5$ under the *assumption* that $\Omega_\Lambda = 0$. Our independent analysis of 10 SNe Ia using the "snapshot" distance method (with which conclusions are drawn from sparsely observed SNe Ia) gives quantitatively similar conclusions [46].

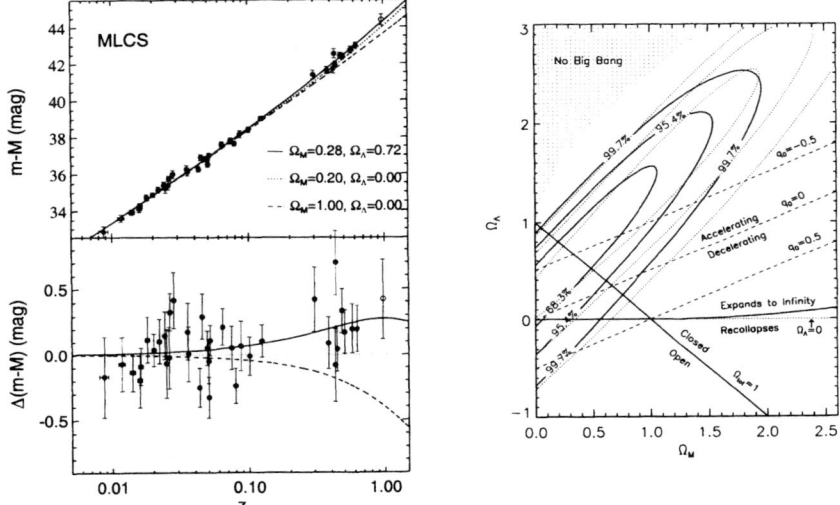

Figure 3 (left): The upper panel shows the Hubble diagram for the low-redshift and high-redshift SN Ia samples with distances measured from the MLCS method; see [48]. Overplotted are three world models: "low" and "high" Ω_M with $\Omega_\Lambda = 0$, and the best fit for a flat universe ($\Omega_M = 0.28$, $\Omega_\Lambda = 0.72$). The bottom panel shows the difference between data and models from the $\Omega_M = 0.20$, $\Omega_\Lambda = 0$ prediction. Except for SN 1997ck (*open symbol*; $z = 0.97$), which lacks spectroscopic confirmation and was excluded from the fit, only the 9 best-observed high-redshift SNe Ia are shown. The average difference between the data and the $\Omega_M = 0.20$, $\Omega_\Lambda = 0$ prediction is 0.25 mag.

Figure 4 (right): Joint confidence intervals for (Ω_M,Ω_Λ) from SNe Ia [48]. The solid contours are results from the MLCS method applied to 10 well-observed SN Ia light curves, together with the snapshot method [46] applied to 6 incomplete SN Ia light curves. The dotted contours are for the same objects excluding SN 1997ck ($z = 0.97$). Regions representing specific cosmological scenarios are illustrated.

Our next results, obtained from a total of 16 high-z SNe Ia, were announced at a conference in February 1998 [47] and formally published by Riess et al. [48] in September 1998. The Hubble diagram (from a refined version of the MLCS method [48]) for the 10 best-observed high-z SNe Ia is given in Figure 3, while Figure 4 illustrates the derived confidence contours in the Ω_M–Ω_Λ plane. We confirm our previous suggestion that Ω_M is low. Even more exciting, however, is our conclusion that Ω_Λ is *nonzero* at the 3σ statistical confidence level. With the MLCS method applied to the full set of 16 SNe Ia, our formal results are $\Omega_M = 0.24 \pm 0.10$ if $\Omega_{\text{total}} = 1$, or $\Omega_M = -0.35 \pm 0.18$ (unphysical) if $\Omega_\Lambda = 0$. If we demand that $\Omega_M = 0.2$, then the best value for Ω_Λ is 0.66±0.21. These conclusions do not change significantly if only the 9 best-observed SNe Ia are used (Figure 3; $\Omega_M = 0.28 \pm 0.10$ if $\Omega_{\text{total}} = 1$). The $\Delta m_{15}(B)$ method yields similar results; if anything, the case for a positive cosmological constant strengthens. (For brevity, in this paper we won't quote the $\Delta m_{15}(B)$ numbers; see [48] for details.) From an essentially independent set of 42 high-z SNe Ia (only 2 objects in common), the SCP obtains almost identical results [49]. This suggests that neither team has made a large, simple blunder!

Recently, we have calibrated an additional sample of 9 high-z SNe Ia, including several observed with *HST*. Preliminary analysis suggests that the new data are entirely consistent with the old results, thereby strengthening their statistical significance. Figure 5 shows the tentative Hubble diagram; full details will be published elsewhere.

Though not drawn in Figure 4, the expected confidence contours from measurements of the angular scale of the first acoustic peak of the cosmic microwave background radiation (CMBR) are nearly perpendicular to those provided by SNe Ia (e.g., [50,51]); thus, the two techniques provide complementary information. A stunning result was already available by mid-1998 from existing measurements [52,55]: our analysis of the data in [48] demonstrates that $\Omega_M + \Omega_\Lambda = 0.94 \pm 0.26$, when the SN and CMBR constraints are combined [54] (see also [55,56], and others). As shown in Figure 6, the confidence contours are nearly circular, instead of highly eccentric ellipses as in Figure 4. Just a few days before the Second Tropical Workshop, the more accurate and precise results of the BOOMERANG collaboration were announced [57], and shortly thereafter the MAXIMA collaboration distributed their very similar findings [58,59]; the TOCO measurements [60] are also relevant. The bottom line is that we appear to live in a flat universe: $\Omega_{\text{total}} = 1$. Combined with the supernova results, the evidence for nonzero Ω_Λ is strong. Making the argument even more compelling is the fact that various studies of clusters of galaxies (see summary by [61]) show that $\Omega_M \approx 0.3$, so if the CMBR results are correct, one is led to the independent conclusion that $\Omega_\Lambda > 0$. We eagerly look forward to future CMBR measurements of even greater precision.

The dynamical age of the Universe can be calculated from the cosmological parameters. In an empty Universe with no cosmological constant, the dynamical age is simply the "Hubble time" (i.e., the inverse of the Hubble constant); there is no deceleration. SNe Ia yield $H_0 = 65 \pm 2$ km s^{-1} Mpc^{-1} (statistical uncertainty only), and a Hubble time of 15.1 ± 0.5 Gyr. For a more complex cosmology, integrating the velocity of the expansion from the current epoch ($z = 0$) to the beginning ($z = \infty$) yields an expression for the dynamical age. As shown in detail by Riess et al. [48], we obtain a value of $14.2^{+1.0}_{-0.8}$ Gyr using the likely range for $(\Omega_M, \Omega_\Lambda)$ that we measure. (The precision is so high because our experiment is sensitive to roughly the *difference* between Ω_M and Ω_Λ, and the dynamical age also varies in approximately this way.) Including the *systematic* uncertainty of the Cepheid distance scale, which may be up to 10%, a reasonable estimate of the dynamical age is 14.2 ± 1.7 Gyr.

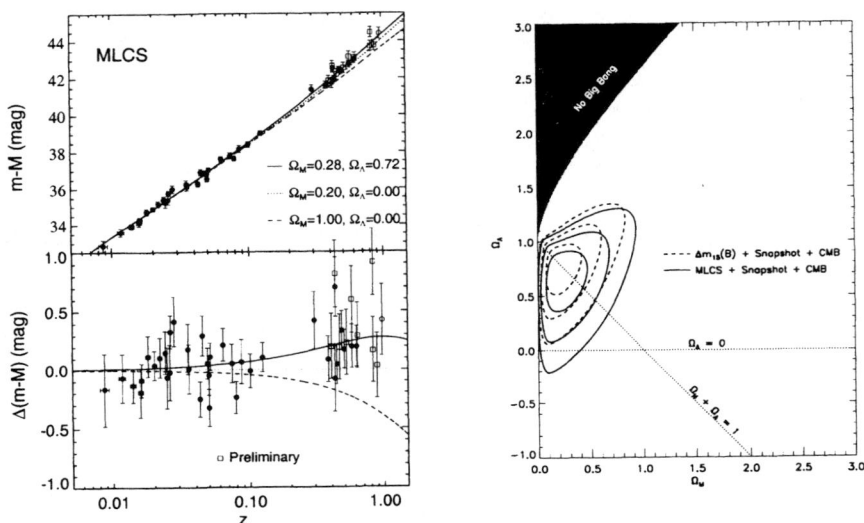

Figure 5 (left): As in Figure 3, the upper panel shows the Hubble diagram for the low-z and high-z SN Ia samples. Here, we include preliminary analysis of 9 additional SNe Ia *(open squares)*. The bottom panel shows the difference between data and models from the $\Omega_M = 0.20$, $\Omega_\Lambda = 0$ prediction.

Figure 6 (right): The HZT's combined constraints from SNe Ia (Figure 3) and the position of the first acoustic peak of the CMB angular power spectrum, based on data available in mid-1998; see [54]. The contours mark the 68%, 95.4%, and 99.7% enclosed probability regions. Solid curves correspond to results from the MLCS method, while dotted ones are from the $\Delta m_{15}(B)$ method; all 16 SNe Ia in [48] were used.

This result is consistent with ages determined from various other techniques such as the cooling of white dwarfs (Galactic disk > 9.5 Gyr [62]), radioactive dating of stars via the thorium and europium abundances (15.2 ± 3.7 Gyr [63]), and studies of globular clusters (10–15 Gyr, depending on whether *Hipparcos* parallaxes of Cepheids are adopted [64,65]). Evidently, there is no longer a problem that the age of the oldest stars seems greater than the dynamical age of the Universe.

DISCUSSION

High-redshift SNe Ia are observed to be dimmer than expected in an empty Universe (i.e., $\Omega_M = 0$) with no cosmological constant. A cosmological explanation for this observation is that a positive vacuum energy density accelerates the expansion. Mass density in the Universe exacerbates this problem, requiring even more vacuum energy. For a Universe with $\Omega_M = 0.2$, the average MLCS distance moduli of the well-observed SNe are 0.25 mag larger (i.e., 12.5% greater distances) than the prediction from $\Omega_\Lambda = 0$. The average MLCS distance moduli are still 0.18 mag bigger than required for a 68.3% (1σ) consistency in a universe with $\Omega_M = 0.2$ and without a cosmological constant. The derived value of q_0 is -0.75 ± 0.32, implying that the expansion of the Universe is accelerating. If Ω_Λ really is constant, then at least the region of the Universe we have observed ($z \lesssim 0.8$) will expand eternally. Under the simplifying assumption of global homogeneity and isotropy, the entire Universe will behave in this manner.

A very important point is that the dispersion in the peak luminosities of SNe Ia ($\sigma = 0.15$ mag) is low after application of the MLCS method of [28,48]. With 16 SNe Ia, our effective uncertainty is $0.15/4 \approx 0.04$ mag, less than the expected difference of 0.25 mag between universes with $\Omega_\Lambda = 0$ and 0.76 (and low Ω_M); see Figure 3. Systematic uncertainties of even 0.05 mag (e.g., in the extinction) are significant, and at 0.1 mag they dominate any decrease in statistical uncertainty gained with a larger sample of SNe Ia. Thus, our conclusions with only 16 SNe Ia are already limited by systematic uncertainties, *not* by statistical uncertainties — but of course the 9 new objects further strengthen our case.

Here we explore the major possible systematic effects that could invalidate our results. Of those that can be quantified at the present time, none appears to reconcile the data with $\Omega_\Lambda = 0$, though further work is necessary to verify this. Additional details can be found in [38] and especially [48].

Evolution

Perhaps the most obvious possible culprit is *evolution* of SNe Ia over cosmic time, due to changes in metallicity, progenitor mass, or some other factor. If the peak luminosity of SNe Ia were lower at high redshift, then the case for $\Omega_\Lambda > 0$ would weaken. Conversely, if the distant explosions are more powerful, then the case for acceleration strengthens. Theorists are not yet sure what the sign of the effect will be, if it's present at all; different assumptions lead to different conclusions [66–70].

Of course, it is very difficult to obtain an independent measure of the peak luminosity of high-z SNe Ia, and hence to directly test for luminosity evolution. However, we can more easily determine whether *other* observable properties of low-z and high-z SNe Ia differ. If they are all the same, it is more probable that the peak luminosity is constant as well — but if they differ, then the peak luminosity might also be affected (e.g., [66]). Drell *et al.* [71], for example, argue that there are reasons to suspect evolution, because the average properties of existing samples of high-z and low-z SNe Ia seem to differ (e.g., the high-z SNe Ia are more uniform).

The local sample of SNe Ia displays a weak correlation between light-curve shape (or luminosity) and host galaxy type, in the sense that the most luminous SNe Ia with the broadest light curves only occur in late-type galaxies. Both early-type and late-type galaxies provide hosts for dimmer SNe Ia with narrower light curves [31]. The mean luminosity difference for SNe Ia in late-type and early-type galaxies is ~ 0.3 mag. In addition, the SN Ia rate per unit luminosity is almost twice as high in late-type galaxies as in early-type galaxies at the present epoch [72]. These results may indicate an evolution of SNe Ia with progenitor age. Possibly relevant physical parameters are the mass, metallicity, and C/O ratio of the progenitor [66].

We expect that the relation between light-curve shape and luminosity that applies to the range of stellar populations and progenitor ages encountered in the late-type and early-type hosts in our nearby sample should also be applicable to the range we encounter in our distant sample. In fact, the range of age for SN Ia progenitors in the nearby sample is likely to be *larger* than the change in mean progenitor age over the 4–6 Gyr lookback time to the high-z sample. Thus, to first order at least, our local sample should correct our distances for progenitor or age effects.

We can place empirical constraints on the effect that a change in the progenitor age would have on our SN Ia distances by comparing subsamples of low-redshift SNe Ia believed to arise from old and young progenitors. In the nearby sample, the mean difference between the distances for the early-type (8 SNe Ia) and late-type hosts (19 SNe Ia), at a given redshift, is 0.04 ± 0.07 mag from the MLCS method. This difference is consistent with zero. Even if the SN Ia progenitors evolved from one population at low redshift to the other at high redshift, we still would not explain the surplus in mean distance of 0.25 mag over the $\Omega_\Lambda = 0$ prediction.

Moreover, it is reassuring that initial comparisons of high-redshift SN Ia spectra appear remarkably similar to those observed at low redshift. For example, the spectral characteristics of SN 1998ai ($z = 0.49$) appear to be essentially indistinguishable from those of normal low-redshift SNe Ia; see Figure 7. In fact, the most obviously discrepant spectrum in this figure is the second one from the top, that of SN 1994B ($z = 0.09$); it is intentionally included as a "decoy" that illustrates the degree to which even the spectra of nearby, relatively normal SNe Ia can vary. Nevertheless, it is important to note that a dispersion in luminosity (perhaps 0.2 mag) exists even among the other, more normal SNe Ia shown in Figure 7; thus, our spectra of SN 1998ai and other high-redshift SNe Ia are not yet sufficiently good for independent, *precise* determinations of luminosity from spectral features [73]. Many of them, however, are sufficient for ruling out other supernovae types (Figure 8), or for identifying gross peculiarities such as those shown by SNe 1991T and 1991bg; see Coil et al. [74].

We can help verify that the SNe at $z \approx 0.5$ being used for cosmology do not belong to a subluminous population of SNe Ia by examining restframe I-band light curves. Normal, nearby SNe Ia show a pronounced second maximum in the I band about a month after the first maximum and typically about 0.5 mag fainter (e.g.,

[75,18]). Subluminous SNe Ia, in contrast, do not show this second maximum, but rather follow a linear decline or show a muted second maximum [15]. As discussed by Riess et al. [76], some evidence for the second maximum is seen from our existing J-band (restframe I-band) data on SN 1999Q ($z = 0.46$); see Figure 9. However, better data on more SNe Ia are needed to confirm the effect.

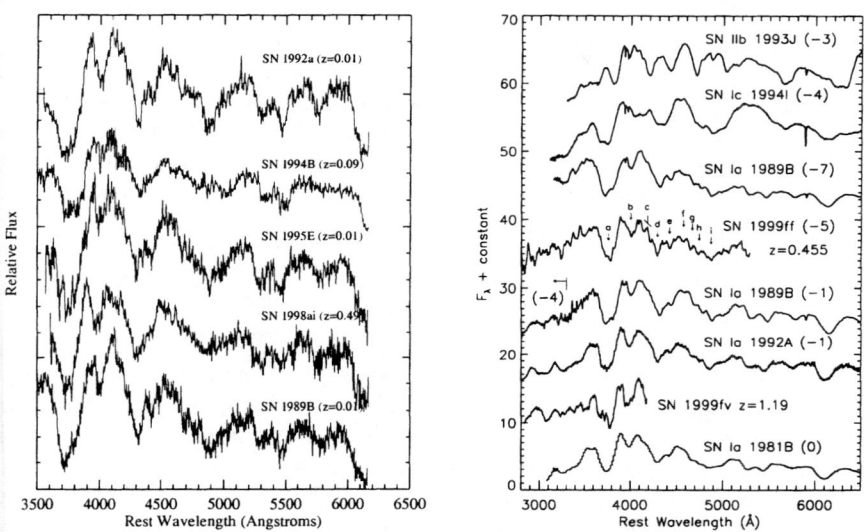

Figure 7 (left): Spectral comparison (in f_λ) of SN 1998ai ($z = 0.49$; Keck spectrum) with low-redshift ($z < 0.1$) SNe Ia at a similar age (~ 5 days before maximum brightness), from [48]. The spectra of the low-redshift SNe Ia were resampled and convolved with Gaussian noise to match the quality of the spectrum of SN 1998ai. Overall, the agreement in the spectra is excellent, tentatively suggesting that distant SNe Ia are physically similar to nearby SNe Ia. SN 1994B ($z = 0.09$) differs the most from the others, and was included as a "decoy."

Figure 8 (right): Heavily smoothed spectra of two high-z SNe (SN 1999ff at $z = 0.455$ and SN 1999fv at $z = 1.19$; quite noisy below ~ 3500 Å) are presented along with several low-z SN Ia spectra (SNe 1989B, 1992A, and 1981B), a SN Ib spectrum (SN 1993J), and a SN Ic spectrum (SN 1994I); see [1] for a discussion of spectra of various types of SNe. The date of the spectra relative to B-band maximum is shown in parentheses after each object's name. Specific features seen in SN 1999ff and labeled with a letter are discussed by Coil et al. [74]. This comparison shows that the two high-z SNe are most likely SNe Ia.

Another way of using light curves to test for possible evolution of SNe Ia is to see whether the rise time (from explosion to maximum brightness) is the same for high-z and low-z SNe Ia; a difference might indicate that the peak luminosities are also different [66]. We recently measured the risetime of nearby SNe Ia, using data from KAIT, the Beijing Astronomical Observatory (BAO) SN search, and a few amateur astronomers [77]. Though the exact value of the risetime is a function of peak luminosity, for typical low-z SNe Ia we find 20.0 ± 0.2 days. We pointed out [78] that this differs by 5.8σ from the *preliminary* risetime of 17.5 ± 0.4 days reported in conferences by the SCP [79–81]. However, a more thorough analysis of the SCP data [82] shows that the high-z uncertainty of ± 0.4 days that the SCP originally reported was much too small because it did not account for systematic effects. The revised discrepancy with the low-z risetime is about 2σ or less. Thus, the apparent difference in risetimes might be insignificant. Even if the difference is real, however, its relevance to the peak luminosity is unclear; the light curves may differ only in the first few days after the explosion, and this could be caused by small variations in conditions near the outer part of the exploding white dwarf that are inconsequential at the peak.

Extinction

Our SN Ia distances have the important advantage of including corrections for interstellar extinction occurring in the host galaxy and the Milky Way. Extinction corrections based on the relation between SN Ia colors and luminosity improve distance precision for a sample of nearby SNe Ia that includes objects with substantial extinction [28]; the scatter in the Hubble diagram is much reduced. Moreover, the consistency of the measured Hubble flow from SNe Ia with late-type and early-type hosts (see above) shows that the extinction corrections applied to dusty SNe Ia at low redshift do not alter the expansion rate from its value measured from SNe Ia in low dust environments.

In practice, our high-redshift SNe Ia appear to suffer negligible extinction; their $B - V$ colors at maximum brightness are normal, suggesting little color excess due to reddening. Riess, Press, & Kirshner [83] found indications that the Galactic ratios between selective absorption and color excess are similar for host galaxies in the nearby ($z \leq 0.1$) Hubble flow. Yet, what if these ratios changed with lookback time (e.g., [84])? Could an evolution in dust grain size descending from ancestral interstellar "pebbles" at higher redshifts cause us to underestimate the extinction? Large dust grains would not imprint the reddening signature of typical interstellar extinction upon which our corrections rely.

However, viewing our SNe through such gray interstellar grains would also induce a *dispersion* in the derived distances. Using the results of Hatano, Branch, & Deaton [85], Riess *et al.* [48] estimate that the expected dispersion would be 0.40 mag if the mean gray extinction were 0.25 mag (the value required to explain the measured MLCS distances without a cosmological constant). This is significantly larger than the 0.21 mag dispersion observed in the high-redshift MLCS distances. Furthermore, most of the observed scatter is already consistent with the estimated *statistical* errors, leaving little to be caused by gray extinction. Nevertheless, if we assumed that *all* of the observed scatter were due to gray extinction, the mean shift in the SN Ia distances would only be 0.05 mag. With the observations presented here, we cannot rule out this modest amount of gray interstellar extinction.

Gray *intergalactic* extinction could dim the SNe without either telltale reddening or dispersion, if all lines of sight to a given redshift had a similar column density of absorbing material. The component of the intergalactic medium with such uniform coverage corresponds to the gas clouds producing Lyman-α forest absorption at low redshifts. These clouds have individual H I column densities less than about 10^{15} cm^{-2} [86]. However, they display low metallicities, typically less than 10% of solar. Gray extinction would require larger dust grains which would need a larger mass in heavy elements than typical interstellar grain size distributions to achieve a given extinction. It is possible that large dust grains are blown out of galaxies by radiation pressure, and are therefore not associated with Lyman-α clouds [87].

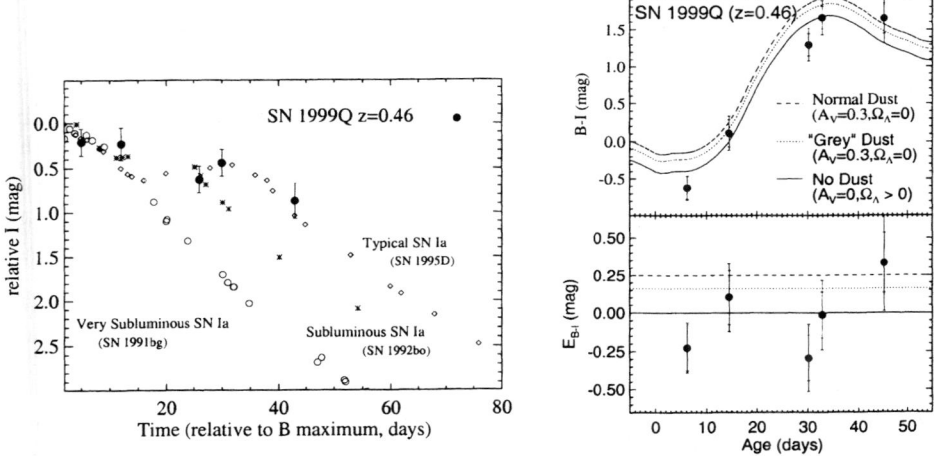

Figure 9 (left): Restframe I-band (observed J-band) light curve of SN 1999Q ($z = 0.46$, 5 solid points; [76]), superposed on the I-band light curves of several nearby SNe Ia. Subluminous SNe Ia exhibit a less prominent second maximum than do normal SNe Ia.

Figure 10 (right): Color excess, E_{B-I}, for SN 1999Q and different dust models [76]. The data are most consistent with no dust and $\Omega_\Lambda > 0$.

But even the dust postulated by Aguirre [84,87,88] is not *completely* gray, having a size of about 0.1 μm. We can test for such nearly gray dust by observing high-redshift SNe Ia over a wide wavelength range to measure the color excess it would introduce. If $A_V = 0.25$ mag, then $E(U - I)$ and $E(B - I)$ should be 0.12–0.16 mag [84,87]. If, on the other hand, the 0.25 mag faintness is due to Λ, then no such reddening should be seen. This effect is measurable using proven techniques; so far, with just one SN Ia (SN 1999Q; Figure 10), our results favor the no-dust hypothesis to better than 2σ [76], but more work along these lines is certainly warranted.

Suppose, though, that for some reason the dust is *very* gray, or our color measurements are not sufficiently precise to rule out Aguirre's (or other) dust. If the cumulative amount of gray dust along the line of sight grows linearly with increasing redshift, we expect that the deviation of the SN Ia peak apparent magnitude from the low-Ω_M, zero-Λ model (Figure 3) will continue growing, to first order (Figure 11). If, on the other hand, the observed faintness of high-z SNe Ia is a consequence of positive Λ, the deviation should actually begin to *decrease* at $z \approx 0.8$ (Figure 11). In essence, we are looking so far back in time that the Λ effect becomes small compared with Ω_M; the Universe is decelerating at that epoch. Thus, we are embarking on a campaign to find and monitor $z = 0.8$–1.2 SNe Ia. Given the expected uncertainties (Figure 11), a sample of 10–20 SNe Ia should give a good statistical result.

Note that this test also applies to other systematic effects that grow monotonically with redshift, as may be expected of possible evolution of the white dwarf progenitors (e.g., [66,67]), or gravitational lensing [89]. Indeed, this is our most decisive test to distinguish between Λ and systematic effects. Unless evolution of dust, or of the progenitors, or of the lenses is fixed in such a way as to mimic the effects of Λ (e.g., [71]), our claim of $\Omega_\Lambda > 0$ will become much more convincing if the deviation of apparent magnitude decreases in the expected manner. Such a turnover (Figure 11) can be considered the "smoking gun" of Λ.

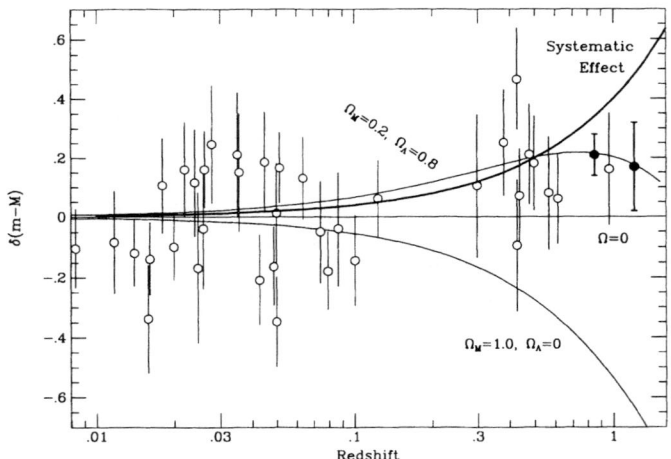

Figure 11: The HZT SN Ia data from Figure 3 *(open circles)* are plotted relative to an empty universe *(horizontal line)*. The two faint curves are the best-fitting Λ model, and the $\Omega_M = 1$ ($\Omega_\Lambda = 0$) model. The darker curve shows a systematic bias that increases linearly with z and is consistent with our $z = 0.5$ data. The expected observational uncertainties of hypothetical SNe Ia at redshifts of 0.85 and 1.2 are shown *(filled circles)*.

CONCLUSIONS

The luminosity distances of the high-redshift Type Ia supernovae studied by the High-z Supernova Search Team exceed the prediction of a low mass-density ($\Omega_M \approx 0.2$) universe by about 0.25 mag. A cosmological explanation is provided by a positive cosmological constant at roughly the 3σ confidence level, with the prior belief that $\Omega_M \geq 0$. We also find that the expansion of the Universe is currently accelerating ($q_0 \leq 0$, where $q_0 \equiv \Omega_M/2 - \Omega_\Lambda$). The independent results of the Supernova Cosmology Project are fully consistent with these conclusions. Moreover, recent precise measurements of the cosmic microwave background radiation strongly suggest that the Universe is flat ($\Omega_{\text{total}} = \Omega_M + \Omega_\Lambda = 1$); hence, if $\Omega_M \approx 0.3$ (as suggested by many studies, such as of clusters of galaxies), then about 70% of the energy density of the Universe must consist of vacuum energy whose precise nature and evolution are unknown (but definitely not radiation, normal matter, or dark matter). Using the best current values of the Hubble constant, Ω_M, and Ω_Λ, we find that the dynamical age of the Universe is 14.2 ±1.7 Gyr, including systematic uncertainties in the Cepheid distance scale used for the host galaxies of three nearby SNe Ia. This value is comparable to the derived ages of globular star clusters.

Though the consistent results from the microwave background experiments are reassuring, we are in the process of testing as exhaustively as possible all systematic biases that could be affecting the SN Ia results. For example, qualitative comparisons of spectra of low-z and high-z SNe Ia do not reveal obvious differences, and quantitative tests are in progress. Moreover, the restframe I-band light curves of low-z SNe Ia and a single measured high-z SN Ia look similar, as do their broadband colors. The risetimes of low-z and high-z SNe Ia may differ a little, but the statistical significance of this result is not high, and in any case the early part of the light curve may have little bearing on the peak luminosity. Further tests are in progress. Compelling evidence for acceleration may come in the next few years from a comparison of the peak apparent brightness of $z \gtrsim 0.8$ SNe Ia with the predictions of various models; the signature of nonzero Λ is quite distinct from that of dust, SN evolution, or other effects that grown with redshift.

ACKNOWLEDGMENTS

We thank all of our collaborators in the HZT for their contributions to this work. A.V.F.'s supernova research at U.C. Berkeley is supported by NSF grants AST-9417213 and AST-9987438, and by grants GO-7505 and GO-8177 from the Space Telescope Science Institute, which is operated by the Association of Universities for Research in Astronomy, Inc., under NASA contract NAS 5-26555. A.V.F. is grateful to the meeting organizers for travel funds.

REFERENCES

1. Filippenko, A. V., *ARAA*, **35**, 309 (1997b).
2. Branch, D., & Tammann, G. A., *ARAA*, **30**, 359 (1992).
3. Branch, D., *ARAA*, **36**, 17 (1998).
4. Branch, D., & Miller, D. L., *ApJ*, **405**, L5 (1993).
5. Riess, A. G., *et al.*, *AJ*, **114**, 722 (1997).
6. Branch, D., Fisher, A., & Nugent, P., *AJ*, **106**, 2383 (1993).
7. Vaughan, T. E., Branch, D., Miller, D. L., & Perlmutter, S., *ApJ*, **439**, 558 (1995).
8. Sandage, A., *et al.*, *ApJ*, **460**, L15 (1996).
9. Saha, A., *et al.*, *ApJ*, **486**, 1 (1997).
10. Filippenko, A. V., in *Thermonuclear Supernovae*, ed. P. Ruiz-Lapuente, *et al.* (Dordrecht: Kluwer), p. 1 (1997a).
11. van den Bergh, S., & Pazder, J., *ApJ*, **390**, 34 (1992).
12. Sandage, A., & Tammann, G. A., *ApJ*, **415**, 1 (1993).
13. Filippenko, A. V., *et al.*, *ApJ*, **384**, L15 (1992b).
14. Phillips, M. M., *et al.*, *AJ*, **103**, 1632 (1992).
15. Filippenko, A. V., *et al.*, *AJ*, **104**, 1543 (1992a).
16. Leibundgut, B., *et al.*, *AJ*, **105**, 301 (1993).
17. Turatto, M., *et al.*, *MNRAS*, **283**, 1 (1996).
18. Suntzeff, N., in *Supernovae and Supernova Remnants*, ed. R. McCray & Z. Wang (Cambridge: Cambridge Univ. Press), p. 41 (1996).
19. Li, W. D., *et al.*, in *Cosmic Explosions*, ed. S. S. Holt & W. W. Zhang (New York: American Inst. Physics), p. 91 (2000a).
20. Pskovskii, Yu. P., *Sov. Astron.*, **21**, 675 (1977).
21. Pskovskii, Yu. P., *Sov. Astron.*, **28**, 658 (1984).
22. Branch, D., *ApJ*, **248**, 1076 (1981).
23. Phillips, M. M., *ApJ*, **413**, L105 (1993).
24. Hamuy, M., *et al.*, *AJ*, **109**, 1 (1995).
25. Hamuy, M., *et al.*, *AJ*, **112**, 2398 (1996b).
26. Tripp, R., *A&A*, **325**, 871 (1997).
27. Riess, A. G., Press, W. H., & Kirshner, R. P., *ApJ*, **438**, L17 (1995).
28. Riess, A. G., Press, W. H., & Kirshner, R. P., *ApJ*, **473**, 88 (1996a).
29. Tripp, R., *A&A*, **331**, 815 (1998).
30. Riess, A. G., *et al.*, in preparation (2000b).
31. Hamuy, M., *et al.*, *AJ*, **112**, 2391 (1996a).
32. Branch, D., Romanishin, W., & Baron, E., *ApJ*, **465**, 73; erratum **467**, 473 (1996).
33. Li, W. D., *et al.*, in *Cosmic Explosions*, ed. S. S. Holt & W. W. Zhang (New York: American Inst. Physics), p. 103 (2000b).
34. Filippenko, A. V., *et al.*, in preparation (2000a).
35. Goobar, A., & Perlmutter, S., *ApJ*, **450**, 14 (1995).
36. Norgaard-Nielsen, H., *et al.*, *Nature*, **339**, 523 (1989).
37. Perlmutter, S., *et al.*, *ApJ*, **483**, 565 (1997).
38. Schmidt, B. P., *et al.*, *ApJ*, **507**, 46 (1998).
39. Perlmutter, S., *et al.*, *IAUC* 6270 (1995).

40. Suntzeff, N., et al., *IAUC* 6490 (1996b).
41. Kim, A., Goobar, A., & Perlmutter, S., *PASP*, **108**, 190 (1996).
42. Leibundgut, B., et al., *ApJ*, **466**, L21 (1996).
43. Goldhaber, G., et al., in *Thermonuclear Supernovae*, ed. P. Ruiz-Lapuente, et al. (Dordrecht: Kluwer), p. 777 (1997).
44. Filippenko, A. V., et al., in preparation (2000b).
45. Garnavich, P., et al., *ApJ*, **493**, L53 (1998a).
46. Riess, A. G., Nugent, P. E., Filippenko, A. V., Kirshner, R. P., & Perlmutter, S., *ApJ*, **504**, 935 (1998a).
47. Filippenko, A. V., & Riess, A. G., *Physics Reports*, **307**, 31 (1998).
48. Riess, A. G., et al., *AJ*, **116**, 1009 (1998b).
49. Perlmutter, S., et al., *ApJ*, **517**, 565 (1999).
50. Zaldarriaga, M., Spergel, D. N., & Seljak, U. *ApJ*, **488**, 1 (1997).
51. Eisenstein, D. J., Hu, W., & Tegmark, M. *ApJ*, **504**, L57 (1998).
52. Hancock, S., Rocha, G., Lazenby, A. N., & Gutiérrez, C. M., *MNRAS*, **294**, L1 (1998).
53. Lineweaver, C. H., & Barbosa, D., *ApJ*, **496**, 624 (1998).
54. Garnavich, P., et al., *ApJ*, **509**, 74 (1998b).
55. Lineweaver, C. H., *ApJ*, **505**, L69 (1998).
56. Efstathiou, G., et al., *MNRAS*, **303**, L47 (1999).
57. de Bernardis, P., et al., *Nature*, **404**, 955 (2000).
58. Hanany, S., et al., *ApJ*, submitted, astro-ph/0005123 (2000).
59. Balbi, A., et al., *ApJ*, submitted, astro-ph/0005124 (2000).
60. Miller, A. D., et al., *ApJ*, **524**, 1 (1999).
61. Bahcall, N. A., Ostriker, J. P., Perlmutter, S., & Steinhardt, P. J., *Science*, **284**, 1481 (1999).
62. Oswalt, T. D., Smith, J. A., Wood, M. A., & Hintzen, P., *Nature*, **382**, 692 (1996).
63. Cowan, J. J., McWilliam, A., Sneden, C., & Burris, D. L., *ApJ*, **480**, 246 (1997).
64. Gratton, R. G., Fusi Pecci, F., Carretta, E., Clementini, G., Corsi, C. E., & Lattanzi, M., *ApJ*, **491**, 749 (1997).
65. Chaboyer, B., Demarque, P., Kernan, P. J., & Krauss, L. M., *ApJ*, **494**, 96 (1998).
66. Höflich, P., Wheeler, J. C., & Thielemann, F. K., *ApJ*, **495**, 617 (1998).
67. Umeda, H., et al., *ApJ*, **522**, L43 (1999).
68. Domínguez, I., Höflich, P., Straniero, O., & Wheeler, J., in *Future Directions of Supernovae Research: Progenitors to Remnants* (Assergi, Italy), astro-ph/9905047 (1999).
69. Yungelson, L. R., & Livio, M., **ApJ, 528**, 108 (2000).
70. Nomoto, K., Umeda, H., Hachisu, I., Kato, M., Kobayashi, C., & Tsujimoto, T., in *Type Ia Supernovae: Theory and Cosmology*, ed. J. Truran & J. Niemeyer (Cambridge: Cambridge Univ. Press), in press, astro-ph/9907386 (2000).
71. Drell, P. S., Loredo, T. J., & Wasserman, I., *ApJ*, **530**, 593 (2000).
72. Cappellaro, E., et al., *A&A*, **322**, 431 (1997).
73. Nugent, P., Phillips, M., Baron, E., Branch, D., & Hauschildt, P., *ApJ*, **455**, L147 (1995).
74. Coil, A. L., et al., submitted (2000).

75. Ford, C. H., et al., *AJ*, **106**, 1101 (1993).
76. Riess, A. G., et al., *ApJ*, **536**, 62 (2000).
77. Riess, A. G., et al., *AJ*, **118**, 2675 (1999b).
78. Riess, A. G., Filippenko, A. V., Li, W. D., & Schmidt, B. P., *AJ*, **118**, 2668 (1999a).
79. Goldhaber, G., et al., *BAAS*, **30**, 1325 (1998a).
80. Goldhaber, G., et al., in *Gravity: From the Hubble Length to the Planck Length*, SLAC Summer Institute (Stanford, CA: SLAC) (1998b).
81. Groom, D. E., *BAAS*, **30**, 1419 (1998).
82. Aldering, G., Knop, R., & Nugent, P. *AJ*, **119**, 2110 (2000).
83. Riess, A. G., Press, W. H., & Kirshner, R. P., *ApJ*, **473**, 588 (1996b).
84. Aguirre, A. N., *ApJ*, **512**, L19 (1999a).
85. Hatano, K., Branch, D., & Deaton, J., *ApJ*, **502**, 177 (1998).
86. Bahcall, J. N., et al., *ApJ*, **457**, 19 (1996).
87. Aguirre, A. N., *ApJ*, **525**, 583 (1999b).
88. Aguirre, A., & Haimin, Z., *ApJ*, **525**, 583 (1999).
89. Wambsganss, J., Cen, R., & Ostriker, J. P., *ApJ*, **494**, 29 (1998).

What's new in the pulsar world?

D. R. Lorimer
dunc@naic.edu

Arecibo Observatory, HC Box 53995
Arecibo PR 00612

Abstract. We review the latest results and prospects for pulsar astronomy in the new millennium. Some of the most exciting discoveries and their impact on our understanding of radio pulsars are discussed as well as speculations for the future.

I AN INTRODUCTION TO PULSAR ASTRONOMY

Pulsar astronomy began in dramatic and serendipitous circumstances back in 1967 when graduate student Jocelyn Bell and her advisor Anthony Hewish first discovered "pieces of scruff" on a chart recorder output (see Fig. 1) from an array of dipoles strung across a field near Cambridge to study interplanetary scintillation.

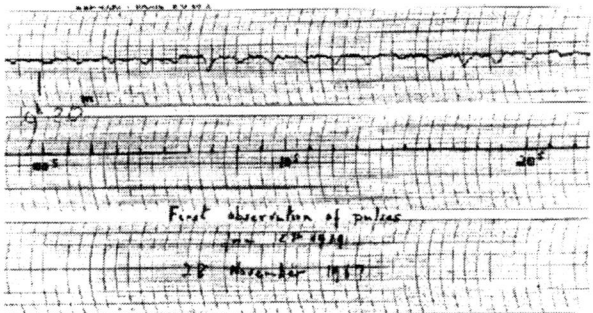

FIGURE 1. The original pen-chart recording for the first pulsar discovery (Hewish et al. 1968).

Over the three decades that have elapsed since the discovery of the first pulsar, now known as PSR B1919+21[1], over 1200 pulsars have been discovered as a result of systematic surveys of the disk of our Galaxy, its Globular Cluster system and also the Magellanic clouds. Although we shall be concerned with the most recent of these discoveries in this review, it is appropriate, by way of introduction, to

---

[1] Pulsars are conveniently named with a PSR prefix followed by their celestial coordinates

summarize some of the basic facts that underpin many of the interpretations of the observational data and review various formation models proposed to explain the rich diversity of objects observed so far.

Rapidly rotating, highly magnetized neutron stars are the only known class of astrophysical objects that can single-handedly explain the basic pulsar phenomenon (Gold 1968, Pacini 1968). The interested reader is referred to Lyne & Smith (1998) for a review of the shortcomings of other scenarios proposed shortly after the Cambridge discoveries. The basic model is shown schematically in Fig. 2.

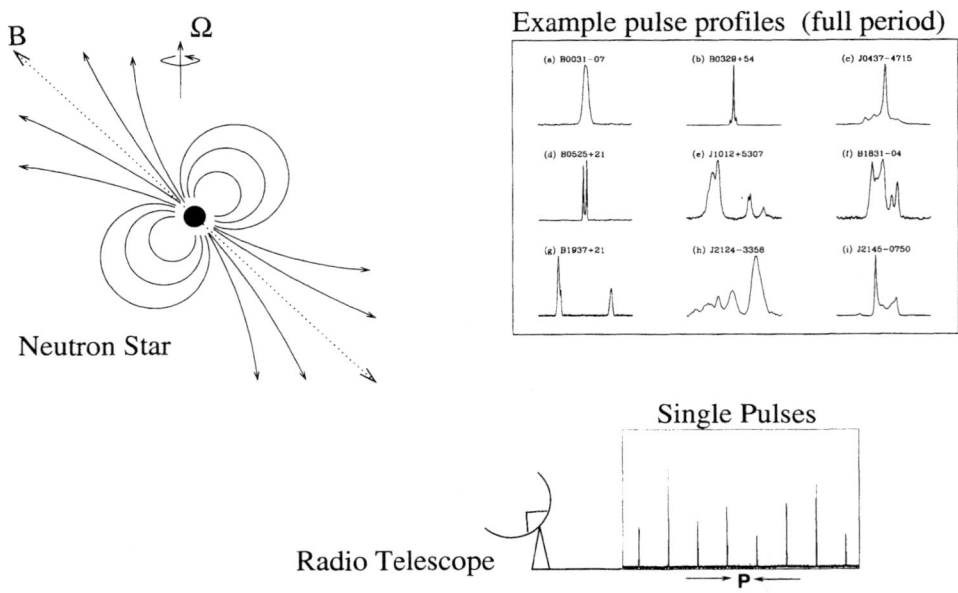

FIGURE 2. The rotating neutron star model. Inset: integrated pulse profiles.

As the neutron star spins, charged particles are accelerated out along magnetic field lines in the magnetosphere. This acceleration causes the particles to emit electromagnetic radiation, most readily detected at radio frequencies as a sequence of pulses produced as the magnetic axis crosses the observer's line of sight each rotation. The time between pulses is the rotation period P of the neutron star.

Pulsars are weak radio sources. Mean flux densities, usually quoted in the literature at a radio frequency of 400 MHz, vary between 1 mJy and 1 Jy (1 Jy = 10^{-26} W m^{-2} Hz^{-1}). This means that the addition of many thousands of pulses is required in order to produce a discernible profile. A remarkable fact is that, although the individual pulses vary quite dramatically from pulse to pulse, at any particular observing frequency the integrated profile is very stable. The pulse profile can thus be thought of as a finger print of the emission beam. A number of examples are shown in Fig. 2 where each profile represents 360 degrees of rotational phase.

Like most things in life, the emission does not come for free. It takes place at the expense of the neutron star's rotational kinetic energy — one of the key predictions of the Gold/Pacini theory. Measurements of the rate of increase in pulse period ($\dot{P}$) through pulsar timing techniques are in excellent agreement with this idea. The distribution of pulse periods and period derivatives is shown in Fig. 3. Amongst other things, the diagram demonstrates clearly the distinction between the "normal pulsars" ($P \sim 0.5$ s and $\dot{P} \sim 10^{-15}$ s/s and populating the "island" of points) and the "millisecond pulsars" ($P \sim 3$ ms and $\dot{P} \sim 10^{-20}$ s/s and occupying the lower left part of the diagram).

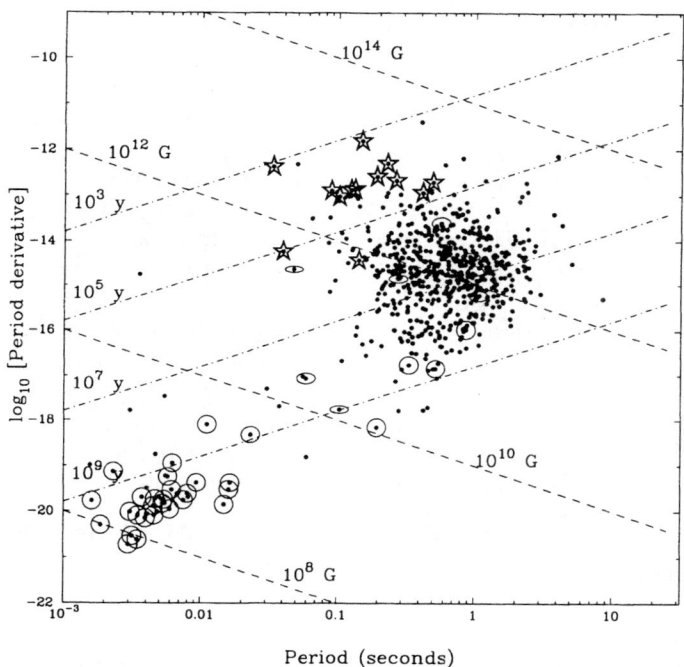

FIGURE 3. The $P - \dot{P}$ diagram. Each pulsar is represented by a black dot. Pulsars known to be members of binary systems are highlighted by a circle (for low-eccentricity orbits) or an ellipse (for elliptical orbits). Pulsars associated with supernova remnants are highlighted by the stars.

From measurements of pulsar spindown we can actually estimate the magnetic field strength and age of the neutron star. The simplest model for this treats the neutron star as a classical dipole in a vacuum (Pacini 1967). For the "canonical neutron star" of mass 1.4 M$_\odot$, radius 10 km and moment of inertia 10^{45} g cm^2 (see for example Shapiro & Teukolsky 1983), it can be shown that the surface dipole magnetic field strength in units of 10^{12} Gauss, $B_{12} \simeq (P_s \dot{P}_{-15})^{1/2}$, where P_s is the period in seconds and $\dot{P}_{-15} = 10^{15} \dot{P}$. Integrating this first-order differential equation over time, assuming a constant magnetic field, results in an expression for the age of the pulsar $t = \tau [1 - (P_0/P)^2]$, where $\tau = P/2\dot{P}$ is the so-called "characteristic age" and P_0 is the period of the pulsar at $t = 0$. Under the assumption that the neutron star has slowed down significantly since birth ($P_0 \ll P$), the characteristic age τ is a good approximation to the true age t. Lines of constant B and τ are drawn on Fig. 3. From these, we infer typical magnetic fields and ages of 10^{12} G and 10^7 yr for the normal pulsars and 10^8 G and 10^9 yr for the millisecond pulsars.

A very important additional difference between normal and millisecond pulsars is binarity. Orbital companions are much more commonly observed around millisecond pulsars ($\sim 80\%$ of the observed sample) than the normal pulsars ($\lesssim 1\%$). The high precision available from pulsar timing measurements yield information on the masses of orbiting companions which, when supplemented by observations at other wavelengths, tell us a great deal about their nature. The present sample of orbiting companions are either white dwarfs, main sequence stars, or other neutron stars. Binary pulsars with low-mass companions ($\lesssim 0.5$ M$_\odot$ — predominantly white dwarfs) usually have millisecond spin periods and essentially circular orbits: $10^{-5} \lesssim e \lesssim 10^{-1}$. Measurements of the "cooling ages" of the white dwarfs (see e.g. van Kerkwijk 1996) provide further evidence that millisecond pulsars have typical ages of a few Gyr. The binary pulsars with high-mass companions ($\gtrsim 1$ M$_\odot$ — neutron stars or main sequence stars) have larger spin periods ($\gtrsim 30$ ms) and are in much more eccentric orbits: $0.2 \lesssim e \lesssim 0.9$.

We now review the various end-products that are implied by standard models for the formation and evolution of single and binary radio pulsars which basically assume that every neutron star in the disk of our Galaxy was formed during the core-collapse phase of a supernova explosion of a massive star. The simplest scenario is shown in Fig. 4 and begins in the final moments of the life of a single massive star that is about to explode as a supernova. The neutron star formed during the

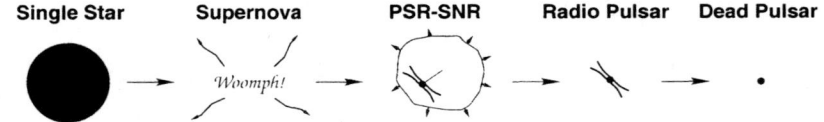

FIGURE 4. Cartoon showing the evolutionary sequence of a single neutron star.

core collapse will receive an impulsive kick (see e.g. Spruit & Phinney 1998) if the explosion is not symmetric and, as a result, begin to move away from the center of the explosion. In the meantime, the outer layers of the star are expanding into the surrounding space at velocities of up to 10,000 km s^{-1}. The result is a pulsar–supernova remnant association (PSR-SNR) which may be visible as a pair for up to 10^5 yr after the explosion. Eventually, the expanding shell becomes so diffuse that it is no longer visible as a supernova remnant. The pulsar, on the other hand, may produce radio emission for a further 10^7 yr or more as it gradually spins down to longer periods. Presently, the longest period for a radio pulsar is 8.5 s (Young, Manchester & Johnston 1999). Ultimately the rotational energy of the neutron star is insufficient to induce pair production in the magnetosphere and, as a result of this, radio emission ceases.

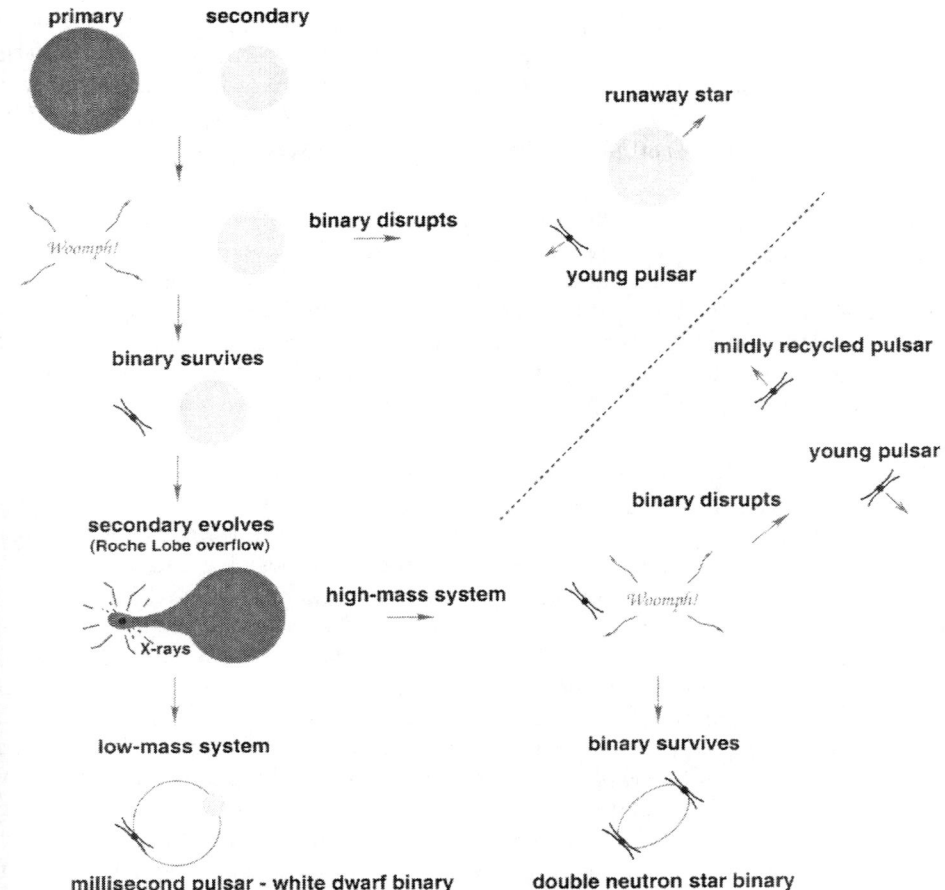

FIGURE 5. Cartoon showing various evolutionary scenarios involving binary pulsars.

The presently favored model to explain the formation of the various types of binary systems has been developed over the years by a number of authors (Bisnovatyi-Kogan & Komberg 1974; Flannery & van den Heuvel 1975; Smarr & Blandford 1976; Alpar et al. 1982). The model is sketched in Fig. 5 and can be qualitatively summarized as follows: starting with a binary star system, the neutron star is formed during the supernova explosion of the initially more massive star (the primary) which has an inherently shorter main sequence lifetime. From the virial theorem, it follows that the binary system gets disrupted if more than half the total pre-supernova mass is ejected from the system during the explosion. In addition, the fraction of surviving binaries is affected by the magnitude and direction of the impulsive kick velocity the neutron star receives at birth (Hills 1983; Bailes 1989). Those binary systems that disrupt produce a high-velocity isolated neutron star and an OB runaway star (Blaauw 1961). Rather like a young student waking up after a wild party, the isolated pulsar has no recollection of previous events — in this case, the binary system it once belonged to and behaves from the moment of release as a single pulsar discussed above. The high binary disruption probability explains why so few normal pulsars are observed with orbiting companions.

For those few binaries that remain bound, and where the companion star is sufficiently massive to evolve into a giant and overflow its Roche lobe, the old spun-down neutron star can gain a new lease of life as a pulsar by accreting matter and therefore angular momentum from its companion (Alpar et al. 1982). The term "recycled pulsar" is often used to describe such objects. During this accretion phase, the X-rays liberated by heating the material falling onto the neutron star mean that such a system is expected to be visible as an X-ray binary system.

Two classes of X-ray binaries relevant to binary and millisecond pulsars exist: neutron stars with high-mass or low-mass companions. The high-mass companions are massive enough to explode themselves as a supernova, producing a second neutron star. If the binary system is lucky enough to survive this explosion, it ends up as a double neutron star binary. The classic example is PSR B1913+16, a 59-ms radio pulsar with a characteristic age of $\sim 10^8$ yr which orbits its companion every 7.75 hr (Hulse & Taylor 1975). In this formation scenario, PSR B1913+16 is an example of the older, first-born, neutron star that has subsequently accreted matter from its companion. Presently, there are no clear observable examples of the second-born neutron star in these systems. This is probably reasonable when one realises that the observable lifetimes of recycled pulsars are much larger than normal pulsars. Double neutron star binary systems are very rare in the Galaxy (see e.g. Curran & Lorimer 1995) — another indication that the majority of binary systems get disrupted when one of the components explodes as a supernova. Systems disrupted after the supernova of the secondary form a midly-recycled isolated pulsar and a young pulsar (formed during the explosion of the secondary).

By definition, the companions in the low-mass X-ray binaries evolve and transfer matter onto the neutron star on a much longer time-scale, spinning the star up to periods as short as a few ms (Alpar et al. 1982). This model has recently gained strong support from the detection of Doppler-shifted 2.49-ms X-ray pulsations from the transient X-ray burster SAX J1808.4–3658 (Wijnands & van der Klis 1998; Chakrabarty & Morgan 1998). At the end of the spin-up phase, the secondary sheds its outer layers to become a white dwarf in orbit around a rapidly spinning millisecond pulsar. A number of binary millisecond pulsars now have compelling optical identifications of the white dwarf companion. The existence of solitary millisecond pulsars in the Galactic disk (which comprise just under 20% of all Galactic millisecond pulsars) cannot easily be explained in the context of this model and alternative formation scenarios need to be developed.

II PULSAR SEARCHING — PAST AND PRESENT

Having gotten a flavor for the diversity of the potentially observable pulsar population, we now turn to a brief history of pulsar searching before concentrating on the latest results from a number of pulsar search efforts going on around the world.

A The pioneering years of pulsar searching: 1967-1982

One means of summarizing the history of progress in pulsar searches is the cumulative number distribution in Fig. 6. This shows, amongst other things, where landmark discoveries of exciting pulsars have been made, and their impacts on subsequent searches. Perhaps the most famous discovery of all of these is the original binary pulsar, PSR B1913+16, discovered by Hulse & Taylor in 1974 in a survey of part of the galactic plane visible from the giant 305–m Arecibo radio telescope (Hulse & Taylor 1975).

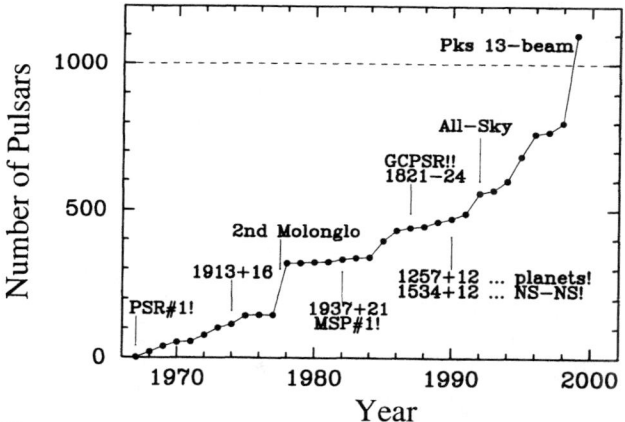

FIGURE 6. The cumulative number of pulsars known as a function of time. This diagram captures some of the "thrill of the chase", as well as the continued motivation to initiate new searches following significant technological advances and conceptual innovations.

This relativistic binary system, which consists of a pair of neutron stars, has been shown to be a truly remarkable natural laboratory for tests of general relativity (Taylor & Weisberg 1982, 1989) — certainly more than ample reward for the initial outlay of telescope time in the survey and subsequent off-line processing. Fascinating accounts of the discovery and study of this system can be found in the 1993 Nobel lectures given by Hulse and Taylor.

In 1977 Dick Manchester and colleagues used the Molonglo radio telescope in Canberra to survey the sky with declinations $-85° \leq \delta \leq +20°$ (~ 8.4 sr). This monumental effort (Manchester et al. 1978) discovered no less than 155 pulsars, thereby doubling the observed pulsar population known at that time (the

1977/8 leap in Fig. 6). The new discoveries included one more binary system PSR B0820+02 — an ordinary 0.865 s pulsar in a wide orbit around a white dwarf star (Manchester et al. 1980). Together with the results of contemporary searches using northern hemisphere telescopes (Davis et al. 1977; Damashek et al. 1978), a fairly well-understood sample of pulsars over the whole sky was available for studying the statistical properties of the Galactic population (Phinney & Blandford 1981; Vivekanand & Narayan 1981, Lyne et al. 1985). These studies suggested a total population of $\sim 10^5$ active radio pulsars in the Galaxy.

B The middle ages... advent of the millisecond pulsars

Data acquisition and processing limitations during the 1970s and early 1980s limited the sampling rates of surveys conducted during this era to $\gtrsim 20$ ms. The corresponding sensitivity to short-period objects, such as the 33 ms pulsar in the Crab nebula, was thus far from ideal suggesting that the true population of short-period pulsars was being underrepresented in the observed sample. This was dramatically confirmed in 1982 with the discovery by Backer et al. of "the millisecond pulsar" B1937+21 (Backer et al. 1982). With a period of 1.56 ms and a corresponding rotation frequency over 20 times larger than the Crab, this remarkable object is still the most rapidly rotating neutron star known to man. Subsequent timing observations of B1937+21 soon showed that it is an extremely stable celestial clock on time-scales $\sim$ years (Davis et al. 1985) having a host of astrophysical applications including the detection of long-period gravitational waves from the early Universe (Bertotti et al. 1983). In addition, PSR B1937+21 is a bright source, having a mean luminosity roughly 240 times that of the original Cambridge pulsar, B1919+21. It thus seemed natural to suppose that B1937+21 was just the tip of the iceberg of a larger population of rapidly rotating neutron stars missed by previous searches.

The large increase in computing power and data storage requirements in a search for millisecond pulsars meant that most early search efforts in the 1980s had only limited sensitivity to millisecond pulsars. This problem is highlighted by the fact that only 4 millisecond pulsars were found in the galactic disk prior to 1990 which, in turn, hampered early attempts to determine their galactic population (Kulkarni & Narayan 1988). Surveys during this period were, however, very successful at discovering young pulsars along the Galactic plane (Clifton et al. 1992; Johnston et al. 1992) and also discovered a number of interesting binary systems.

Following initially unsuccessful searches of globular clusters prior to the discovery of PSR B1937+21 (Seiradakis & Graham 1980), search systems armed with faster sampling rates returned to globular clusters — where low-mass X-ray binaries, the probable progenitors of millisecond pulsars, were already known to exist (Clark 1975). This "targeted search" approach circumvents the need to cover large areas of sky, since each cluster can be observed with one telescope pointing, thereby greatly reducing the total amount of data to process. Searches soon proved fruitful, with the discovery of PSR B1821-24, a 3.1 ms pulsar in the globular cluster M28 (Lyne et al. 1987). Surveys of other clusters have since been very successful, discovering over 20 millisecond pulsars. Notable highlights were the discovery of an eclipsing binary system in Terzan 5 (Lyne et al. 1990), 11 millisecond pulsars in 47 Tucanae (Manchester et al. 1991; Robinson et al. 1995), and a neutron star–neutron star binary in M15, B2127+11C (Prince et al. 1991; Anderson 1992).

C All-sky surveys for millisecond pulsars in the 1990s

Advances in low-cost computing power and data storage capabilities towards the end of the 1980s meant that a return to galactic disk surveys with much improved sensitivity was possible. The breakthrough was made by Wolszczan (1991) in a search of just 200 deg^2 of sky away from the galactic plane during an upgrade period at Arecibo. The survey found the millisecond pulsar planetary system B1257+12 (Wolszczan & Frail 1992), and yet another neutron star–neutron star binary system PSR B1534+12 (Wolszczan 1991). A statistical analysis by Johnston & Bailes (1991) demonstrated that a large number of millisecond pulsars would be found by an all-sky search of similar sensitivity to Wolszczan's survey.

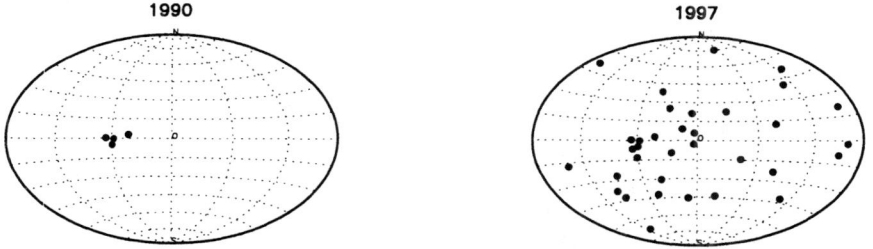

FIGURE 7. Hammer-Aitoff projections showing the known Galactic disk millisecond pulsar population in 1990 (left) and in 1997 (right).

Large-area searches began in earnest in the early 1990s at Parkes (Manchester et al. 1996; Lyne et al. 1998), Jodrell Bank (Nicastro et al. 1995), Arecibo (Thorsett et al. 1993; Foster et al. 1995; Ray et al. 1995; Camilo et al. 1996), and Green Bank (Sayer, Nice & Taylor 1997). With the exception of the Arecibo surveys, these have now been completed. The great success of these surveys can be seen by the sharp rise in Fig. 6 during the mid 1990s, and also in Fig. 7 which compares the sky distributions circa 1990 with the present situation in which the sources are much more uniformly distributed on the sky. Over 30 millisecond pulsars have so far been discovered by these searches as well as many more low-luminosity ordinary pulsars. Notable highlights include PSRs J0437–4715 (Johnston et al. 1993) and J1713+0747 (Foster et al. 1993) — bright, nearby millisecond pulsars which are already proving to be more stable clocks than PSR B1937+21. A future application of such an "array" of clocks will be as a very sensitive detector of long-period gravitational waves (Foster & Backer 1990).

D The current revolution in pulsar hunting... multibeams!

After so many exciting discoveries in 30 years of pulsar searching, perhaps the most striking feature of Fig. 6 is that the gradient of the curve is currently steeper than it has ever been! Clearly then, pulsar astronomy is currently enjoying its most productive phase to date and we now turn to the latest results.

Most of the new discoveries are a result of a multibeam system currently in use at the Parkes 64-m telescope in New South Wales, Australia. With 13×20-cm 25-K receivers on the sky (Fig. 8), along with $13 \times 2 \times 288$-MHz filterbanks, the telescope is presently making major contributions in a number of different pulsar search projects. In its main use for a Galactic plane survey (Lyne et al. 2000), the system achieves a sensitivity of 0.15 mJy in 35 min and covers about one square degree of sky per hour of observing — a standard that is far beyond the present capabilities of any other observatory. As a result, the Parkes telescope now surpasses even Arecibo as a pulsar finding instrument at L-band (20 cm).

FIGURE 8. The 13-element multibeam system being hoisted up into the focus of the 64-m radio telescope at Parkes (see: www.atnf.csiro.au). This new system dramatically improves the sensitivity of large-area surveys carried out by pulsar and spectral line astronomers.

The staggering total of over 550 new pulsars has come from an analysis of over half the total data. On completion, the survey will have covered out to $b = \pm 5°$ along most of the Galactic plane visible from the Parkes observatory and is likely to find well over another 150 pulsars in addition to the new discoveries so far. Such a large haul is resulting in significant numbers of interesting individual objects: several of the new pulsars are observed to be spinning down at high rates, suggesting that they are young objects with large magnetic fields (Camilo et al. 2000a). The characteristic age for the 400-ms pulsar J1119−6127, for example, is only 1.6 kyr. Another member of this group is the 4-s pulsar J1814−1744. These two pulsars have the largest inferred dipole magnetic field strengths for any pulsar so far known (4.1 and 5.5×10^{13} G respectively) and challenge some models which do not predict any radio emission much above 4×10^{13} G. In addition J1814−1744 has similar spin properties to the so-called "anomalous X-ray pulsar" 1E 2259+586

(Fahlman & Gregory 1981) which is known to be radio quiet. It is clear that these is much to be learned from these and other young pulsars from the survey. On completion of the survey, a detailed statistical analysis of the population of young pulsars discovered will undoubtably help us to better constrain the birth properties of pulsars, particularly the distributions of initial spin period and magnetic field strengths which are at present poorly understood.

The multibeam system has not only been finding young, distant pulsars close to the Galactic plane. Edwards et al. (in prep. see also astro-ph/9911221) have been using the same system to search intermediate Galactic latitudes ($5° \leq |b| \leq 15°$). The discovery of 8 short-period pulsars during this search, not to mention 50 long-period objects, strongly supports a recent suggestion by Toscano et al. (1998) that L-band searches are an excellent means of finding relatively distant millisecond pulsars. Full details of the search results should be publicly-available by the end of 2000. It is, however, already clear that this novel experiment is likely to significantly enhance our knowledge of the Galactic population of millisecond pulsars.

TABLE 1. Parameters for Parkes multibeam binary pulsars. From left to right, the parameters listed (when known) are pulsar period, characteristic age, inferred magnetic field strength, orbital period, eccentricity and minimum inferred companion mass.

PSR J	P (ms)	τ_c (10^6 y)	$\log B$ $\log$ (G)	P_b (d)	Ecc.	Min. M_c ($M_\odot$)
1232−6501	88.28	1400	10.0	1.863	0.00	0.15
1904+04	71.09	–	–	15.750	0.04	0.23
1810−2005	32.82	4000	9.3	15.012	0.00	0.29
1453−58	45.25	–	–	12.422	0.00	0.88
1435−60	9.35	–	–	1.355	0.00	0.90
1811−1736	104.18	950	10.1	18.779	0.83	0.87
1141−6545	393.90	1.45	12.1	0.198	0.17	1.01
1740−3052	570.31	0.36	12.6	231.039	0.58	11.07

A number of the new discoveries from the inner-plane survey have orbiting companions, the parameters of these systems are summarized in Table 1 which has been adapted from the compilation presented by (Manchester et al. 2000). Several low-eccentricity systems are known where the likely companions are white dwarf stars. What is interesting about these new discoveries is the range of spin periods of the pulsars in the binary systems: 9–90 ms. The spin periods are generally longer than previously observed and suggests that these neutron stars have undergone a shorter phase of accretion and spin up from their companions than the "classical" low-eccentricity binary millisecond pulsars described in §I. This in turn suggests a wider range of companion masses in the progenitor systems which produce more massive white dwarfs after an abbreviated period of accretion.

Among the three eccentric binary systems discovered by the survey so far is a probable double neutron star system J1811−1736 (Lyne et al. 2000). Unlike the original binary pulsar, B1913+16 which will merge due to gravitational wave emission in another 110 Myr, this system has a much wider orbit and as such will not merge for another 10^{12} yr. The contribution to the cosmic merger rate of double neutron star systems from binaries similar to J1811−1736 is, therefore, rather small. It is, however, clear that this new system, and others which will undoubtably come from this survey, will teach us much about the still poorly-understood population of double neutron star systems.

Currently, the shortest orbital period system discovered in the multibeam survey is PSR J1141−6545 (Kaspi et al. 2000). This 4.75-hr binary is in a moderate-eccentricity orbit where the likely companion is a massive white dwarf star. The orbital parameters are such that this system will coalesce due to the emission of gravitational radiation within 1.5 Gyr. The fact that J1141−6545 may have a characteristic age of just over 1 Myr implies that the likely birth-rate of such objects could be significant. Accurate determinations of the neutron star and companion mass for this system should be possible over the next few years of regular timing measurements. An exciting additional possibility is the significant degree of geodetic precession of the pulsar spin axis caused by the perturbing effects of the companion's gravitational field (Kaspi et al. 2000). This effect should also be detectable over the coming years as a change in the pulse shape due to the changing impact angle between the radio beam and our line of sight.

The most massive binary system yet from either of the multibeam surveys is J1740−3052, whose orbiting companion must be at least 11 $M_\odot$! Recent optical observations (Manchester et al. 2000) reveal a K-supergiant as being the likely companion star in this system. Combined radio and optical observations of the pulsar and companion star during periastron passage currently underway should tell us a great deal about the interaction between the pulsar wind and the outer layers of the supergiant. Given the existence of such high-mass systems in the Galaxy, surely it is only a matter of time before a radio pulsar will be found orbiting a stellar-mass black hole — one of the holy grails of pulsar astronomy.

One place to look for a pulsar orbiting a black hole is in the cores of dense globular clusters where the high stellar density produces an ideal environment for stellar encounters and exchange of binary companions. As such, it is not inconceivable that a millisecond pulsar could be found orbiting a stellar mass black hole in a globular cluster. Such a system would be the next step beyond PSR B1913+16 as a laboratory for testing strong-field gravity. While no such system has so far been found, the field of pulsar searches in globular clusters has received a much-needed kick from a deep search of the cluster 47 Tucanae by Camilo et al. (2000b). This survey, which makes use of just the central beam in the 13-element array, employs long (5-hr) integration times and computationally-intensive "acceleration search" techniques to look for short-period binary systems that are already known to exist in this and other clusters (as described in the previous discussion, §II B).

In addition to the 11 pulsars previously known in 47 Tucanae (Manchester et al. 1991; Robinson et al. 1995), this new search has resulted in the discovery of a further 11 new pulsars so far. All of the 9 pulsars published by Camilo et al. (2000b) are members of binary systems, many of which were only detectable with the aid of the acceleration searching. Pulse profiles for 20 of the known pulsars are shown in Fig. 9. It is important to note that not all these pulsars are detectable on a given observing day. The flux densities of each of the pulsars varies considerably due to interstellar scintillation — constructive and destructive fringing events that occur as the pulses propagate through the turbulent interstellar medium between us and the cluster. This effects occasionally amplifies nominally weak and otherwise undetectable pulsars above the survey detection threshold.

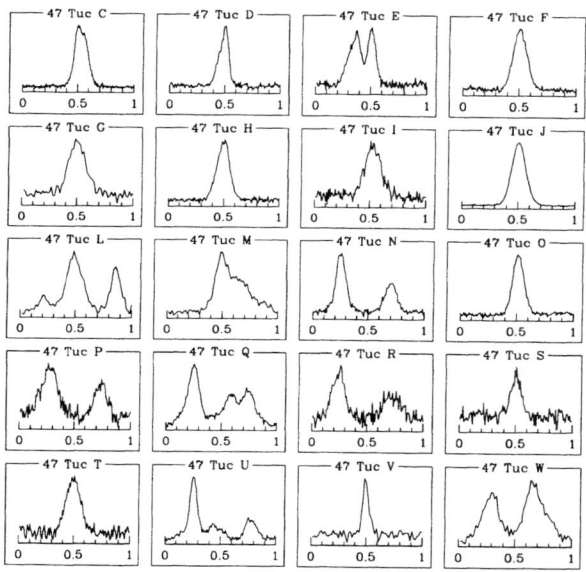

FIGURE 9. 20 millisecond pulsars in one telescope pointing! The compilation of new and previously-known pulsars published by Camilo et al. (2000b).

Based on current observations, Camilo et al. (2000b) estimate the total population of millisecond pulsars in this cluster to be over 200! The range of spin periods for the presently-observed sample is 2–8 ms: this is a strong indication that the recycling process is extremely effective in globular clusters. Over 60% of the known pulsars in 47 Tucanae are members of binary systems. Currently, the shortest orbital period pulsar known is for 47 Tuc R, a 3.48-ms pulsar in a 96-min orbit around a low-mass (0.03 $M_\odot$) companion star. This is in fact the shortest orbital period of any known radio pulsar binary. Its existence, together with the 11-min orbit observing in the low-mass X-ray binary 1820-303 in the globular cluster NGC 6624 (Stella Pridhorsky & White 1987), strongly suggests that there may be a substantial population of even shorter period radio pulsar binaries in 47 Tucanae and other globular clusters. Searches for these elusive objects continues.

An additional bonus of the new observations of 47 Tucanae has been their use for timing the new and previously known pulsars in the cluster. Freire et al. (2000) have now obtained accurate parameters for 15 of the pulsars. These new data (spin, astrometric and binary parameters) provide a wealth of information on the individual pulsars in the cluster as well as the physical properties of 47 Tucanae. The radial distribution of pulsars in this cluster out to a few core radii seems to be consistent with an isothermal sphere. 47 Tucanae is a cluster that has not yet gone through a core collapse phase and is supporting itself by "burning" binaries (thereby releasing kinetic energy) in the core. Among the questions posed is why there are so many short-period binaries in 47 Tucanae but relatively few low-mass X-ray binaries. In addition, there is a distinct absence of long-period ($\gtrsim$ 3 day) binaries. Presumably, the latter objects get quickly disrupted during exchange interactions, which may in turn result in short-period binaries and solitary millisecond pulsars. Fossil evidence for such interactions may be the small, but significant, eccentricities measurable for some of the binaries in 47 Tucanae.

III FUTURE GOALS AND DIRECTIONS

In this review, I have tried to give a flavor for the current status of pulsar research which is being revolutionized by a number of high-sensitivity searches which continue to probe new areas of the neutron-star parameter space. Presently, however, we are aware of only about 1% of the total active pulsar population in our Galaxy. Thus, in terms of the exotic systems that exist, we have almost certainly not seen all of the pulsar zoo. Present and future multibeam surveys (e.g. Arecibo and the SKA) could result in many interesting discoveries. Possible examples are:

- A dual-line binary pulsar, i.e. a double neutron star system in which both components are observable as radio pulsars. The additional clock in such a binary system would be most valuable in further tests of strong-field gravity.

- A radio pulsar with a black-hole companion would undoubtably also be a fantastic laboratory for studying gravity in the strong-field regime.

- A sub-millisecond pulsar. The original millisecond pulsar, B1937+21, rotating at 642 Hz is still the most rapidly rotating neutron star known. Do kHz neutron stars exist? Searches now have sensitivity to such objects and a discovery of even one would constrain the equation of state of matter at high densities.

- A binary system in which the neutron star is in the process of transforming from an X-ray-emitting neutron star to a millisecond radio pulsar.

The phenomenal timing stability of radio pulsars leads naturally to a large number of applications, including their use as natural detectors of gravitational radiation (Foster & Backer 1990). Long-term timing experiments currently underway appear to have tremendous potential to further constrain and perhaps detect the gravitational wave background within the next decade.

REFERENCES

Alpar M. A., Cheng A. F., Ruderman M. A., Shaham J., 1982, Nat, 300, 728
Anderson S. B., 1992, PhD thesis, California Institute of Technology
Backer D. C., Kulkarni S. R., Heiles C., Davis M. M., Goss W. M., 1982, Nat, 300, 615
Bailes M., 1989, ApJ, 342, 917
Bertotti B., Carr B. J., Rees M. J., 1983, MNRAS, 203, 945
Bisnovatyi-Kogan G. S., Komberg B. V., 1974, Sov. Astron., 18, 217
Blaauw A., 1961, Bull. Astr. Inst. Netherlands, 15, 265
Camilo F., Nice D. J., Shrauner J. A., Taylor J. H., 1996, ApJ, 469, 819
Camilo F., Kaspi V. M., Lyne A. G., Manchester R. N., Bell J. F., McKay N. P. F., Crawford F., 2000a, ApJ, in press (astro-ph/0004330)
Camilo F., Lorimer D. R., Freire P., Lyne A. G., Manchester R. N., 2000b, ApJ, 535, 975
Camilo F., Nice D. J., Taylor J. H., 1996, ApJ, 461, 812
Chakrabarty D., Morgan E. H., 1998, Nat, 394, 346
Clark G. W., 1975, ApJ, 199, L143
Clifton T. R., Lyne A. G., Jones A. W., McKenna J., Ashworth M., 1992, MNRAS, 254, 177
Curran S. J., Lorimer D. R., 1995, MNRAS, 276, 347
Damashek M., Taylor J. H., Hulse R. A., 1978, ApJ, 225, L31
Davies J. G., Lyne A. G., Seiradakis J. H., 1977, MNRAS, 179, 635
Davis M. M., Taylor J. H., Weisberg J. M., Backer D. C., 1985, Nat, 315, 547
Fahlman G. G., Gregory P. C., 1981, Nat, 293, 202
Flannery B. P., van den Heuvel E. P. J., 1975, A&A, 39, 61
Foster R. S., Backer D. C., 1990, ApJ, 361, 300
Foster R. S., Cadwell B. J., Wolszczan A., Anderson S. B., 1995, ApJ, 454, 826
Foster R. S., Wolszczan A., Camilo F., 1993, ApJ, 410, L91
Freire P., Camilo F., Lorimer D. R., Lyne A. G., Manchester R. N., D'Amico N., 2000, MNRAS, in press
Gold T., 1968, Nat, 218, 731
Hewish A., Bell S. J., Pilkington J. D. H., Scott P. F., Collins R. A., 1968, Nat, 217, 709
Hills J. G., 1983, ApJ, 267, 322
Hulse R. A., Taylor J. H., 1975, ApJ, 195, L51
Hulse R. A., 1994, Reviews of Modern Physics, 66, 699
Johnston S., Bailes M., 1991, MNRAS, 252, 277
Johnston S., Lyne A. G., Manchester R. N., Kniffen D. A., D'Amico N., Lim J., Ashworth M., 1992, MNRAS, 255, 401
Johnston S. et al., 1993, Nat, 361, 613
Kaspi V. M. et al., 2000, ApJ, in press (astro-ph/0005214)
Kulkarni S. R., Narayan R., 1988, ApJ, 335, 755

Lyne A. G., Smith F. G., 1998, Pulsar Astronomy, 2nd ed. Cambridge University Press, Cambridge
Lyne A. G., Brinklow A., Middleditch J., Kulkarni S. R., Backer D. C., Clifton T. R., 1987, Nat, 328, 399
Lyne A. G. et al., 1990, Nat, 347, 650
Lyne A. G. et al., 1998, MNRAS, 295, 743
Lyne A. G. et al., 2000, MNRAS, 312, 698
Lyne A. G., Manchester R. N., Taylor J. H., 1985, MNRAS, 213, 613
Manchester R. N., Lyne A. G., Taylor J. H., Durdin J. M., Large M. I., Little A. G., 1978, MNRAS, 185, 409
Manchester R. N., Newton L. M., Cooke D. J., Lyne A. G., 1980, ApJ, 236, L25
Manchester R. N., Lyne A. G., Robinson C., D'Amico N., Bailes M., Lim J., 1991, Nat, 352, 219
Manchester R. N. et al., 1996, MNRAS, 279, 1235
Manchester R. N. et al., 2000, in Kramer M., Wex N., Wielebinski R., eds, Pulsar Astronomy – 2000 and Beyond. p. 49, (astro-ph/9911319)
Nicastro L., Lyne A. G., Lorimer D. R., Harrison P. A., Bailes M., Skidmore B. D., 1995, MNRAS, 273, L68
Pacini F., 1967, Nat, 216, 567
Pacini F., 1968, Nat, 219, 145
Phinney E. S., Blandford R. D., 1981, MNRAS, 194, 137
Prince T. A., Anderson S. B., Kulkarni S. R., Wolszczan W., 1991, ApJ, 374, L41
Ray P. S. et al., 1995, ApJ, 443, 265
Robinson C. R., Lyne A. G., Manchester A. G., Bailes M., D'Amico N., Johnston S., 1995, MNRAS, 274, 547
Sayer R. W., Nice D. J., Taylor J. H., 1997, ApJ, 474, 426
Seiradakis J. H., Graham D. A., 1980, A&A, 85, 353
Shapiro S. L., Teukolsky S. A., 1983, Black Holes, White Dwarfs and Neutron Stars. The Physics of Compact Objects. Wiley–Interscience, New York
Smarr L. L., Blandford R., 1976, ApJ, 207, 574
Spruit H., Phinney E. S., 1998, Nat, 393, 139
Stella L., Priedhorsky W., White N. E., 1987, ApJ, 312, L17
Taylor J. H., Weisberg J. M., 1982, ApJ, 253, 908
Taylor J. H., Weisberg J. M., 1989, ApJ, 345, 434
Taylor J. H., 1994, Reviews of Modern Physics, 66, 711
Thorsett S. E., Deich W. T. S., Kulkarni S. R., Navarro J., Vasisht G., 1993, ApJ, 416, 182
van Kerkwijk M. H., 1996, in Johnston S., Walker M. A., Bailes M., eds, Pulsars: Problems and Progress, IAU Colloquium 160. ASAP, San Francisco, p. 489
Vivekanand M., Narayan R., 1981, J. Astrophys. Astr., 2, 315
Wijnands R., van der Klis M., 1998, Nat, 394, 344
Wolszczan A., Frail D. A., 1992, Nat, 355, 145
Wolszczan A., 1991, Nat, 350, 688
Young M. D., Manchester R. N., Johnston S., 1999, Nat, 400, 848

SNAP: Supernova / Acceleration Probe An Experiment to Measure the Properties of the Accelerating Universe

P. Nugent
for the SNAP Collaboration[1]

Lawrence Berkeley National Laboratory Berkeley, CA 94720

Abstract. A ~2-meter satellite telescope with a 1-square-degree optical imager, a small near-IR imager, and a three-arm near-UV-to-near-IR spectrograph can discover over 2000 Type Ia supernovae in a year at redshifts between $z = 0.1$ and 1.7, and follow them with high-signal-to-noise calibrated light-curves and spectra. The resulting data set can determine the cosmological parameters with precision: mass density Ω_M to ± 0.02, vacuum energy density Ω_Λ to ± 0.05, and curvature Ω_k to ± 0.06. The data set can test the nature of the "dark energy" that is apparently accelerating the expansion of the universe. In particular, a cosmological constant dark energy can be differentiated from alternatives including a range of "quintessence" dynamical scalar-field models, by measuring the ratio of the dark energy's pressure to its density to ± 0.05 over a range of redshifts. The large numbers of supernovae across a wide range of redshifts are necessary but not sufficient to accomplish these goals; the controls for systematic uncertainties are primary drivers of the design of this space-based experiment. These systematic and statistical controls cannot be obtained with other ground-based and/or space-based telescopes, either currently in construction or in planning stages.

INTRODUCTION

In the past few decades the study of cosmology has taken some of its first major steps as an empirical science, combining concepts and tools from astrophysics and particle physics. The most recent of these results have already brought surprises. The universe's expansion is apparently accelerating rather than decelerating as

---

[1] SNAP Collaboration: G. Aldering, S. Deustua, W. Edwards, B. Frye, D. Groom, S. Holland, D. Kasen, R. Knop, R. Lafever, M. Levi, P. Nugent, S. Perlmutter, K. Robinson (LBNL), D. Curtis, G. Goldhaber, J. R. Graham, S. Harris, P. Harvey, H. Heetderks, A. Kim, M. Lampton, R. Lin, D. Pankow, C. Pennypacker, A. Spadafora, G. F. Smoot (UC Berkeley), P. Astier, J.F. Genat, D. Hardin, J.- M. Levy, R. Pain, K. Schamahneche (IN2P3), A. Baden, J. Goodman, G. Sullivan (U. Maryland), R. Ellis, M. R. Metzger (CalTech), D. Huterer (U. Chicago), A. Fruchter (STScI), C. Bebek (Cornell U.), L. Bergstrom, A. Goobar (U. Stockholm), I. Hook (U. Edinburgh), C. Lidman (ESO), J. Rich (CEA/DAPNIA), A. Mourao (Inst. Superior Tecnico,Lisbon)

expected due to gravity. This implies that the simplest model for the universe – flat and dominated by matter – appears not to be true, and that our current fundamental physics understanding of particles, forces, and fields is likely to be incomplete.

The most clear evidence for this surprising conclusion comes from the recent supernova measurements of changes in the universe's expansion rate that directly show the acceleration. These measurements indicate the presence of a new, mysterious energy component that can cause acceleration. This conclusion is supported by current measurements of the mass density of the universe, when taken together with current Cosmic Microwave Background measurements or inflationary theory.

To address this new puzzle and begin to establish a solid cosmological picture, we propose a satellite experiment to carry out a definitive supernova study that will determine the values of the cosmological parameters and may unveil the unidentified accelerating energy. Here, we will show that this experiment addresses these fundamental science questions with a necessary level of statistical and systematic rigor that cannot be matched by plausible alternatives, whether on the ground or in space [1].

This proposed supernova measurement will play a key role in the larger set of cosmological measurement approaches expected to yield results over the next decade. (This proposed satellite will also use some of these other approaches as part of its science mission.) Together these measurements will complement and cross-check our understanding of the cosmological model of the universe. Since the supernova approach is arguably the most direct and least model dependent, we expect it to provide a touchstone for this concordance of measurement results. Moreover, since this experiment is sensitive to the redshift range in which the accelerating energy is dominant, it will provide a nearly unique window on the properties of this entity of fundamental physics.

This experiment seizes upon the many recent advances in instrumentation and space technology to explore fundamental questions about the nature of our universe.

SCIENTIFIC MOTIVATION AND BACKGROUND

A Simple, Direct Approach to the Cosmological Parameters

Type Ia supernovae (SNe Ia) provide simple cosmological measurement tools. Each one is a strikingly similar explosion event whose physics can be analyzed in some detail from its intensity and spectrum as it brightens and fades. Most observed SNe Ia have nearly the same peak luminosity, and the variations that do exist can be correlated with other observables and hence calibrated to 5% in distance [2,3]. The variation-corrected peak brightness (magnitude) is then a measure of the distance to the supernova.

The wavelengths of the photons from the supernova are stretched— "redshifted"— in exact proportion to the stretching of the universe during the period that the photon travels to us. Thus the comparison of SN Ia redshifts and magnitudes provides a particularly straightforward measurement of the changing rate of expansion of the universe: the apparent magnitude indicates the distance and hence time back to the supernova explosion, while the redshift measures the total relative expansion of the universe since that time.

This satellite project is designed to establish a Hubble-diagram (redshift vs. magnitude) plot dense with supernova events looking back over two-thirds the age of the universe. With such a history of the expansion of the universe we can determine the contributions of decelerating and accelerating energies—mass density Ω_M, vacuum energy density Ω_Λ, and/or other yet-to-be-studied "dark energies"—as the expansion rate changes over time.

This is an extremely transparent methodology. Almost everyone, even non-scientists, can appreciate and perhaps critique every step. Aside from the basic cosmological equations, there is no model dependence in this empirically-based method, and it is sensitive to only a few parameters of cosmology so there is no fit required in a large-dimensional parameter space. (Conversely, this method of course does not help determine these other parameters, except by narrowing down the whole phase space, as discussed below.) This transparency is an unusual and important feature of this particular very fundamental measurement.

The Current Results: Questions Answered and Posed by an Accelerating Universe

The cosmological results from the magnitude/redshift measurements of a few score SNe Ia already present surprises and puzzles [4,5]. Most striking is the indication that we live in an accelerating universe, which must be dominated by a positive cosmological constant or other vacuum energy whose pressure is negative and large. The very simplest cosmological model, the Einstein-de Sitter ($\Omega_M = 1$) universe, which is flat and has zero cosmological constant, is strongly inconsistent with the data. Of the two arguably next-simplest models, only the flat model with the cosmological constant, Λ, fits the data, while the low-mass open universe with zero Λ does not. (All of these statements can be made with very strong statistical confidence; even stretching the range of imagined systematic uncertainties, it is very difficult to fit the data without a cosmological constant in a flat universe.)

These current results immediately raise important questions. Although the data indicate that an accelerating dark energy density—perhaps the cosmological constant—has overtaken the decelerating mass density, they do not tell us the actual magnitude of either one. These two density values are two of the fundamental parameters that describe the constituents of our universe, and determine its geometry and destiny. The proposed satellite project is designed to obtain sufficient magnitude-redshift data for a large enough range of redshifts ($0.1 < z < 1.7$) that these absolute densities can each be determined to unprecedented accuracy

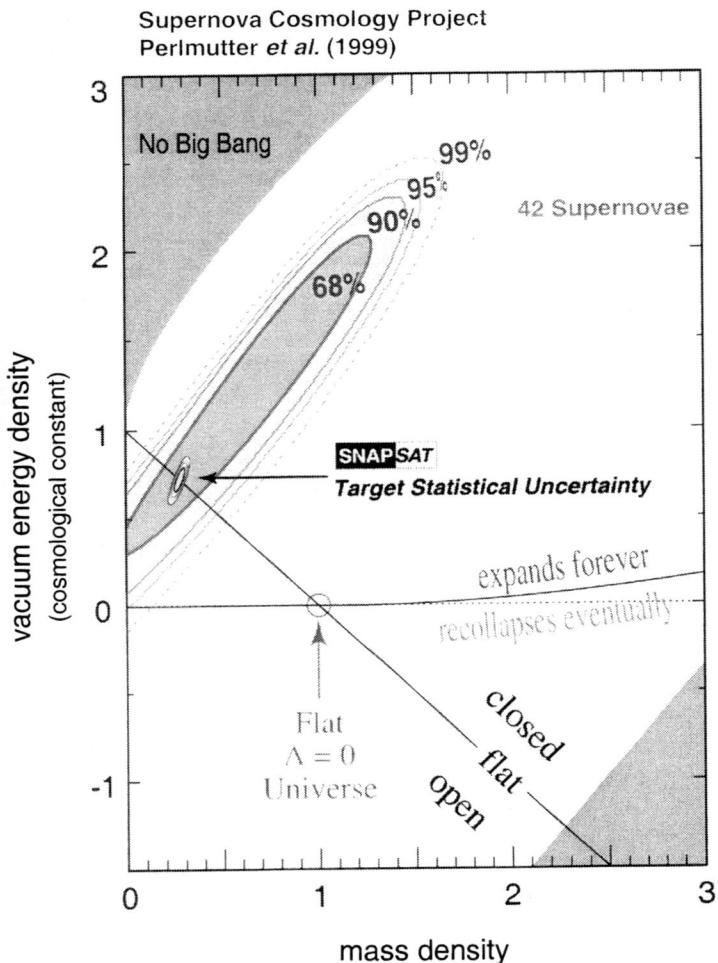

FIGURE 1. 68%, 90%, and 99% confidence regions in the Ω_M–Ω_Λ plane from the 42 distant SNe Ia in [4]. These results rule out a simple flat, [$\Omega_M = 1$, $\Omega_\Lambda = 0$] cosmology. They further show strong evidence (probability > 99%) for $\Omega_\Lambda > 0$. Also shown is the expected confidence region from the SNAP satellite for an $\Omega_M = 0.28$ flat universe.

(see Figure 1). Taken together, the sum of these energy densities then provides a measurement of the curvature of the universe.

The current data also do not tell us the nature of the dark energy; all we know is that it must have a sufficiently negative pressure to cause the universe's expansion to accelerate. Our one long-known physical model for the dark energy, the vacuum energy density that Einstein called "the cosmological constant," presents difficult theoretical problems. Why, for example, is the vacuum energy density so small when compared to the natural energy scales of the particles and fields that would be expected to account for it: the values that are consistent with the current SN Ia results are 10^{120} times smaller than the Planck scale. Moreover, why would a vacuum energy density that remains constant throughout history turn out just now to be within a factor of two or three of the mass energy density, which has fallen by many orders of magnitude since the Big Bang?

In response to these theoretical problems, several alternative physical models have been proposed as candidates for the dark energy. These models can generally be characterized by their equation of state, $p = w\rho$ (the speed of light, c, is set to unity). The ratio of pressure to density, w, can be constant or time-varying depending on the model, and has a constant value of -1 in the case of the cosmological constant. The current SN Ia data allow some crude constraints on the alternative dark energy models, since not all equations of state fit the data. With the proposed satellite project we can begin to study these alternative dark energy models in some detail, by determining w to much higher accuracy and by studying it over a range of redshifts.

The existence of a negative-pressure vacuum energy density is in remarkable concordance with combined galaxy cluster measurements [6], which are sensitive to Ω_M, and current CMB results [7,8], which are sensitive to the curvature Ω_k (see Figure 2). Two of these three independent measurements and standard Inflation would have to be in error to make the cosmological constant (or dark energy) unnecessary in the cosmological models. If this were, in fact, to be the case, a definitive accounting of the systematic uncertainties for the supernova measurements would be particularly crucial, and any new cosmological models would still require the basic product of the SNAP mission, a history of the scale of the universe.

Scientific Goals of SNAP

The primary scientific objective of this mission is to measure important cosmological parameters with low statistical and systematic errors. Assuming that the dark energy is the cosmological constant, this experiment can simultaneously determine mass density Ω_M to accuracy of 0.02, cosmological constant energy density Ω_Λ to 0.05 and curvature $\Omega_k = 1 - \Omega_M - \Omega_\Lambda$ to 0.06.

The proposed experiment is one of very few that can study the dark energy directly, and test a cosmological constant against alternative dark energy candidates. Assuming a flat universe with mass density Ω_M and a dark energy component with

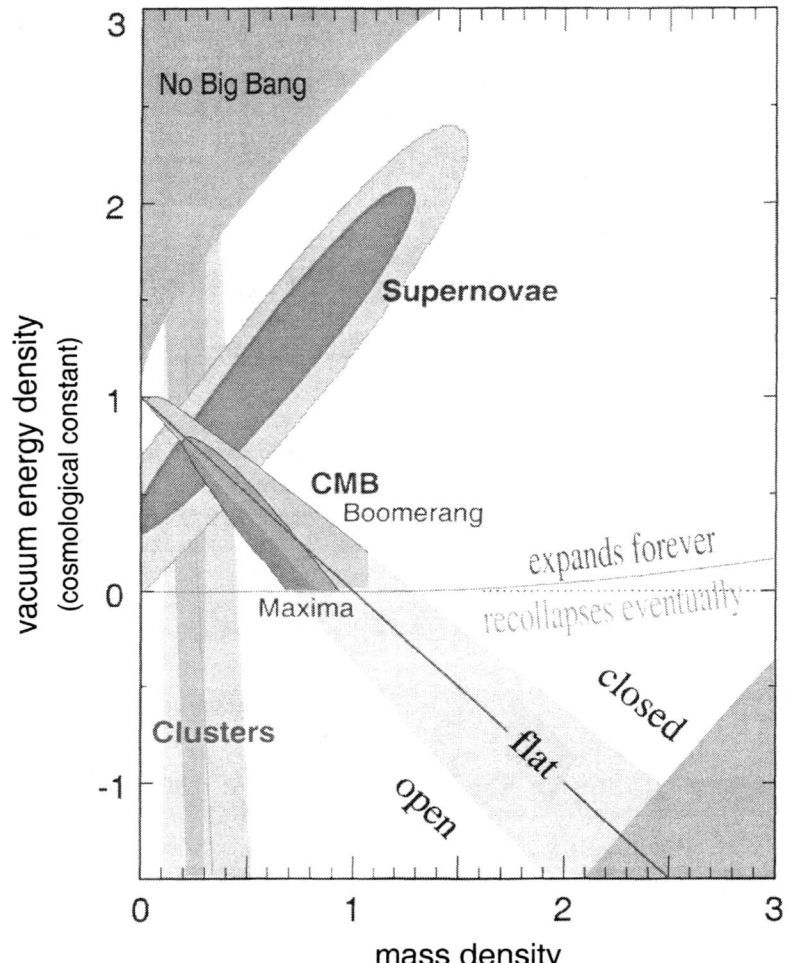

FIGURE 2. There is strong evidence for the existence of a cosmological vacuum energy density. Plotted are Ω_M–Ω_Λ confidence regions for current SN, galaxy cluster, and CMB results. Their consistent overlap is a strong indicator for dark energy.

a non-evolving equation of state, the proposed experiment will be able to measure the equation-of-state ratio w with accuracy of 0.05, at least a factor of five better than the best planned cosmological probes. With such a strong constraint on w we will be able to differentiate between the cosmological constant and such theoretical alternatives as "topological defect" models and a range of dynamical scalar-field ("quintessence") particle-physics models (see Figure 3). Moreover, with data of such high quality one can relax the assumption of the constant equation of state, and begin simple tests of its variation with redshift. These determinations would directly shed light on physics at high energy/small scale and physics of the early universe.

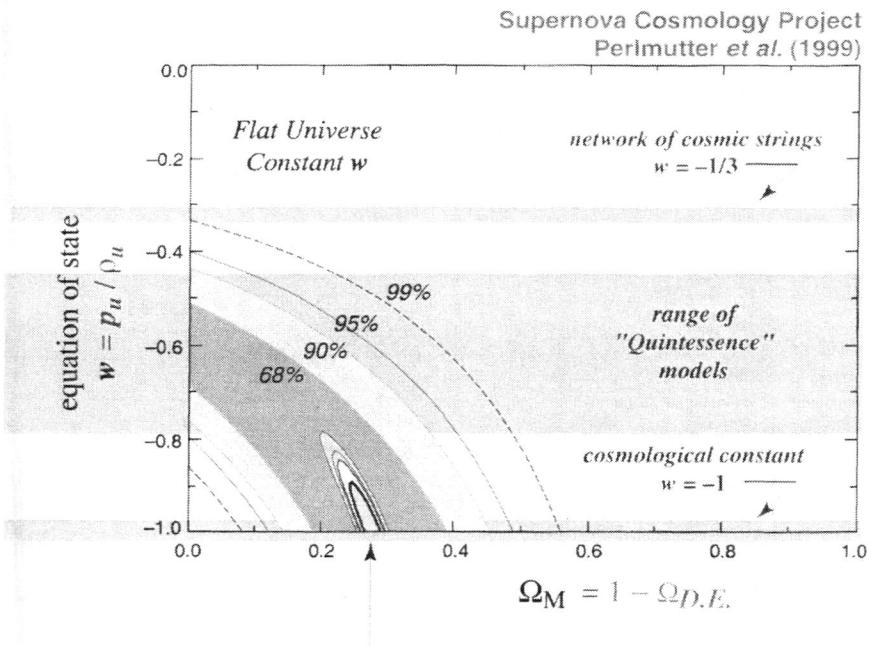

FIGURE 3. Best-fit 68%, 90%, 95%, and 99% confidence regions in the Ω_M–w plane for an additional energy density component, Ω_w, characterized by an equation-of-state $w = p/\rho$. (If this energy density component is Einstein's cosmological constant, Λ, then the equation of state is $w = p_\Lambda/\rho_\Lambda = -1$.) Also shown is the expected confidence region allowed by SNAP.

It is important to add that these SN Ia results are not the only available cosmological measurements, nor will they be at the time of the proposed satellite. The estimates of the mass density from large-scale structure (LSS) surveys and cluster evidence are constantly improving. The MAP and Planck satellite experiments are expected to give high-precision fits of ~ 11 cosmological and model-dependent parameters, both before and after the proposed satellite's SN Ia measurements. Perhaps surprisingly, these supernova measurements will provide stronger constraints on Ω_M and Ω_Λ than those expected from either LSS or CMB measurements, and constraints on curvature Ω_k that are comparable with those expected from MAP and Planck. The important cosmological test will be the cross comparison of these and other fundamental measurements — and it is even possible that cosmology will next progress when we discover that they do not agree. In any case, it will be all of these measurements fit simultaneously, that will provide us with our best understanding of the cosmology of the universe; the final results can be as much as an order of magnitude better than the constraints from any one measurement approach.

To accomplish these goals, it is not sufficient simply to discover and study more supernovae and more distant supernovae. The current SN Ia data set already has statistical uncertainties that are only a factor of two larger than the identified systematic uncertainties. There are also several additional proposed systematic effects that might confound attempts at higher precision, in particular the possibilities of "grey dust" or systematic shifts in the population of SN Ia host galaxy environments. Addressing each of these systematic concerns requires a major leap forward in the supernova measurement techniques, and has driven us to the satellite experiment describe here [1].

PROPOSED EXPERIMENT

Instrumentation

The baseline proposed satellite experiment is based on a simple, dedicated combination of a 1.8- to 2.0-meter telescope, a 1-square-degree imager, a 1-square-arcminute near-IR imager, and a "three-arm" near-UV-to-near-IR spectrograph. The 1-square-degree wide field is obtained with a three-mirror telescope, and a feedback loop based on fast-readout chips on the focal plane to stabilize the image.

The wide-field imager is completely filled with a CCD mosaic for the "optical" wavelengths between 0.3 and 1.0 microns. The near-IR imager will be a single HgCdTe detector to obtain images of specific targets in the wavelengths between 1.0 and 1.7 microns.

The spectrograph uses dichroic beam-splitters to send the light into two optical arms (0.3 – 0.6 μm and 0.55 – 1.0 μm) and one near-IR arm (0.95 – 1.7 μm). Each of the three arms employs an "integral field unit" (IFU) to obtain an effective image of a 2" by 2" field, split into 0.07" by 0.07" regions that are each individually sent to the spectrograph to obtain a flux at each position and wavelength (sometimes called a three-dimensional "data cube"). In operation, these integral field units will allow simultaneous spectroscopy of a supernova target and its surrounding galactic environment; the 2" by 2" field of view also removes any requirement for precise positioning of a supernova target in a traditional spectrograph slit. This point is particularly important for absolute flux calibration, because all of the supernova light is collected with the integral field units. The spectrograph is thus designed to allow the spectra to be used to obtain photometry in any "synthetic" filter band that one chooses.

Observation Strategy and Baseline Data Package

This instrumentation will be used with a simple, predetermined observing strategy designed to monitor a 20-square-degree region of sky near the north and south ecliptic poles, discovering and following supernovae that explode in that region. Every field will be visited frequently enough with sufficiently long exposures that at any given redshift up to $z = 1.7$ every supernova will be discovered within, on average, two restframe days of explosion. Every supernova at $z < 1.2$ will be followed as it brightens and fades, while at $z > 1.2$ there will be sufficient numbers of supernovae that it will only be necessary (and possible) to follow a subsample to obtain comparable numbers of supernovae.

The wide-field imager makes it possible to find and follow approximately 2000 SNe Ia in a year. The 1.8- to 2.0-meter aperture of the mirror, along with high throughput instruments, allow this dataset to extend to redshift $z = 1.7$.

This prearranged observing strategy will provide a uniform, standardized, calibrated dataset for each supernova, allowing for the first time comprehensive comparisons across complete sets of supernovae. The standardized dataset will have the following measurements that will address, and often eliminate, each of the statistical and systematic uncertainties that have been identified or proposed.

- A light curve sampled at frequent, standardized epochs that extends from ~2 restframe days to ~80 restframe days after explosion.

- Multiple color measurements, including optical and near-IR bands, at key epochs on the light curve.

- Spectrum at maximum light, extending from 0.3 μm to 1.7 μm.

- Final reference images and spectra to enable clean subtraction of host galaxy light.

The quality of these measurements is as important as the time and wavelength coverage, so we require:

- Control over signal-to-noise ratio for these photometry and spectroscopy measurements, to permit comparably high statistical significance for supernovae over a wide range of redshifts.

- Control over calibration for these photometry and spectroscopy measurements, with constant monitoring data collected to ensure that cross-instrument and cross-wavelength calibration remains stable over time.

Note that to date not one single SN Ia has ever been observed with this complete set of measurements, either from the ground or in space, and only a handful have a dataset that is comparably thorough. With the observing strategy proposed here, *every one* of $\sim$2000 followed SN Ia will have this complete set of measurements.

In addition to this minimum-required-dataset, a still more extensive set of observations will be performed for a randomly selected subset of SNe Ia (with more at lower redshifts and fewer at higher redshifts). These additional observations will include:

- A time series of spectra, sampled frequently over the entire 80 restframe days of the observed light curve.
- Multiple filter-band light curves. (These are not necessary when the time series of spectra is obtained, since this provides synthetic-filter photometry.)

CONTROL OF STATISTICAL AND SYSTEMATIC UNCERTAINTIES

The satellite instrumentation and observation strategy is designed to provide comprehensive control of the previously identified or proposed sources of uncertainty. The completeness of the resulting dataset will make it possible to monitor the physical properties of each supernova explosion, allowing studies of effects that have *not* been previously identified or proposed.

At present, the identified systematic uncertainty is over half the size of the statistical uncertainty; this would provide the "floor" on the proposed measurement uncertainty, if it were not improved. However, almost every one of the sources of identified systematics is due to limitations of the previous (and even planned NGST baseline SN program) measurements. The dataset described here removes these limitations so that the relevant effects can be measured and the previous systematic uncertainties now become controllable *statistical* uncertainties.

Previously Identified Sources of Systematic Uncertainty

In Table 1, we summarize the identified sources of systematic error, and give the uncertainty that each contributed to previous measurements. With the proposed satellite experiment, each of these effects can either be measured so that it can become part of the statistical error budget, or else bounded (the target overall systematic uncertainty is kept below $\sim$0.02 magnitudes, so that it will contribute comparably to the final statistical uncertainties). The final column of the table summarizes the observations required to reach this target systematic uncertainty.

These previously identified sources of systematic uncertainty are each discussed in more detail in the full proposal [1].

Systematic	Current ground-based δM	SNAP requirement to satisfy $\delta M < 0.02$
Malmquist bias	0.04	Detection of every supernova 3.8 magnitudes below peak in the target redshift range
K-Correction and Cross-Filter Calibration	0.025	Spectral time series of representative SN Ia and cross-wavelength relative flux calibration
Non-SN Ia Contamination	< 0.05	Spectrum for every supernova at maximum covering the rest frame Si II 6250Å feature
Milky Way Galaxy extinction	< 0.04	SDSS & SIRTF observations; SNAP spectra of Galactic subdwarfs
Gravitational lensing by clumped mass	< 0.06	Average out the effect with large statistics with ~ 75 SNe Ia per 0.03 redshift bin. SNAP microlensing measurements.
Extinction by "ordinary" dust outside the Milky Way	0.03	Cross-wavelength calibrated spectra to observe wavelength dependent absorption

TABLE 1. Listed are the main systematic errors in the measurement of the cosmological parameters. Their contribution to magnitude uncertainties in the current analyzed data set is tabulated, along with the observational requirements needed to reduce those uncertainties to $\delta M < 0.02$

Proposed Sources of Systematic Uncertainty

Extinction by Proposed "Gray Dust":

Models of "gray dust" have been proposed to evade detection by the usual measurements of reddening [9]. However, even grey dust cannot remain completely invisible, since it will re-emit absorbed light and contribute to the far-infrared (FIR) background. Current SCUBA observations indicate that FIR emission from galaxies is close enough to account for all the FIR background. Deeper SCUBA and SIRTF observations should tighten the constraints on the amount of gray dust allowed.

Another tell-tale observation will allow us to independently detect and measure gray dust. The physical models so far proposed have dust grains that are large enough that they dim blue and red light equally, however the near-IR light ($\sim$1.2 μm) is less affected. The same technique can therefore be used to measure this dust as would be used to measure the "ordinary" dust, by extending the broad-wavelength measurements into the near-IR. This will measure dimming due to proposed large-grain gray dust out to $z = 0.5$, and this proposed systematic uncertainty, too, can become part of our statistical error budget.

Current space-based observations of existing supernovae are already being used in this way to test if gray dust in a non-accelerating universe can mimic the effects of an accelerating universe at $z = 0.5$. Results show that the observed color excess is too small to be compatible with the 30% opacity of gray dust needed in a $\Lambda = 0$ universe to be consistent with observations. Our proposed satellite measurements would improve greatly on these first results and allow detection and measurement of much smaller gray-dust opacity.

Requirement: Cross-wavelength calibrated spectra, at controlled SN-explosion epochs, that extend to rest-frame 1.2 μm.

In principle, gray dust models can be constructed that would evade these broad-wavelength measurements, either because the "gray dust" does not exist closer than $z = 0.5$ or because the dust grains are even larger than first proposed and thus absorb light equally at 0.4 μm and 1.2 μm. (Such larger grain sizes are strongly disfavored by other astrophysical constraints, however.) Even these more contrived dust models can be measured by the proposed dataset because of its large redshift range: at redshifts beyond $z = 1.4$ models with dust would be distinguished from cosmological models with no dust but with Λ at the 50 standard-deviation level.

Requirement: A redshift distribution that extends to $z \geq 1.5$ for followed SNe Ia.

Proposed Uncorrected Evolution:

Uncorrected "evolution" has also been proposed as a potential source of systematic uncertainty [10]. Supernova behavior may depend on properties of its progenitor star or binary-star system. The distribution of these stellar properties is likely to change over time—"evolve"—in a given galaxy, and over a set of galaxies.

As galaxies age, generation after generation of stars complete their life-cycles, enriching the galactic environment with heavy elements (the abundance of these elements is termed "metallicity"). In a given generation of stars, the more massive ones will complete their life cycles sooner, so the distribution of stellar masses will also change over time. Such statistical changes in the galactic environments are expected to affect the typical properties of supernova-progenitor stars, and hence the details of the triggering and evolution of the supernova explosions. Even the SNe Ia might be expected to show some differences that reflect the galactic environment in which their progenitor stars exploded, even though they are triggered under very similar physical conditions every time (as mass is slowly added to a white dwarf star until it approaches the Chandrasekhar limit).

Evidence for such galactic-environment driven differences among SNe Ia has in fact already been seen among nearby, low-redshift supernovae [2]. The range of intrinsic SN Ia luminosities seen in spiral galaxies differs from that seen in elliptical galaxies. So far, it appears that the differences that have been identified are well calibrated by the SN Ia light curve width-luminosity relation. The standard supernova analyses thus already are correcting for a luminosity effect due to galactic-environment-distribution evolution. There are likely to be additional, more subtle effects of changes in the galactic environment and shifts in the progenitor star population, although it is not clear that these effects would change the peak luminosity of the SNe Ia. The proposed satellite experiment is designed to provide sufficient data to measure these second-order effects, which might be collectively called "proposed uncorrected evolution."

In this discussion it is important to recognize that each individual galaxy begins its life at a different time since the Big Bang, at a different absolute time. Even today, there are newly formed, "young," first-generation galaxies present that have not yet gone through the life cycles of their high-mass stars, nor yet produced significant heavy element abundance. Thus at any given redshift there will be a large range of galactic environments present and the supernovae will correspondingly exhibit a large range of progenitor-star ages and heavy-element abundances. (This is why we can currently observe and correct an evolutionary range of SNe Ia using only low-redshift, nearby SNe Ia.) It is only the relative distribution of these environment ages that will change with universal clock time. By identifying matching sets of supernova that come from essentially the same progenitor stars in the same galactic environments, but across a wide variety of redshifts, we can then perform the cosmological measurements using SNe Ia in the same evolutionary state. This only requires that the SN Ia sample sizes are sufficiently large and varied at each redshift that we can find matching examples in sufficient quantities.

We have identified a series of key supernova features that respond to differences in the underlying physics of the supernova (see the full proposal for a complete list and detailed description [1]). By measuring all of these features for each supernova we can tightly constrain the physical conditions of the explosion, making it possible to recognize sets of supernovae with matching initial conditions. The current theoretical models of SN Ia explosions are not sufficiently complete to predict the precise luminosity of each supernova, but they are able to give the rough relationships between changes in the physical conditions of the supernovae (such as opacity, metallicity, fused nickel mass, and nickel distribution) and changes in their peak luminosities. We can therefore give the approximate accuracy needed for the measurement of each feature to ensure that the physical condition of each set of supernovae is well enough determined so that the range of luminosities for those supernovae is well below the systematic uncertainty bound ($\sim 2\%$ in total).

In addition to these features of the supernovae themselves, we will also study the host galaxy of the supernova. We can measure the host galaxy luminosity, colors, morphology, type, and the location of the supernova within the galaxy, even at redshifts $z \sim 1.7$. These observations are not possible from the ground.

Requirements	Addresses and Resolves
Detection of every supernova 3.8 magnitudes below peak for $z \leq 1.5$	• Rise time measurement • Eliminates Malmquist Bias
SNe Ia at $0.3 \leq z \leq 1.7$	• Statistics and lever-arm for the precision measurement of Ω_M, Ω_Λ • Detection of Gray Dust • Detection of SN Ia evolution
~ 75 SNe Ia per 0.03 redshift bin	• Statistics and lever-arm for the precision measurement of w • The effect of gravitational lensing by clumped mass is averaged out
Well sampled light-curves between ~ 2 restframe days to ~ 80 restframe days after explosion	• Determination of the peak magnitude of each SN Ia • Determination of the light-curve shape of each SN Ia • Detection of SN Ia evolution
Multiple IR and optical color measurements at key epochs	• Determination of extinction for each SN Ia • Confirmation of the light-curve shape of each SN Ia
Spectrum for every supernova at maximum covering the rest frame Si II 6250Å feature and that extend from rest frame UV to 1.2μm	• Eliminates non-SN Ia contamination • Measures extinction due to "ordinary" dust outside the Milky Way • Spectral feature – peak magnitude relation
Spectral time series of representative SN Ia with cross-wavelength relative flux calibration	• Determine K-corrections • Allow cross-filter comparisons • Detection of Gray Dust • Detection of SN Ia evolution

TABLE 2. Observational requirements to ensure various statistical and systematic errors each contribute uncertainties of $\delta M < 0.02$. The particular sources of error that each requirement addresses are also listed.

WHY A NEW SATELLITE?: DESIGN REQUIREMENTS AND GROUND- OR SPACE-BASED ALTERNATIVES

The science goals that we have described drive the design requirements of this experiment. The target statistical uncertainties are closely matched to the target systematic uncertainties, so that the numbers of supernovae, their redshift range, and the quality and comprehensive nature of the dataset of measurements for each supernova all together can achieve the stated cosmological measurements.

In particular, the mirror aperture is about as small as it can be before spectroscopy at the requisite resolution is no longer zodiacal-light-noise limited. A smaller mirror design would quickly degrade the achievable signal-to-noise of the spectroscopy measurements, and drastically reduce the number of supernovae followed. The field of view for the optical imager has been optimized to obtain the follow-up photometry of multiple supernovae simultaneously; a smaller field would require multiple pointings of the telescope and again would greatly reduce the number of supernovae that could be followed. The three-arm spectrograph covers precisely the wavelength range necessary to capture, over the entire target redshift range, the Si II 6250 Å feature that both identifies the SNe Ia and provides a key measurement of the explosion physics to identify each supernova's evolutionary state. In general, more than one critical design requirement has driven each of these instrument choices; for example, the wavelength range of the spectrograph also is required to measure the effects of any "gray dust" on the supernova magnitudes.

Although calculations based on Poisson noise would indicate that ground-based telescopes of sufficiently-large aperture can compete with a smaller space-based telescope, it ignores the difficulty inherent in obtaining accurate photometry of faint sources overwhelmed by a foreground $\sim 10^5 \times$ brighter (typical for red/NIR observations for SNe of $z > 0.7$). Achieving photometry with an accuracy of 2% on such a source requires the foreground to be uniform and stable to 2×10^{-7} on small scales. This level of accuracy cannot be achieved from the ground. Furthermore, H_2O absorption will decimate the Si 6250Å and other other key spectral features.

Given this inherent limitation of ground-based observations, a comparison with plausible ground- and space-based alternatives makes it particularly clear why this satellite design is required to achieve the science. Simply finding the supernovae near their explosion date from the ground is the first challenge, even for an entirely dedicated 8-meter telescope with a special-purpose 9-square-degree imager. To detect SNe Ia within ~ 2 restframe days of explosion (as required for the risetime measurement) the photometry must extend to 3.8 magnitudes below peak with a signal-to-noise of 10. From the ground, with its bright sky and atmospheric seeing, this limits the search to redshifts less than $z = 0.6$—and fewer than 300 SNe Ia per year would be measured. If one begins to degrade the experiment by removing this risetime measurement's control on systematics, the next key requirement is a measurement of the plateau phase of the light-curve, approximately 2.8 magnitudes below peak, which would limit ground-based searches to redshifts less than $z =$

0.7. Finally, if we give up this plateau-measurement control on systematics, the fundamental measurement requirement is 2% photometry at peak (to determine the color and thus reddening) and 15 days after peak (to determine light-curve width). From the ground, even this minimal dataset is only obtainable to redshifts less than $z = 0.75$. (See Ref. [1] for a more complete discussion of the comparison with ground-based alternatives.)

Using the existing Hubble Space Telescope or even the planned Next Generation Space Telescope (NGST) does not improve the ability to discover these supernovae, since neither telescope has a wide-field camera. With the 8-meter NGST's 16-square-arcminute field of view, it would require tens of years of full-time searching to obtain a comparable sample of SNe Ia in the target redshift range. The NGST does have a quite useful supernova program planned, but all at higher redshifts than this project, and without the extensive controls on systematic uncertainties that we require. This NGST program is aimed at different science, since it is not possible to study the "dark energy" at redshifts much beyond $z \sim 1.2$, when the universe had smaller scale and the matter-density dominated.

One might wonder if the NGST could be used simply to follow up the spectroscopy of the supernovae discovered with this telescope. This would be possible, but it is a rather wasteful use of the 8-meter's capabilities; most of time for over half a year would be spent simply slewing the NGST from supernova to supernova, with the shutter open for only a small fraction of the time. (A coordinated wide-field ground-based search with NGST follow-up would suffer this same problem and further add the disadvantages of discovering the higher-redshift supernova late after explosion.)

OVERVIEW OF FEASIBILITY

The essential elements of the project's feasibility have already been studied. We were able to establish many of the baseline design feasibility issues by reference to other satellite missions that have successfully flown, or are currently being built.

- We made a top-down cost estimate based on other similar satellite designs and costs.

- We performed a study of orbit options and found several options that allowed a workable combination of launch vehicle, mass-to-orbit, thermal control, cosmic-ray load, continuous observing duty cycle, telemetry rates, and power budget.

- We have baselined a three-mirror anastigmat telescope design which provides a diffraction limited wide field of view with minimum obscuration. We are also likely to adopt a flight-proven lightweight glass mirror technology.

- Pointing requirements can be met two ways: (1) using feedback from the focal plane detectors to the spacecraft attitude control system, or (2) using a fast-steering mirror. Both are legacy technologies developed for earth-observing satellites. This image-stabilizing option avoids the need to maintain a precisely stable spacecraft.

- At University of California at Berkeley and at Lawrence Berkeley National Laboratory's microfabrication facility, we have built and tested high-resistivity CCDs that provide greater than 90% quantum efficiency up to 1 μm, and are at least ten times more radiation hard than conventional CCDs. This fabrication process has now been transferred to a high-volume commercial vendor, and two fabrication runs are currently in their final stages of processing.

- For some years, much larger CCD and silicon strip arrays have been routinely built by the high energy physics community and operated in comparably inaccessible locations, where they are exposed to high radiation levels.

- We have conducted extensive simulation and modeling of the science reach and performance of various observing strategies and instrument trade-offs.

SUMMARY OF OTHER MAJOR SCIENCE

The dataset of images and spectra obtained with this wide-field imager and three-arm spectrograph can address other important science goals with very little additional effort in data collection or in the instrument specifications. Although these science goals will not be discussed in detail here, it is important in particular to note that we can obtain complementary measurements of the cosmological parameters with completely independent measurement methods.

Weak lensing.

Because the observation strategy observes the same patches of sky repeatedly over a year of supernova observations, a very deep, high-resolution image can be added together from thousands of images taken at every orientation of the spacecraft. This is an ideal way to look for weak-lensing elongations of distant galaxies, since the optical distortions of the image will be small and well characterized. Such images of several dozen square-degree fields can constrain the cosmological parameters in a manner complementary to the SN Ia measurements, with different systematics.

Type II supernovae.

As we discover and follow the SNe Ia, we will also discover and have the option of following SNe II. While these supernovae are not of predictable luminosity, they are close enough matches to a black body that their luminosities can be determined from the size and temperature of their photospheres, along with a fit to any spectral deviations from black body. (Since our experiment provides a very tight constraint on the date of explosion and the velocity of the expanding gas, the size of the photosphere will be easy to determine.) Most SNe II are about six times fainter than the prototypical SNe Ia, so most will not be studied with early detections beyond redshifts $z \sim 0.5$. However since SNe II are much more frequent than SNe Ia, we can afford to study the brightest few percent and this will extend the SN II study beyond $z = 1$. The sources of systematic uncertainty for these SNe II measurements would generally be different from the SNe Ia systematic uncertainties.

There is also important science to be gained from this project that is not aimed specifically at the cosmological models. It is clear, for example, that the final set of very deep, wide field images would become a resource for all of astrophysics, as the Hubble Deep Fields have been.

CONCLUSIONS

The fundamental questions and surprising discoveries of recent years make this a fascinating new era of empirical cosmology. This proposed satellite project presents a unique opportunity to extend this exciting work and advance our understanding of the universe. The origin and destiny of the universe have intrigued humanity for at least as long as there are written records. We live at a time when we can begin to find answers.

REFERENCES

1. This article is drawn from the full SNAP proposal available at the SNAP World Wide Web Page: *http://snap.LBL.gov*
2. M. Hamuy et al., AJ **112**, 2391 (1996).
3. A. G. Riess, W. H. Press, and R. P. Kirshner, ApJ **473**, 88 (1996).
4. S. Perlmutter et al., ApJ **517**, 565 (1999).
5. A. Riess et al., AJ **116**, 1009 (1998).
6. N. A. Bahcall, J. P. Ostriker, S. Perlmutter, and P. J. Steinhardt, Science **284**, 1481+ (1999).
7. A. E. Lange et al., astro-ph/0005004 (2000).
8. A. Balbi et al., astro-ph/0005124 (2000).
9. A. N. Aguirre and Z. Haiman, astro-ph **9907039** (1999).
10. H. Umeda et al., ApJ **513**, 861 (1999).

CP, *CPT*, AND *B* PHYSICS

CP Violation – A Brief Review[1]

Jonathan L. Rosner

Enrico Fermi Institute and Department of Physics
University of Chicago, Chicago, IL 60637 USA

Abstract. Some past, present, and future aspects of CP violation are reviewed. The discrete symmetries C, P, and T are introduced with an example drawn from Maxwell's Equations. The history of the discovery of CP violation in the kaon system is described briefly, and brought up-to-date with a review of recent results on kaon decays. The candidate theory of CP violation, based on phases in the Cabibbo-Kobayashi-Maskawa (CKM) matrix, will be tested by studies of B mesons, both in decays to CP eigenstates and in "direct" decays; we will soon learn a great deal more about whether the CKM picture is self-consistent. Future measurements are noted and some brief remarks are made about the "other" manifestation of CP violation, the baryon asymmetry of the Universe.

I INTRODUCTION

Fundamental discrete symmetries have provided both guidance and puzzles in our evolving understanding of elementary particle interactions. The discrete symmetries C (charge inversion), P (parity, or space reflection), and T (time reversal) are preserved by strong and electromagnetic processes, but violated by weak decays. For a brief period of several years, it was thought that the products CP and T were preserved by all processes, but that belief was shattered with the discovery of CP violation in neutral kaon decays in 1964 [1]. The product CPT seems to be preserved, as is expected in local Lorentz-invariant quantum field theories [2].

Since 1973 we have had a candidate theory of CP violation [3], based on phases in the coupling constants describing the weak charge-changing transitions of quarks. These couplings are described by the unitary 3×3 *Cabibbo-Kobayashi-Maskawa* (CKM) [3,4] matrix. This theory has survived a qualitative test with the establishment of direct CP violation in neutral kaon decays [5,6]. It is well on its way to being tested in a wealth of B decay processes. Will these tests be passed? What are the implications in either case? What will we learn about the "other" manifestation of CP asymmetry in nature, the baryon asymmetry of the Universe? This brief review is devoted to these questions.

---

[1] This work was supported in part by the United States Department of Energy under Grant No. DE FG02 90ER40560

TABLE 1. Behavior of Maxwell's equations under discrete symmetries.

Equation	P	T	C	CPT
$\nabla \cdot \mathbf{E} = 4\pi\rho$	+	+	−	−
$\nabla \cdot \mathbf{B} = 0$	−	−	−	−
$\nabla \times \mathbf{B} - \frac{1}{c}\frac{\partial \mathbf{E}}{\partial t} = \frac{4\pi}{c}\mathbf{j}$	−	−	−	−
$\nabla \times \mathbf{E} + \frac{1}{c}\frac{\partial \mathbf{B}}{\partial t} = 0$	+	+	−	−

In Section II we introduce the discrete symmetries P, T, and C by the example of Maxwell's equations. Section III is devoted to the history and present status of CP violation and related phenomena in kaon decays, while Section IV deals with results and prospects for B mesons. Some future measurements are discussed in Section V. The baryon number of the Universe and its relation to CP violation are treated briefly in Section VI, while Section VII concludes.

II DISCRETE SYMMETRIES

Maxwell's equations in vacuum provide a convenient framework for illustrating the action of discrete symmetries, since each term in each equation must transform similarly.

Under P, we have $\mathbf{E}(\mathbf{x},t) \to -\mathbf{E}(-\mathbf{x},t)$, $\mathbf{B}(\mathbf{x},t) \to \mathbf{B}(-\mathbf{x},t)$, $\nabla \to -\nabla$, $\mathbf{j}(\mathbf{x},t) \to -\mathbf{j}(-\mathbf{x},t)$, i.e., electric fields change in sign while magnetic fields do not, and currents change in direction. Under time reversal, $\mathbf{E}(\mathbf{x},t) \to \mathbf{E}(\mathbf{x},-t)$, $\mathbf{B}(\mathbf{x},t) \to -\mathbf{B}(\mathbf{x},-t)$, $\partial/\partial t \to -\partial/\partial t$, $\mathbf{j}(\mathbf{x},t) \to -\mathbf{j}(\mathbf{x},-t)$, i.e., magnetic fields change in sign while electric fields do not, since directions of currents are reversed. Under C, $\mathbf{E}(\mathbf{x},t) \to -\mathbf{E}(\mathbf{x},t)$, $\mathbf{B}(\mathbf{x},t) \to -\mathbf{B}(\mathbf{x},t)$, $\rho(\mathbf{x},t) \to -\rho(\mathbf{x},t)$, $\mathbf{j}(\mathbf{x},t) \to -\mathbf{j}(\mathbf{x},t)$, i.e., both electric and magnetic fields change sign, since their sources ρ and $\mathbf{j}$ change sign. Finally, under CPT, space and time are inverted but electric and magnetic fields retain their signs: $\mathbf{E}(\mathbf{x},t) \to \mathbf{E}(-\mathbf{x},-t)$, $\mathbf{B}(\mathbf{x},t) = \mathbf{B}(-\mathbf{x},-t)$.

The behavior of the Maxwell equations under P, T, C, and CPT is summarized in Table 1. Each term behaves as shown. It is interesting that a fundamental term in the Lagrangian behaving as $\mathbf{E} \cdot \mathbf{B}$, while Lorentz covariant, violates P and T. The strong suppression of such a term (as evidenced by the small value of the neutron electric dipole moment) is known as the *strong CP problem* [7], and, although of fundamental importance, will not be discussed further here.

III CP SYMMETRY FOR KAONS

A $K \to \pi\pi$ decays

While some neutral particles (such as γ, Z^0, and π^0) are equal to their antiparticles, others (such as the neutron) are not. The K^0, discovered in cosmic radiation in the late 1940's [8], is one such particle. It is characterized by an additive quantum number $S = 1$, *strangeness*, introduced [9] in order to explain its strong production (which conserves strangeness) and weak decay (which does not). The antiparticle of K^0, the $\overline{K^0}$, has $S = -1$. Since strangeness is violated in decays, one must appeal to discrete symmetries to describe the linear combinations of K^0 and $\overline{K^0}$ corresponding to states of definite mass and lifetime. These states are

$$K_1 = \frac{K^0 + \overline{K^0}}{\sqrt{2}} \quad , \quad K_2 = \frac{K^0 - \overline{K^0}}{\sqrt{2}} \quad . \tag{1}$$

The K_1 is permitted to decay to $\pi\pi$ and thus should be short-lived, while the K_2 is forbidden to decay to $\pi\pi$, must instead decay to 3π, $\pi\ell\nu_\ell$, etc., and thus will be longer-lived. Indeed, the short-lived neutral kaon ($\sim K_1$) lives for only 0.089 ns, while the long-lived neutral kaon ($\sim K_2$) lives for 52 ns, nearly a factor of 600 longer.

The original argument by Gell-Mann and Pais [10], based in 1955 on C and P conservation, was recast in 1957 in terms of the product CP [11], to correspond to the newly formulated CP-invariant theory of the weak interactions. The K^0 and $\overline{K^0}$ have spin zero. A spin-zero final state of $\pi\pi$ has CP eigenvalue equal to $+1$. Thus, if CP is conserved, it is the CP-even linear combination of K^0 and $\overline{K^0}$ which decays to $\pi\pi$. With a phase convention such that $CP|K^0\rangle = |\overline{K^0}\rangle$, this is just the combination K_1. The Gell-Mann–Pais proposal was soon confirmed [12] by the discovery of the predicted long-lived particle corresponding to K_2.

Similar behavior is encountered in many cases of degenerate systems, such as two coupled pendula [13] or a drum-head in its first excited state. In the latter case, the drum has two degenerate modes, each with one nodal line corresponding to a diameter, which will be orthogonal to one another if the corresponding nodal lines are perpendicular to each other. Consider two equally valid bases:

- (B1) Diagonal nodal lines point to the upper right (R) and the upper left (L).

- (B2) The nodal lines are horizontal (H) and vertical (V).

We can draw the analogy $R \leftrightarrow K^0$, $L \leftrightarrow \overline{K^0}$. Suppose, now, that a fly alights on the bottom edge of the drum head, such that it sits on the nodal line of the V mode. Then the modes V and H are split from one another. The mode $H = (R + L)/\sqrt{2}$ which couples to the fly will shift in mass and lifetime. .It is analogous to K_1 and the fly is analogous to the $\pi\pi$ system. The mode V is unaffected by the fly. It is analogous to K_2.

In 1964, Christenson, Cronin, Fitch, and Turlay [1], using a spark chamber exposed to a beam of long-lived neutral kaons, found that these particles indeed *did* decay to $\pi\pi$. For many years this phenomenon could be described in terms of a single parameter ϵ, such that the states of definite mass and lifetime become

$$K_1 \to K_S \text{ ("short")} \simeq K_1 + \epsilon K_2 \,, \quad K_2 \to K_L \text{ ("long")} \simeq K_2 + \epsilon K_1 \,, \qquad (2)$$

with $|\epsilon| \simeq 2 \times 10^{-3}$, and $\text{Arg}(\epsilon) \simeq \pi/4$. Confirmation of this description was provided by the rate asymmetry in the decays $K_L \to \pi^{\pm} \ell^{\mp} \nu_\ell$, which measures Re ϵ. But what is the source of ϵ?

One possibility was suggested almost immediately by Wolfenstein [14]: A new "superweak" $|\Delta S = 2|$ interaction could mix $K^0 = d\bar{s}$ and $\overline{K^0} = s\bar{d}$ (where d and s denote quarks) without any other observable consequences. This theory would imply, for example, that no difference in the ratio of CP-violating and CP-conserving amplitudes would arise when comparing $\pi^+\pi^-$ and $\pi^0\pi^0$ final states.

A new opportunity for generating not only ϵ but other CP-violating effects as well arises when there are at least three quark families, as first proposed by Kobayashi and Maskawa [3]. Loop diagrams inducing the transition $d\bar{s} \leftrightarrow s\bar{d}$ involving internal lines of W^+W^- and u,c,t quarks and antiquarks can lead to $\epsilon \neq 0$ when the coupling constants are complex. With three quark families, one cannot redefine phases of quarks so that all the couplings are real. Some other consequences of the Kobayashi-Maskawa theory will be mentioned presently.

The time-dependence of the two-component K^0 and $\overline{K^0}$ system is governed by a 2×2 *mass matrix* $\mathcal{M}$ (for reviews see [15]):

$$i\frac{\partial}{\partial t} \begin{bmatrix} K^0 \\ \overline{K^0} \end{bmatrix} = \mathcal{M} \begin{bmatrix} K^0 \\ \overline{K^0} \end{bmatrix} \,, \qquad (3)$$

where $\mathcal{M} = M - i\Gamma/2$, and M and Γ are Hermitian matrices. The eigenstates are, approximately,

$$K_S \simeq K_1 + \epsilon K_2 \,, \quad K_L \simeq K_2 + \epsilon K_1 \,, \qquad (4)$$

corresponding to the eigenvalues $\mu_{S,L} = m_{S,L} - i\gamma_{S,L}/2$, with

$$\epsilon \simeq \frac{\text{Im}(\Gamma_{12}/2) + i\,\text{Im}\,M_{12}}{\mu_S - \mu_L} \,. \qquad (5)$$

Using both data and the magnitude of CKM matrix elements one can show [15] that the second term dominates. Since the mass difference $m_L - m_S$ and width difference $\gamma_S - \gamma_L$ are nearly equal, the phase of $\mu_L - \mu_S$ is about $\pi/4$, so that the phase of ϵ is also $\pi/4$ (mod π).

It is easy to emulate the *CP-conserving* neutral kaon system in table-top demonstrations of systems with two degenerate states, such as the pair of coupled pendula mentioned above [13]. The demonstration of CP violation is harder, requiring systems that emulate $\text{Im}(M_{12}) \neq 0$ or $\text{Im}(\Gamma_{12}) \neq 0$. One can couple two identical resonant circuits "directionally" to each other so that the energy fed from circuit 1 to circuit 2 differs from that fed in the reverse direction [16]. Devices with this property utilize Faraday rotation of the plane of polarization of radio-frequency waves. More recently, it was realized [17] that this asymmetric coupling is inherent in the equations of motion of a spherical (or "conical") pendulum in a rotating coordinate system, giving rise to the precession of the plane of oscillation of the Foucault pendulum. In either case the analogy actually deepens the mystery of CP violation, since the CP-violating effect is imposed, so to speak, "from the outside," using a magnetic field in the case of directional couplers or a rotating coordinate frame in the case of the Foucault pendulum.

To return to the CKM matrix, we have the following parameterization suggested by Wolfenstein [18]:

$$V \equiv \begin{bmatrix} V_{ud} & V_{us} & V_{ub} \\ V_{cd} & V_{cs} & V_{cb} \\ V_{td} & V_{ts} & V_{tb} \end{bmatrix} = \begin{bmatrix} 1 - \frac{\lambda^2}{2} & \lambda & A\lambda^3(\rho - i\eta) \\ -\lambda & 1 - \frac{\lambda^2}{2} & A\lambda^2 \\ A\lambda^3(1 - \rho - i\eta) & -A\lambda^2 & 1 \end{bmatrix}, \quad (6)$$

where $\lambda = \sin\theta_C \simeq 0.22$ describes strange particle decays. Here θ_C is the Gell-Mann–Lévy–Cabibbo [4,19] angle, originally introduced to preserve the universal strength of the hadronic weak current. The unitarity of the CKM matrix, $V^\dagger = V^{-1}$, is the modern way of implementing this requirement.

We learn $|V_{cb}| = A\lambda^2 \simeq 0.039 \pm 0.003$ from the dominant decays of b quarks, which are to charmed quarks [20]. (We have expanded errors somewhat in comparison with those quoted in some reviews [21]. The dominant source of error in many cases is theoretical.) Similarly, charmless b decays give $|V_{ub}/V_{cb}| = 0.090 \pm 0.025 = \lambda(\rho^2 + \eta^2)^{1/2}$, leading to a constraint on $\rho^2 + \eta^2$.

As a result of the unitarity of the CKM matrix, the quantities $V^*_{ub}/A\lambda^3 = \rho + i\eta$, $V_{td}/A\lambda^3 = 1 - \rho - i\eta$, and 1 form a triangle in the (ρ, η) plane (Fig. 1). The angles opposite these sides are, respectively, $\beta = -\text{Arg}(V_{td})$, $\gamma = \text{Arg}(V^*_{ub})$, and $\alpha = \pi - \beta - \gamma$. We still do not have satisfactory limits on the angle γ (equivalently, on the magnitude of the side V_{td}) of this "unitarity triangle." Further information comes from the following constraints (see [22] for more details):

1. The magnitude of ϵ constrains mainly the imaginary part of V^2_{td}, which is proportional to $\eta(1 - \rho)$, since the top quark dominates the loop diagram giving rise to K^0–$\overline{K^0}$ mixing. A correction due to charmed quarks changes the 1 to 1.44, with the result $\eta(1.44 - \rho) = 0.51 \pm 0.18$.

2. We have taken the amplitude for mixing of the neutral B^0 meson with its antiparticle $\overline{B^0}$ to be $\Delta m_d = 0.473 \pm 0.016$ ps^{-1} [23]. The subscript d denotes the light quark in the B^0. Taking the matrix element of the four-quark operator inducing the relevant $\bar{b}d \leftrightarrow \bar{d}b$ transition to be $f_B\sqrt{B_B} = 200 \pm 40$ MeV, we find a constraint on $|V_{td}|$ which amounts to $|1 - \rho - i\eta| = 1.01 \pm 0.21$.

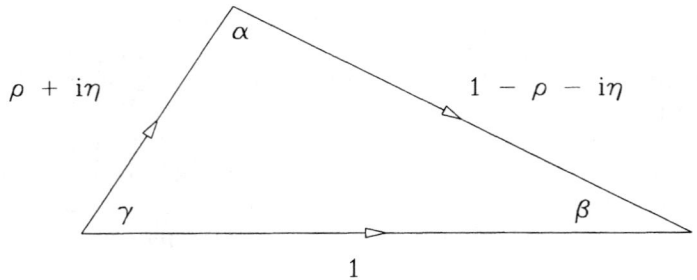

FIGURE 1. Unitarity triangle for CKM elements. Here $\rho + i\eta = V_{ub}^*/A\lambda^3$; $1 - \rho - i\eta = V_{td}/A\lambda^3$.

TABLE 2. Ranges of angles in the unitarity triangle.

Angle	Expression		Degrees	ρ	η
α	$\pi - \beta - \gamma$	Min	72	−0.01	0.30
		Max	113	0.25	0.27
β	$\tan^{-1}[\rho/(1-\eta)]$	Min	17	−0.01	0.30
		Max	31	0.29	0.43
γ	$\tan^{-1}(\eta/\rho)$	Min	48	0.25	0.27
		Max	92	−0.01	0.30

3. We have used the following lower limit for mixing of the strange b meson $B_s = \bar{b}s$ with its antiparticle: $\Delta m_s > 14.3$ ps^{-1} (95% c.l.) [23,24]. Since the relevant CKM elements (including $|V_{ts}| = A\lambda^2$) are fairly well known, this result serves mainly to constrain the combination of hadronic parameters $f_{B_s}\sqrt{B_{B_s}}$ and hence, through the assumption $[f_{B_s}\sqrt{B_{B_s}}]/[f_B\sqrt{B_B}] < 1.25$ [25], yields the bound $|V_{ts}/V_{td}| > 4.3$ or $|1 - \rho - i\eta| < 1.05$.

The resulting limits on (ρ, η) are a roughly rectangular region bounded on the left by $|1 - \rho - i\eta| < 1.05$, on the top and bottom by $0.3 < (\rho^2 + \eta^2)^{1/2} < 0.52$, and on the right by $|1 - \rho - i\eta| > 0.8$. Only a small region is excluded by the bound arising from the parameter ϵ: $\eta(1.44 - \rho) > 0.33$. Even without this bound, the case of real CKM matrix elements ($\eta = 0$), i.e., a superweak origin for ϵ, is disfavored. The boundaries of this region give rise to the minimum and maximum values of α, β, γ shown in Table 2. These bounds imply

$$-0.71 < \sin 2\alpha < 0.59, \quad 0.59 < \sin 2\beta < 0.89, \quad 0.54 < \sin^2 \gamma < 1 \qquad (7)$$

for quantities which are measurable in B decays (see below). The allowed values of (ρ, η) are $\simeq (0.14 \pm 0.15, 0.38 \pm 0.13)$.

The Kobayashi-Maskawa theory predicts small differences in CP-violating decays to pairs of charged and neutral pions. These arise in the following way.

TABLE 3. Experimental values for Re(ϵ'/ϵ).

Experiment	Reference	Value ($\times 10^{-4}$)	$\Delta\chi^2$
Fermilab E731	[29]	7.4 ± 5.9	3.97
CERN NA31	[30]	23.0 ± 6.5	0.35
Fermilab E832	[5]	28.0 ± 4.1	4.65
CERN NA48	[6]	14.0 ± 4.3	1.44
Average		19.2 ± 4.6	$\sum = 10.4$

1. "Tree" amplitudes are governed by $\bar{s} \to \bar{u}ud$. Since this subprocess has three nonstrange quarks in the final state, it contributes to both $\Delta I = 1/2$ and $\Delta I = 3/2$ transitions, and hence to both $I_{\pi\pi} = 0$ and $I_{\pi\pi} = 2$ final states. The corresponding CKM matrix elements are real, so these amplitudes do not have a weak phase.

2. "Penguin" amplitudes involve a transition $\bar{s} \to \bar{d}$ with internal W and u, c, t lines and emission or absorption of a gluon. The subprocess has only one nonstrange quark in the final state so it contributes only to $\Delta I = 1/2$ transitions and hence only to the $I_{\pi\pi} = 0$ final state. Because of the presence of all three $Q = 2/3$ quarks in internal lines, these amplitudes have a weak phase.

As a consequence of the different isospin structure and weak phases of the tree and penguin amplitudes, the $I_{\pi\pi} = 0$ and $I_{\pi\pi} = 2$ amplitudes thus acquire different weak phases, leading to a small difference from unity of the ratio

$$R \equiv \frac{\Gamma(K_L \to \pi^+\pi^-)/\Gamma(K_S \to \pi^+\pi^-)}{\Gamma(K_L \to \pi^0\pi^0)/\Gamma(K_S \to \pi^0\pi^0)} = 1 + 6\,\text{Re}\frac{\epsilon'}{\epsilon} \quad , \tag{8}$$

where ϵ' is related to the imaginary part of the ratio of the $I_{\pi\pi} = 2$ and $I_{\pi\pi} = 0$ amplitudes. The ratio ϵ'/ϵ acquires an important term proportional to the CKM parameter η from the penguin contribution. This term is partially canceled by an "electroweak penguin" in which the gluon mentioned above is replaced by a virtual photon or Z, whose isospin-dependent couplings to quarks induce $\Delta I = 3/2$ contributions. ϵ'/ϵ is expected to be nearly real. Its magnitude was estimated by one group [26] to be a few parts in 10^4, with a broad and somewhat asymmetric probability distribution extending from slightly below zero to above 2×10^{-3}. Some other estimates, discussed in Refs. [27], permit higher values.

The most recent experiments on Re(ϵ'/ϵ) are summarized in Table 3. A scale factor [28] of 1.86 is included in the error of the average to account for the large spread in quoted results. The value of ϵ'/ϵ is non-zero, with a magnitude in the ballpark of estimates based on the Kobayashi-Maskawa theory. The fact that it is larger than some theoretical estimates is not a serious problem, given that we still cannot account reliably for the large enhancement of $\Delta I = 1/2$ amplitudes with respect to $\Delta I = 3/2$ amplitudes in *CP-conserving* $K \to \pi\pi$ decays.

B Other rare kaon decays

A CP- or T-violating angular asymmetry in $K_L \to \pi^+\pi^-e^+e^-$ has recently been reported [31,32]. With a final state consisting of four distinct particles, using the three independent final c.m. momenta, one can construct a T-odd observable whose presence is signaled by a characteristic distribution in the angle ϕ between the $\pi^+\pi^-$ and e^+e^- planes.

The asymmetry in $\sin\phi\cos\phi$ reported in Ref. [31] is $(13.6 \pm 2.5 \pm 1.2)\%$. It arises from interference between two processes. (1) The K_L decays to $\pi^+\pi^-$ with an amplitude ϵ. This process is CP-violating. One of the pions then radiates a virtual photon which internally converts to e^+e^-. (2) The CP-odd state K_2 can decay directly to $\pi^+\pi^-\gamma$ via a weak magnetic dipole transition. This process is CP-conserving.

The decay $K_L \to \mu^+\mu^-\gamma$ has recently been studied with sufficiently high statistics to permit a greatly improved measurement of the virtual-photon form factor in $K_L \to \gamma^*\gamma$ [33]. This measurement is useful in estimating the long-distance contribution to the real part of the amplitude in $K_L \to \gamma^{(*)}\gamma^{(*)} \to \mu^+\mu^-$, which in turn allows one to limit the short-distance contribution to $K_L \to \mu^+\mu^-$. Since this contribution involves loops with virtual W's and u, c, t quarks, useful bounds on CKM matrix elements can be placed. Preliminary results [33] indicate $\rho > -0.2$, the best limit so far from any process involving kaons.

Several neutral-current processes involving $K \to \pi +$ (lepton pair) can shed further light on the Kobayashi-Maskawa theory of CP violation [34].

1. The decay $K^+ \to \pi^+\nu\bar{\nu}$ is sensitive primarily to $|V_{td}|$, with a small charm correction, and so constrains the combination $|1.4 - \rho - i\eta|$. The predicted branching ratio is roughly

$$\mathcal{B}(K^+ \to \pi^+\nu\bar{\nu}) \simeq 10^{-10} \left|\frac{1.4 - \rho - i\eta}{1.4}\right|^2 , \qquad (9)$$

For $0 \le \rho \le 0.3$ one then predicts (see [34]) $\mathcal{B}(K^+ \to \pi^+\nu\bar{\nu}) = (0.8 \pm 0.2) \times 10^{-10}$, with additional uncertainties associated with the charmed quark mass and the magnitude of V_{cb}. A measurement of this branching ratio with an accuracy of 10% is of high priority in constraining (ρ, η) further.

The Brookhaven E787 Collaboration has reported one event with negligible background [35], corresponding to

$$\mathcal{B}(K^+ \to \pi^+\nu\bar{\nu}) = (1.5^{+3.4}_{-1.2}) \times 10^{-10} . \qquad (10)$$

More data are expected from the final stages of analysis of this experiment, as well as from a future version (Brookhaven E949) with improved sensitivity.

2. The decays $K_L \to \pi^0 \ell^+ \ell^-$ are expected to be dominated by CP-violating contributions, both indirect ($\sim \epsilon$) and direct. There is also a CP-conserving "contaminant" from the intermediate state $K_L \to \pi^0 \gamma\gamma$. The direct contribution probes the CKM parameter η. It is expected to be comparable in magnitude to the indirect contribution, and to have a phase of about $\pi/4$ with respect to it. Each contribution (including the CP-conserving one) is expected to correspond to a $\pi^0 e^+ e^-$ branching ratio of a few parts in 10^{12}. However, the decay $K_L \to \pi^0 e^+ e^-$ may be limited by backgrounds in the $\gamma\gamma e^+ e^-$ final state associated with radiation of a photon in $K_L \to \gamma e^+ e^-$ from one of the leptons [36]. Present experimental upper limits (90% c.l.) [37] are

$$\mathcal{B}(K_L \to \pi^0 e^+ e^-) < 5.64 \times 10^{-10}, \quad \mathcal{B}(K_L \to \pi^0 \mu^+ \mu^-) < 3.4 \times 10^{-10}, \quad (11)$$

still significantly above theoretical expectations.

3. The decay $K_L \to \pi^0 \nu \bar{\nu}$ is expected to be due entirely to CP violation, and provides a clean probe of η. Its branching ratio, proportional to $A^4 \eta^2$, is expected to be about 3×10^{-11}. The best current experimental upper limit (90% c.l.) for this process [38] is $\mathcal{B}(K_L \to \pi^0 \nu \bar{\nu}) < 5.9 \times 10^{-7}$, several orders of magnitude above the expected value.

C Is the CKM picture of CP violation right?

Two key tests have been passed so far. The theory has succeeded, albeit qualitatively, in predicting the range $\text{Re}(\epsilon'/\epsilon) = (1 \text{ to } 2) \times 10^{-3}$. Its prediction for the branching ratio for $K^+ \to \pi^+ \nu \bar{\nu}$ is in accord with the experimental rate deduced from the one event observed so far.

One test still to be passed in the decays of neutral kaons is the measurement of the height η of the unitarity triangle through the decay $K_L \to \pi^0 \nu \bar{\nu}$. Prospects for this measurement will be mentioned below. However, in the nearer term, one looks forward to a rich set of effects in decays of particles containing b quarks, particularly the B mesons. To this end, experiments are under way at a number of laboratories around the world.

Asymmetric $e^+ e^-$ collisions are being studied at two "B factories," the PEP-II machine at SLAC with the BaBar detector, and the KEK-B collider in Japan with the Belle detector. By end of April 2000, these detectors were recording about 100 and 60 pb^{-1} of data per day, respectively, and had accumulated about 6 and 2 fb^{-1} of data at the energy of the $\Upsilon(4S)$ resonance, which decays almost exclusively to $B\bar{B}$. The BaBar experiment expects to have about 100 tagged $B^0 \to J/\psi K_S$ decays by this coming summer [39].

Significant further data on $e^+ e^-$ collisions at the $\Upsilon(4S)$ are expected from the Cornell Electron Storage Ring with the upgraded CLEO-III detector. The HERA-b experiment at DESY in Hamburg will study b quark production via the collisions of 920 GeV protons with a fixed target. The CDF and D0 detectors at Fermilab will devote a significant part of their program at Run II of the Tevatron to B physics. In the longer term, one can expect further results on B physics from the general-purpose LHC detectors ATLAS and CMS and the dedicated LHC-b detector at CERN, and possibly the dedicated BTeV detector at Fermilab.

IV CP VIOLATION AND B DECAYS

In constrast to the neutral kaon system, in which the eigenstates of the mass matrix differ in lifetime by nearly a factor of 600, the eigenstates of the corresponding B^0–$\overline{B^0}$ mass matrix are expected to differ in lifetime by at most 10–20% for strange B's [40], and considerably less for nonstrange B's. Thus, instead of studying the properties of mass eigenstates like K_L, one must resort to other means. There are two main avenues of study.

- *Decays to CP eigenstates* $f = \pm \mathrm{CP}(f)$ utilize interference between direct decays $B^0 \to f$ or $\overline{B^0} \to f$ and the corresponding paths involving mixing: $B^0 \to \overline{B^0} \to f$ or $\overline{B^0} \to B^0 \to f$. Final states such as $f = J/\psi K_S$ provide "clean" examples in which one quark subprocess is dominant. In this case one measures $\sin 2\beta$ with negligible corrections. For the final state $\pi^+\pi^-$, one measures $\sin 2\alpha$ only to the extent that the direct decay is dominated by a "tree" amplitude (the quark subprocess $b \to u\bar{u}d$). When contamination from the penguin subprocess $b \to d$ is present (as it is expected to be at the level of several tens of percent), one must measure decays to other $\pi\pi$ states (such as $\pi^\pm \pi^0$ and $\pi^0\pi^0$) to sort out various decay amplitudes [41].

- *"Self-tagging" decays* involve final states f such as $K^+\pi^-$ which can be distinguished from their CP-conjugates $\bar{f}$. A CP-violating rate asymmetry arises if there exist two weak amplitudes a_i with weak phases ϕ_i and strong phases δ_i ($i=1,2$):

$$A(B \to f) = a_1 e^{i(+\phi_1+\delta_1)} + a_2 e^{i(+\phi_2+\delta_2)} \quad ,$$

$$A(\bar{B} \to \bar{f}) = a_1 e^{i(-\phi_1+\delta_1)} + a_2 e^{i(-\phi_2+\delta_2)} \quad . \qquad (12)$$

Note that the weak phase changes sign under CP-conjugation, while the strong phase does not. The rate asymmetry is then

$$\mathcal{A}(f) \equiv \frac{\Gamma(f) - \Gamma(\bar{f})}{\Gamma(f) + \Gamma(\bar{f})} = \frac{2 a_1 a_2 \sin(\phi_1 - \phi_2) \sin(\delta_1 - \delta_2)}{a_1^2 + a_2^2 + 2 a_1 a_2 \cos(\phi_1 - \phi_2) \cos(\delta_1 - \delta_2)} \quad . \qquad (13)$$

Thus the two amplitudes must have different weak *and* strong phases in order for a rate asymmetry to be observable. The CKM theory predicts the weak phases, but no reliable estimates of strong phases in B decays exist. Some ways of circumventing this difficulty will be mentioned.

A Decays to CP eigenstates

The interference between mixing and decay in decays of neutral B mesons to CP eigenstates leads to a term which modulates the exponential decay (see, e.g., [42]):

$$\frac{d\Gamma(t)}{dt} \sim e^{-\Gamma t}(1 \mp \mathrm{Im}\lambda_0 \sin \Delta mt) \quad , \tag{14}$$

where the upper sign refers to B^0 decays and the lower to $\overline{B^0}$ decays. Δm is the mass splitting mentioned earlier, and the factor λ_0 expresses the interference of decay and mixing amplitudes. For $f = J/\psi K_S$, $\lambda_0 = -e^{-2i\beta}$ to a good approximation, while for $f = \pi^+\pi^-$, $\lambda_0 \simeq e^{2i\alpha}$ only to the extent that the effect of penguin amplitudes can be neglected in comparison with the dominant tree contribution.

The time integral of the modulation term is

$$\int_0^\infty dt\, e^{-\Gamma t} \sin \Delta mt = \frac{1}{\Gamma}\frac{x}{1+x^2} \leq \frac{1}{\Gamma}\cdot\frac{1}{2} \quad , \tag{15}$$

where $x \equiv \Delta m/\Gamma$. This expression is maximum for $x = 1$, and 95% of maximum for the observed value $x \simeq 0.72$. It has been fortunate that the B^0 mixing amplitude and decay rate are so well matched to one another.

The CDF Collaboration [43] has learned how to "tag" neutral B mesons at the time of their production and thus to measure the decay rate asymmetry in B^0 ($\overline{B^0}$) $\to J/\psi K_S$. This asymmetry arises from the phase 2β characterizing the two powers of V_{td} in the B^0–$\overline{B^0}$ mixing amplitude. The tagging methods are of two main types. "Opposite-side" methods rely on the fact that strong interactions always produce b and $\bar{b}$ in pairs, so that in order to determine the initial flavor of a decaying B one must find out something about the "other" b-containing hadron produced in association with it, either via the charge of the jet containing it or via the charge of the lepton or kaon it emits when decaying. "Same-side" methods [44] utilize the fact that a B^0 tends to be associated more frequently with a π^+, and a $\overline{B^0}$ with a π^-, somewhere nearby in phase space, whether through the dynamics of fragmentation or through the decays of excited B resonances.

The CDF result is $\sin 2\beta = 0.79^{+0.41}_{-0.44}$. An earlier result from OPAL [45] and a newer result from ALEPH [46], both utilizing B's produced in the decays of the Z^0, can be combined with the CDF value to obtain $\sin 2\beta = 0.91 \pm 0.35$, which exceeds zero at the 99% confidence level [46]. At the 1σ lower limit (0.56) this is very close to the lower bound (0.59) quoted in Table 2.

B "Self-tagging" decays and direct CP violation

An example of direct CP violation can occur in $B^0 \to K^+\pi^-$. One expects two types of contribution to this process: a "tree" amplitude governed by the quark subprocess $\bar{b} \to \bar{u}u\bar{s}$ with CKM factor $V_{ub}^* V_{us}$, and a "penguin" amplitude governed

TABLE 4. Main amplitudes contributing to $B^0 \to K^{(*)+}\pi^-$.

Amplitude	Subprocess	CKM factor	Weak phase
Tree	$\bar{b} \to \bar{u}u\bar{s}$	$V_{ub}^* V_{us}$	γ
Penguin	$\bar{b} \to \bar{s}$	$V_{tb}^* V_{ts}$	π

by the quark subprocess $\bar{b} \to \bar{s}$ with dominant CKM factor $V_{tb}^* V_{ts}$ (since the contribution of the top quark in the internal loop is dominant). These contributions are summarized in Table 4.

Since the tree and penguin amplitudes have a relative weak phase γ (mod π), one can have $\Gamma(B^0 \to K^+\pi^-) \neq \Gamma(\overline{B^0} \to K^-\pi^+)$ as long as the strong phases δ_T and δ_P are different in the tree and penguin amplitudes. However, even if these strong phases do not differ from one another, the ratios of rates for various charge states of $B \to K\pi$ decays can provide separate information on the weak phase γ [47–49] and the strong phase difference $\delta_T - \delta_P$.

One must first deal with electroweak penguins which were also relevant for the interpretation of ϵ'/ϵ. An early suggestion (see the first of Refs. [47]) proposed a way to extract γ from the rates for $B^+ \to (\pi^0 K^+, \pi^+ K^0, \pi^+\pi^0)$ and the charge-conjugate processes. The amplitudes for the first two processes (with appropriate factors of $\sqrt{2}$) form a triangle with an amplitude related to the third process by flavor SU(3) as long as electroweak penguins are negligible, which they are not [50]. It turns out, however [49], that the relevant electroweak penguin's contribution to this process can be calculated, so that sufficiently precise measurements of the rates for the above processes can indeed yield useful information on γ.

The possibility has been raised recently [49,51,52] that the weak phase γ may exceed 90°. Two processes whose rates hint at this constraint are $B^0 \to \pi^+\pi^-$ and $B^0 \to K^{*+}\pi^-$. The former process has a rate which is somewhat smaller than expected, while the rate for the latter is larger than expected.

The amplitudes contributing to $B^0 \to \pi^+\pi^-$ are summarized in Table 5. The relative phase of the tree and penguin amplitudes is $\gamma + \beta = \pi - \alpha$. The two amplitudes will interfere destructively if the final strong phase difference is small (as expected from perturbative QCD estimates, which indeed may be risky), and if $\alpha < \pi/2$. This would tend to favor not-too-positive values of ρ. There is some hint that the interference is indeed destructive. The observed branching ratio [53] $\mathcal{B}(B^0 \to \pi^+\pi^-) = (4.3^{+1.6}_{-1.4} \pm 0.5) \times 10^{-6}$ is less than the value of about 10^{-5} which one would estimate [51] from the tree amplitude alone (e.g., using the observed $B \to \pi e \nu_e$ branching ratio and factorization).

The same types of amplitudes contributing to $B^0 \to K^+\pi^-$ also contribute to $B^0 \to K^{*+}\pi^-$ (see Table 4). As in $B^0 \to K^+\pi^-$, the relative phase between the tree and penguin amplitudes is expected to be $\gamma - \pi$. One thus expects constructive

TABLE 5. Main amplitudes contributing to $B^0 \to \pi^+\pi^-$.

Amplitude	Subprocess	CKM factor	Weak phase
Tree	$\bar{b} \to \bar{u}u\bar{d}$	$V_{ub}^* V_{ud}$	γ
Penguin	$\bar{b} \to \bar{d}$	$V_{tb}^* V_{td}$	$-\beta$

interference between the two amplitudes if the strong phase difference is small and $\gamma > \pi/2$. Indeed, the branching ratio for $B^0 \to K^{*+}\pi^-$ appears to exceed 2×10^{-5}, while the pure "penguin" process $B^+ \to K^+\phi$ has a branching ratio less than 10^{-5}.

A global fit to the above two processes and many others (see the second of Refs. [52]) finds $\gamma = (114^{+24}_{-23})°$, which just grazes the allowed region quoted in Table 2. Since the upper bound on γ in Table 2 is set primarily by the lower limit on B_s–$\bar{B}_s$ mixing, such mixing should be visible in experiments of only slightly greater sensitivity than those performed up to now.

The Tevatron and the LHC will copiously produce both nonstrange and strange neutral B's, decaying to $\pi^+\pi^-$, $K^\pm\pi^\mp$, and K^+K^- [54]. Each of these channels has particular advantages.

- The decays $B^0 \to K^+K^-$ and $B_s \to \pi^+\pi^-$ should be highly suppressed unless these final states are "fed" by rescattering from other channels [55].

- The decays $B^0 \to \pi^+\pi^-$ and $B_s \to K^+K^-$ can yield γ when their time-dependence is measured [56]. The kinematic peaks for these two states overlap significantly, so one must either use particle identification or utilize the vastly different oscillation frequencies for B^0–$\bar{B^0}$ and B_s–$\bar{B}_s$ mixing to distinguish the two final states.

- A recent proposal for measuring γ [57] utilizes the decays $B^0 \to K^+\pi^-$, $B^+ \to K^0\pi^+$, $B_s \to K^-\pi^+$, and the corresponding charge-conjugate processes. The $B^0 \to K^+\pi^-$ and $B_s \to K^-\pi^+$ peaks are well separated from one another and from $B^0 \to \pi^+\pi^-$ and $B_s \to K^+K^-$ kinematically [54].

The proposal of Ref. [57] is based on the observation that $B \to K\pi$ decays involve two types of amplitudes, tree (T) and penguin (P), with relative weak phase γ and relative strong phase δ. The decays $B^+ \to K^0\pi^+$ are expected to be dominated by the penguin amplitude (there is no tree contribution except through rescattering from other final states), so this channel is not expected to display any CP-violating asymmetries. One expects $\Gamma(B^+ \to K^0\pi^+) = \Gamma(B^- \to \bar{K^0}\pi^-)$. This will provide a check of the assumption that rescattering effects can be neglected. A typical amplitude is given by $A(B^0 \to K^+\pi^-) = -[P + Te^{i(\gamma+\delta)}]$, where the signs are associated with phase conventions for states [58].

We now define

TABLE 6. CP-violating asymmetries in decays of B mesons to light quarks.

Mode	Signal events	$\mathcal{A}_{CP}$
$K^+\pi^-$	80^{+12}_{-11}	-0.04 ± 0.16
$K^+\pi^0$	$42.1^{+10.9}_{-9.9}$	-0.29 ± 0.23
$K_S\pi^+$	$25.2^{+6.4}_{-5.6}$	$+0.18 \pm 0.24$
$K^+\eta'$	100^{+13}_{-12}	$+0.03 \pm 0.12$
$\omega\pi^+$	$28.5^{+8.2}_{-7.3}$	-0.34 ± 0.25

$$\left\{ \begin{array}{c} R \\ A_0 \end{array} \right\} \equiv \frac{\Gamma(B^0 \to K^+\pi^-) \pm \Gamma(\overline{B^0} \to K^-\pi^+)}{2\Gamma(B^+ \to K^0\pi^+)} , \qquad (16)$$

$$\left\{ \begin{array}{c} R_s \\ A_s \end{array} \right\} \equiv \frac{\Gamma(B_s \to K^-\pi^+) \pm \Gamma(\overline{B_s} \to K^+\pi^-)}{2\Gamma(B^+ \to K^0\pi^+)} , \qquad (17)$$

and $r \equiv T/P$, $\tilde{\lambda} \equiv V_{us}/V_{ud}$. Then one finds

$$R = 1 + r^2 + 2r\cos\delta\cos\gamma , \quad R_s = \tilde{\lambda}^2 + \left(\frac{r}{\tilde{\lambda}}\right)^2 - 2r\cos\delta\cos\gamma , \qquad (18)$$

$$A_0 = -A_s = -2r\sin\gamma\sin\delta . \qquad (19)$$

The sum of R and R_s allows one to determine r. Then using R, r, and A_0, one can solve for both δ and γ. The prediction $A_s = -A_0$ serves as a check of the flavor SU(3) assumption which gave these relations. An error of $10°$ on γ seems feasible with forthcoming data from Run II of the Tevatron.

The CLEO Collaboration has recently presented some upper limits on CP-violating asymmetries in B decays to light-quark systems [59], based on 9.66 million events recorded at the $\Upsilon(4S)$. With asymmetries defined as

$$\mathcal{A}_{CP} \equiv \frac{\Gamma(\overline{B} \to \bar{f}) - \Gamma(B \to f)}{\Gamma(\overline{B} \to \bar{f}) + \Gamma(B \to f)} , \qquad (20)$$

the results are shown in Table 6. No statistically significant asymmetries have been seen yet. The sensitivity of these results is not yet adequate to probe the maximum predicted values [60] $|\mathcal{A}_{CP}^{K^+\pi}| \leq 1/3$, but is getting close.

V SOME FUTURE MEASUREMENTS

The future of the experimental study of CP violation involves a broad program of experiments with kaons, charmed and B mesons, and neutrinos. We mention just a few of the possibilities.

A Rare kaon decays

Plans are afoot for measurement of the branching ratio for $K_L \to \pi^0 \nu \bar{\nu}$ at the required sensitivity ($\mathcal{B} \simeq 3 \times 10^{-3}$). Experiments are envisioned using both relatively slow kaons at Brookhaven National Laboratory [61] and faster kaons at the Fermilab Main Injector [62]. A Fermilab proposal [63] seeks to accumulate 100 events of $K^+ \to \pi^+ \nu \bar{\nu}$ in order to measure $|V_{td}|$ to a statistical precision of 5% and an overall precision of 10%.

B Charmed mesons

Impressive strides have been taken in the measurement of mass differences and lifetime differences for CP eigenstates of the neutral charmed mesons D^0 [64,65]. No significant effects have been seen yet at the level of a percent or so, but there are tantalizing hints [66]. It would be worth while to follow up these possibilities. Electron-positron colliders, mentioned below, will devote much of their running time to the study of B mesons, but charmed mesons are accumulated as well in such experiments, and the samples of them will increase. Hadronic experiments dedicated to producing large numbers of B's may also have more to say about mixing, lifetime differences, and CP violation for charmed mesons.

C B production in symmetric e^+e^- collisions

Although asymmetric e^+e^- colliders, known as "B-factories," are now starting to take data at an impressive rate, the CLEO Collaboration at the symmetric CESR machine has recently celebrated 20 years of B physics, and is continuing with an active program. It will be able in the CLEO-III program to probe charmless B decays down to branching ratios of 10^{-6}. In so doing, it may be able to detect the elusive $B^0 \to \pi^0 \pi^0$ mode, whose rate will help pin down the penguin amplitude's contribution and permit a determination of the CKM phase α [41].

Other final states of great interest at this level include VP and VV, where P,V denote light pseudoscalar and vector mesons. There is a good chance that direct CP violation may show up in one or more channels if final state phase differences are sufficiently large. The detailed study of angular correlations in VV channels may be able to provide useful information on strong final state phases.

A useful probe of rescattering effects [55], mentioned above, is the decay $B^0 \to K^+K^-$. This decay is expected to have a branching ratio of only a few parts in 10^8 if rescattering is unimportant, but could be enhanced by more than an order of magnitude in the presence of rescattering from other channels.

A challenging channel of fundamental importance is $B^+ \to \tau^+\bar{\nu}_\tau$. The rate for this process will provide information on the combination $f_B|V_{cb}|$. Rare decays which have not yet been seen (such as $B \to X\ell^+\ell^-$ and $B \to X\nu\bar{\nu}$) will probe the effects of new particles in loops.

D B production in asymmetric e^+e^- collisions

The benchmark process for the BaBar and Belle detectors will be the measurement of $\sin 2\beta$ in $B^0 \to J/\psi K_S$. The PEP-II and KEK-B machines utilize asymmetric e^+e^- collisions in order to create a moving reference frame in which the decays of B^0 and $\overline{B^0}$ are separated by a large enough distance for their separation to be detectable. (Each travels only an average distance of 30 μm in the center of mass.) This facilitates both flavor tagging and improvement of signal with respect to background. These machines will make possible a host of time-dependent studies in such decays as $B \to \pi\pi$, $B \to K\pi$, etc., and their impressive luminosities will eventually add significantly to the world's tally of detected B's.

E Hadronic B production

The strange B's cannot be produced at the $\Upsilon(4S)$ which will dominate the attention of e^+e^- colliders for some years to come. Hadronic reactions at high energies will produce copious b's incorporated into all sorts of hadrons: nonstrange, strange, and charmed mesons, and baryons. One looks forward to a measurement of the strange-B mixing parameter $x_s = \Delta m_s/\Gamma_s$. The decays of B_s can provide valueable information on CKM phases and CP violation, as in $B_s \to K^+K^-$ [56]. The width difference of 10–20% expected between the CP-even and CP-odd eigenstates of the B_s system [40] should be visible in the next round of experiments.

F Neutrino studies

The origin of magnitudes and phases in the CKM matrix is intimately connected with the origin of the quark masses themselves, whose physics still eludes us. We will not understand this pattern until we have mapped out a similar pattern for the leptons, a topic to which many other talks in this Workshop are devoted. Our understanding of neutrino masses and mixings will benefit greatly from forthcoming experiments at the Sudbury Neutrino Observatory [67], Borexino [68], K2K [69], and Fermilab (BooNE and MINOS) [70], to name a few.

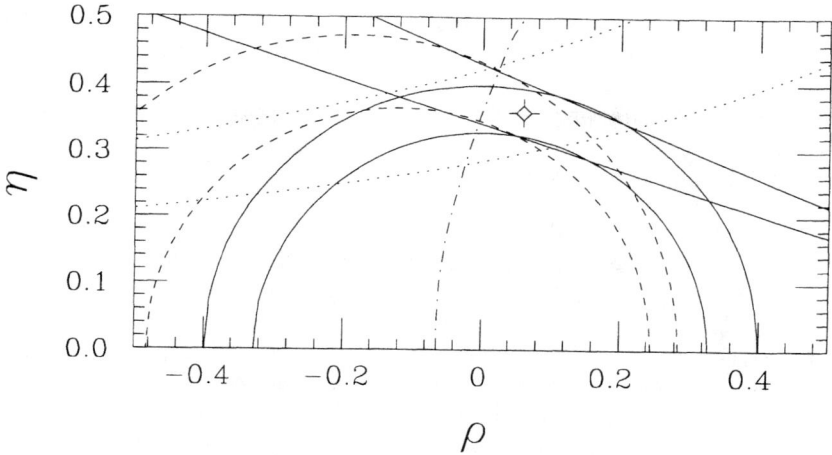

FIGURE 2. Example of a region in the (ρ, η) plane that might be allowed by data in the year 2003. Constraints are based on the following assumptions: $|V_{ub}/V_{cb}| = 0.08 \pm 0.008$ (solid semicircles), $|V_{ub}/V_{td}| = |(\rho - i\eta)/(1 - \rho - i\eta)| = 0.362 \pm 0.036$ based on present data on B^0–$\bar{B}^0$ mixing and a measurement of $B(B^+ \to \tau^+ \nu_\tau)$ to $\pm 20\%$ (dashed semicircles), CP-violating K–$\bar{K}$ mixing as discussed in Sec. 2 but with V_{cb} measured to $\pm 4\%$ (dotted hyperbolae), the bound $x_s > 20$ for B_s^0–$\bar{B}_s^0$ mixing (to the right of the dot-dashed semicircle), and measurement of $\sin 2\beta$ to ± 0.059 (diagonal straight lines). The plotted point, corresponding to $(\rho, \eta) = (0.06, 0.36)$, lies roughly within the center of the allowed region.

G   The (ρ, η) plot in a few years

The (ρ, η) plot might appear as shown in Fig. 2 in a few years [22,71]. We can look forward either to a reliable determination of parameters or to the possibility that one or more experiments give contradictory results, indicating the need for new physics. Such new physics most typically shows up in the form of additional contributions to B^0–$\overline{B^0}$ mixing [72], though it can also show up in decays [73].

VI   BARYON ASYMMETRY

The ratio of the number of baryons n_B to the number of photons n_γ in the Universe is a few parts in 10^{10}, much larger than the corresponding ratio for antibaryons. Shortly after the discovery of CP violation in neutral kaon decays, Sakharov proposed in 1967 [74] three ingredients needed to understand this preponderance of matter over antimatter: (1) an epoch in which the Universe was not in thermal equilibrium, (2) an interaction violating baryon number, and (3) CP (and C) violation. However, one can't explain the observed baryon asymmetry merely by means of the CP violation contained in the CKM matrix. The effects are too small unless some new physics is introduced. Two examples are the following:

- The concept of supersymmetry, in which each particle of spin J has a "superpartner" of spin $J \pm 1/2$, affords many opportunities for introducing new CP-violating phases and interactions which could affect particle-antiparticle mixing [75].

- The presence of neutrino masses at the sub-eV level can signal large Majorana masses for right-hand neutrinos, exceeding 10^{11} GeV [76]. Lepton number (L) is violated by such masses. The violation of L can easily be reprocessed into baryon number (B) violation by $B - L$ conserving interactions at the electroweak scale [77]. New CP-violating interactions must then exist at the high mass scale if lepton number is to be generated there. It is conceivable that these interactions are related to CKM phases, but the link will be very indirect [78]. In any case, if this alternative is the correct one, it will be very important to understand the leptonic analogue of the CKM matrix!

VII CONCLUSIONS

The CKM theory of CP violation in neutral kaon decays has passed a crucial test. The parameter ϵ'/ϵ is nonzero, and has the expected order of magnitude, though exceeding some theoretical estimates. Still to come will be several tests using B mesons, including the observation of a difference in rates between $B^0 \to J/\psi K_S$ and $\overline{B^0} \to J/\psi K_S$. There will be more progress in "tagging" neutral B's, and we can look forward to rich information from measurements of decay rates of charged and neutral B's into a variety of final states.

I see two possibilities for our understanding of CP violation in the next few years. (1) If B decays do not provide a consistent set of CKM phases, we will be led to examine other sources of CP violation. Most of these, in contrast to the CKM theory, predict neutron and electron dipole moments very close to their present experimental upper limits. (2) If, on the other hand, the CKM picture still hangs together after a few years, attention should naturally shift to the next "layer of the onion": the origin of the CKM phases (and the associated quark and lepton masses). It is probably time to start anticipating this possibility, given the resilience of the CKM picture since it was first proposed nearly 30 years ago.

VIII ACKNOWLEDGEMENTS

It is a pleasure to thank Maria Eugenia and Jose Nieves for their wonderful hospitality in San Juan.

REFERENCES

1. J. H. Christenson, J. W. Cronin, V. L. Fitch, and R. Turlay, Phys. Rev. Lett. **13**, 138-140 (1964).
2. J. Schwinger, Phys. Rev. **91**, 713-728 (1953); Phys. Rev. **94**, 1362-1384 (1954); G. Lüders, Kong. Danske Vid. Selsk., Matt-fys. Medd. **28**, No. 5, 1-17 (1954); Ann. Phys. (N.Y.) **2**, 1-15 (1957); W. Pauli, in *Niels Bohr and the Development of Physics*, edited by W. Pauli (Pergamon, New York, 1955), pp. 30-51.
3. M. Kobayashi and T. Maskawa, Prog. Theor. Phys. **49**, 652-657 (1973).
4. N. Cabibbo, Phys. Rev. Lett. **10**, 531-532 (1963).
5. KTeV (Fermilab E832) Collaboration, A. Alavi-Harati *et al.*, Phys. Rev. Lett. **83**, 22 (1999).
6. NA48 Collaboration, G. D. Barr *et al.*, presented at CERN seminar by A. Ceccucci, Feb. 29, 2000 (unpublished). Follow the links on http://www/cern.ch/NA48/ for a copy of the transparencies.
7. For a review see S. M. Barr, in *TASI 94: CP Violation and the Limits of the Standard Model*, Boulder, CO, 29 May – 24 June 1994, edited by J. F. Donoghue (World Scientific, River Edge, NJ, 1995), pp. 87-111.
8. G. D. Rochester and C. C. Butler, Nature **160**, 855-857 (1947).
9. M. Gell-Mann, Phys. Rev. **92**, 833-834 (1953); "On the Classification of Particles," 1953 (unpublished); M. Gell-Mann and A. Pais, in *Proceedings of the 1954 Glasgow Conference on Nuclear and Meson Physics*, edited by E. H. Bellamy and R. G. Moorhouse (Pergamon, London and New York, 1955); M. Gell-Mann, Nuovo Cim. **4**, Suppl. 848-866 (1956); T. Nakano and K. Nishijima, Prog. Theor. Phys. **10**, 581-582 (1953); K. Nishijima, Prog. Theor. Phys. **12**, 107-108 (1954); Prog. Theor. Phys. **13**, 285-304 (1955).
10. M. Gell-Mann and A. Pais, Phys. Rev. **97**, 1387-1389 (1955).
11. T. D. Lee, R. Oehme, and C. N. Yang, Phys. Rev. **106**, 340-345 (1957); L. D. Landau, Nucl. Phys. **3**, 127-131 (1957).
12. K. Lande, E. T. Booth, J. Impeduglia, and L. M. Lederman, Phys. Rev. **103**, 1901-1904 (1956).
13. B. Winstein, in *Festi-Val – Festschrift for Val Telegdi*, ed. by K. Winter (Elsevier, Amsterdam, 1988), pp. 245-265.
14. L. Wolfenstein, Phys. Rev. Lett. **13**, 562-564 (1964).
15. P. K. Kabir, *The CP Puzzle* (Academic Press, New York, 1968); T. P. Cheng and L. F. Li, *Gauge Theory of Elementary Particles* (Oxford University Press, 1984); R. G. Sachs, *The Physics of Time Reversal Invariance* (University of Chicago Press, Chicago, 1988); K. Kleinknecht, in *CP Violation*, edited by C. Jarlskog (World Scientific, Singapore, 1989), pp. 41-104; J. L. Rosner, in *Testing the Standard Model* (Proceedings of the 1990 Theoretical Advanced Study Institute in Elementary Particle Physics, Boulder, Colorado, 3–27 June, 1990), edited by M. Cvetič and P. Langacker (World Scientific, Singapore, 1991), pp. 91-224; B. Winstein and L. Wolfenstein, Rev. Mod. Phys. **63**, 1113-1148 (1992).
16. J. L. Rosner, Am. J. Phys. **64**, 982-985 (1996).
17. J. L. Rosner and S. A. Slezak, Enrico Fermi Institute Report No. EFI 99-51, hep-

ph/9912506, submitted to Am. J. Phys.
18. L. Wolfenstein, Phys. Rev. Lett. **51**, 1945-1947 (1983).
19. M. Gell-Mann and M. Lévy, Nuovo Cim. **16**, 705-725 (1960).
20. For reviews of information on CKM matrix elements see, e.g., F. Gilman, K. Kleinknecht, and Z. Renk, Eur. Phys. J. C **3**, 103-106 (1998); A. F. Falk, hep-ph/9908520, published in *Proceedings of the XIXth International Symposium on Lepton and Photon Interactions*, Stanford, California, August 9–14 1999, edited by J. Jaros and M. Peskin (World Scientific, Singapore, 2000); A. Ali and D. London, preprint hep-ph/0002167.
21. See, e.g., F. Caravaglio, F. Parodi, P. Roudeau, and A. Stocchi, talk given at 3rd Int. Conf. on *B* Physics and CP Violation, Taipei, Taiwan, 3–7 Dec. 1999, preprint hep-ph/0002171.
22. J. L. Rosner, in *Lattice '98* (Proceedings of the XVIth International Symposium on Lattice Field Theory, Boulder, Colorado, 13–18 July 1998), edited by T. DeGrand, C. DeTar, R. Sugar, and D. Toussaint, Nucl. Phys. B (Proc. Suppl.) **73**, 29-42 (1999).
23. Slightly more up-to-date world averages for nonstrange and strange neutral *B* mixing amplitudes may be found in http://www.cern.ch/LEPBOSC/.
24. G. Blaylock, in *Proceedings of the XIXth International Symposium on Lepton and Photon Interactions*, Stanford, California, August 9–14 1999, edited by J. Jaros and M. Peskin (World Scientific, Singapore, 2000).
25. J. L. Rosner, Phys. Rev. D **42**, 3732-3740 (1990).
26. A. J. Buras, M. Jamin, and M. E. Lautenbacher, Phys. Lett. B **389**, 749-756 (1996).
27. See the articles on ϵ'/ϵ by A. J. Buras, S. Bertolini, R. Gupta, G. Martinelli, and W. A. Bardeen in *Kaon Physics*, edited by J. L. Rosner and B. Winstein (University of Chicago Press, 2000).
28. Particle Data Group, C. Caso *et al.*, Eur. Phys. J. C **3**, 1-794 (1998).
29. Fermilab E731 Collaboration, L. K. Gibbons *et al.*, Phys. Rev. Lett. **70**, 1203-1206 (1993); Phys. Rev. D **55**, 6625-6715 (1997).
30. CERN NA31 Collaboration, G. D. Barr *et al.*, Phys. Lett. B **317**, 233-242 (1993).
31. KTeV Collaboration, A. Alavi-Harati *et al.*, Phys. Rev. Lett. **84**, 408-411 (2000).
32. NA48 Collaboration, G. D. Barr *et al.*, presented by S. Wronka in *Kaon Physics* [27].
33. KTeV Collaboration, G. Breese Quinn, Ph. D. Thesis, University of Chicago, May, 2000 (unpublished).
34. G. Buchalla and A. J. Buras, Nucl. Phys. **B548**, 309-327 (1999); G. Buchalla, in *Kaon Physics* [27].
35. Brookhaven E787 Collaboration, S. Adler *et al.*, Phys. Rev. Lett. **84**, 3768-3770 (2000).
36. H. B. Greenlee, Phys. Rev. D **42**, 3724-3731 (1990).
37. Fermilab E-799-II/KTeV Collaboration, A. Alavi-Harati *et al.*, presented by J. Whitmore at Kaon 99 Conference, Chicago, IL, June 21-26, 1999, published in *Kaon Physics* [27]; preprint hep-ex/0001005, and preprint in preparation.
38. Fermilab E-799-II/KTeV Collaboration, A. Alavi-Harati *et al.*, Phys. Rev. D **61**, 072006 (2000).
39. A. J. S. Smith, private communication.
40. M. Beneke, G. Buchalla, and I. Dunietz, Phys. Rev. D **54**, 4419-4431 (1996).

41. M. Gronau and D. London, Phys. Rev. Lett. **65**, 3381-3384 (1990).
42. I. Dunietz and J. L. Rosner, Phys. Rev. D **34**, 1404-1417 (1986).
43. CDF Collaboration, T. Affolder *et al.*, Phys. Rev. D **61**, 072005 (2000).
44. M. Gronau, A. Nippe, and J. L. Rosner, Phys. Rev. D **47**, 1988-1993 (1993); M. Gronau and J. L. Rosner, Phys. Rev. Lett. **72**, 195-198 (1994); Phys. Rev. D **49**, 254-264 (1994).
45. OPAL Collaboration, K. Ackerstaff *et al.*, Eur. Phys. J. C **5**, 379 (1998).
46. ALEPH Collaboration, ALEPH report ALEPH 99-099, CONF 99-054, presented by R. Forty at 3rd Int. Conf. on *B* Physics and CP Violation, Taipei, Taiwan, 3–7 December 1999.
47. M. Gronau, J. Rosner and D. London, Phys. Rev. Lett. **73**, 21-24 (1994); M. Gronau and J. L. Rosner, Phys. Rev. Lett. **76**, 1200-1203 (1996); Phys. Rev. D **57**, 6843-6850 (1998); A. S. Dighe, M. Gronau, and J. L. Rosner, Phys. Rev. D **54**, 3309-3320 (1996); A. S. Dighe and J. L. Rosner, **54**, 4677-4679 (1996); M. Gronau and D. Pirjol, Phys. Lett. B **449**, 321-327 (1999); Phys. Rev. D **61**, 013005 (2000).
48. R. Fleischer, Phys. Lett. B **365**, 399-406 (1996); Phys. Rev. D **58**, 093001 (1998); R. Fleischer and T. Mannel, Phys. Rev. D **57**, 2752-2759 (1998); A. J. Buras, R. Fleischer, and T. Mannel, Nucl. Phys. **B533**, 3-24 (1998); R. Fleischer and A. J. Buras, Eur. Phys. J. C **11**, 93-109 (1999).
49. M. Neubert and J. L. Rosner, Phys. Lett. B **441**, 403-409 (1998); Phys. Rev. Lett. **81**, 5076-5079 (1998); M. Neubert, JHEP **9902**, 014 (1999).
50. N. G. Deshpande and X.-G. He, Phys. Rev. Lett. **74**, 26-29 (1995); **74**, 4099(E) (1995); O. F. Hernández, D. London, M. Gronau, and J. L. Rosner, Phys. Rev. D **52**, 6374-6382 (1995).
51. M. Gronau and J. L. Rosner, Phys. Rev. D **61**, 073008 (2000).
52. X.-G. He, W.-S. Hou, and K.-C. Yang, Phys. Rev. Lett. **83**, 1100-1103 (1999); W.-S. Hou, J. G. Smith, and F. Würthwein, preprint hep-ex/9910014 (unpublished).
53. CLEO Collaboration, D. Cronin-Hennessy *et al.*, Cornell University preprint CLNS 99-1650, hep-ex/0001010 (unpublished).
54. F. Würthwein and R. Jesik, talks for Working Group 1 presented at Workshop on B Physics at the Tevatron – Run II and Beyond, Fermilab, February 2000 (unpublished).
55. B. Blok, M. Gronau, and J. L. Rosner, Phys. Rev. Lett. **78**, 3999-4002 (1997); **79**, 1167 (1997); A. Falk, A. L. Kagan, Y. Nir, and A. A. Petrov, Phys. Rev. D **57**, 4290-4300 (1998); M. Gronau and J. L. Rosner, Phys. Rev. D **57**, 6843-6350 (1998); **58**, 113005 (1998); R. Fleischer, Phys. Lett. B **435**, 221-232 (1998); Eur. Phys. J. C **6**, 451-470 (1999).
56. R. Fleischer, Phys. Lett. B **459**, 306-320 (1999); DESY preprint DESY 00-014, hep-ph/0001253. See also I. Dunietz, Proceedings of the Workshop on *B* Physics at Hadron Accelerators, Snowmass, CO, 1993, p. 83; D. Pirjol, Phys. Rev. D **60**, 054020 (1999).
57. M. Gronau and J. L. Rosner, preprint hep-ph/0003119, to be published in Phys. Lett. B.
58. M. Gronau, O. F. Hernández, D. London and J. L. Rosner, Phys. Rev. D **50**, 4529-4543 (1994).

59. CLEO Collaboration, S. Chen et al., Cornell University preprint CLNS 99-1651A, hep-ex/0001009.
60. M. Gronau and J. L. Rosner, Phys. Rev. D **59**, 113002 (1999).
61. KOPIO Collaboration, in *Rare Symmetry Violating Processes*, proposal to the National Science Foundation, October 1999, and Brookhaven National Laboratory Proposal P926 (unpublished).
62. KAMI Collaboration, Fermilab Proposal P804 (unpublished).
63. CKM Collaboration, Fermilab Proposal P905.
64. CLEO Collaboration, R. Godang et al., Cornell University preprint CLNS 99-1659, hep-ex/0001060, submitted to Phys. Rev. Lett.
65. FOCUS Collaboration, Fermilab E831, J. M. Link et al., FERMILAB-PUB-00-091-E, HEP-EX/0004034, submitted to Phys. Lett. B.
66. These are discussed, for example, by I. I. Bigi and N. G. Uraltsev, University of Notre Dame preprint UND-HEP-00-BIG01, hep-ph/0005089; S. Bergmann et al., preprint hep-ph/0005181.
67. R. van de Water, this Workshop.
68. See the description of this experiment at http://almime.mi.infn.it/
69. M. Vagins, this Workshop.
70. J. Conrad, this Workshop; M. Shaevitz, this Workshop.
71. P. Burchat et al., Report of the NSF Elementary Particle Physics Special Emphasis Panel on B Physics, July, 1998 (unpublished).
72. See, e.g., C. O. Dib, D. London, and Y. Nir, Int. J. Mod. Phys. A **6**, 1253-1266 (1991); M. Gronau and D. London, Phys. Rev. D **55**, 2845-2861 (1997).
73. Y. Grossman and M. P. Worah, Phys. Lett. B **395**, 241-249 (1997); Y. Grossman, G. Isidori, and M. P. Worah, Phys. Rev. D **58**, 057504 (1998).
74. A. D. Sakharov, Pis'ma Zh. Eksp. Teor. Fiz. **5**, 32-35 (1967) [JETP Lett. **5**, 24-27 (1967)].
75. M. P. Worah, Phys. Rev. D **56**, 2010-2018 (1997); Phys. Rev. Lett. **79**, 3810-3813 (1997); M. Carena, M. Quiros, A. Riotto, I. Vilja, and C. E. M. Wagner, Nucl. Phys. **B503**, 387-404 (1997); M. P. Worah, in *Kaon Physics* [27].
76. P. Ramond, this Workshop, and references therein.
77. G. 't Hooft, Phys. Rev. Lett. **37**, 8-11 (1976); M. Fukugita and T. Yanagida, Phys. Lett. B **174**, 45-47 (1986); M. A. Luty, Phys. Rev. D **45**, 455-465 (1992); M. Plümacher, Zeit. Phys. C **74**, 549-559 (1997); W. Buchmüller and M. Plumacher, Phys. Lett. B **389**, 73–77 (1996); **431**, 354-362 (1998); Phys. Rep. **320**, 329–339 (1999).
78. J. L. Rosner, in *The Albuquerque Meeting* (Proceedings of the 8th Meeting, Division of Particles and Fields of the American Physical Society, Aug. 2–6, 1994, The University of New Mexico), edited by S. Seidel (World Scientific, Singapore, 1995), pp. 321-350; M. P. Worah, Phys. Rev. D **53**, 3902-3912 (1996).

Experimental Tests of CPT Invariance

D. Zavrtanik

Nova Gorica Polytechnic, Nova Gorica and
J. Stefan Institute, Ljubljana, Slovenia
(danilo.zavrtanik@ses-ng.si)

for the CPLEAR Collaboration

A. Angelopoulos[1], A. Apostolakis[1], E. Aslanides[11], G. Backenstoss[2], P. Bargassa[13], O. Behnke [17], A. Benelli [9], V. Bertin[11], F. Blanc[7,13], P. Bloch[4], P. Carlson[15], M. Carroll[9], E. Cawley[9], M.B. Chertok[3], M. Danielsson[15], M. Dejardin[14], J. Derre[14], A. Ealet[11], C. Eleftheriadis[16], L. Faravel [7], P. Fassnacht[11], W. Fetscher[17], M. Fidecaro[4], A. Filipčič[10], D. Francis[3], J. Fry[9], E. Gabathuler[9], R. Gamet[9], H.- J. Gerber[17], A. Go[4], , A. Haselden[9], P.J. Hayman[9], F. Henry-Couannier[11], R.W. Hollander[6], K. Jon-And[15], P.-R. Kettle[13], P. Kokkas[4], R. Kreuger[6], R. Le Gac[11], F. Leimgruber[2], I. Mandić[10], N. Manthos[8], G. Marel[14], M. Mikuž[10], J. Miller[3], F. Montanet[11], A. Muller[14], T. Nakada[13], B. Pagels [17], I. Papadopoulos[16], P. Pavlopoulos[2], G. Polivka[2], R. Rickenbach[2], B.L. Roberts[3], T. Ruf[4], M. Schäfer[17], L.A. Schaller[7], T. Schietinger[2], A. Schopper[4], L. Tauscher[2], C. Thibault[12], F. Touchard[11], C. Touramanis[9], C.W.E. Van Eijk[6], S. Vlachos[2], P. Weber[17], M. Wolter[17], D. Zavrtanik[10] and D. Zimmerman[3]

[1] *University of Athens, Greece,* [2] *University of Basle, Switzerland,* [3] *Boston University, USA,* [4] *CERN, Geneva, Switzerland,* [5] *LIP and University of Coimbra, Portugal,* [6] *Delft University of Technology, Netherlands,* [7] *University of Fribourg, Switzerland,* [8] *University of Ioannina, Greece,* [9] *University of Liverpool, UK,* [10] *J. Stefan Inst. and Phys. Dep., University of Ljubljana, Slovenia,* [11] *CPPM, IN2P3-CNRS et Université d'Aix-Marseille II, France,* [12] *CSNSM, IN2P3-CNRS, Orsay, France,* [13] *Paul Scherrer Institut(PSI), Villigen, Switzerland,* [14] *CEA, DSM/DAPNIA, CE-Saclay, France,* [15] *Royal Institute of Technology, Sweden,* [16] *University of Thessaloniki, Greece,* [17] *ETH-IPP Zürich, Switzerland*

Abstract. The CPLEAR experiment at CERN has directly studied matter and antimatter symmetries via the measurement of the time evolution of K^0 and $\bar{K}^0$. The CPT violation parameter $\mathcal{R}e(\delta)$ was directly measured with a precision of a few 10^{-4} while $\mathcal{I}m(\delta)$ is determined from the Bell-Steinberger relation, with a precision of 10^{-5}. The mass and decay-width equality between the K^0 and $\bar{K}^0$ were tested down to the level of 10^{-18} GeV.

Introduction

According to our present knowledge of weak interactions, the three discrete symmetries C, P and T are not exact symmetries in our Universe. The same applies to the combined symmetry CP. This fact has been experimentally well established and well accomodated in the Standard Model, but its origin is not yet understood. Within the framework of a local field theory, of Lorentz invariance and of the usual spin-statistics requirements, the triple product of C, P and T symmetries represents an exact symmetry expressed by the CPT theorem [1]. CPT invariance, being a natural consequence of local quantum field theory, warranties the equality of lifetimes and masses of particles and antiparticles. It is conceivable, however, that a small violation of CPT symmetry could occur in extensions of non-local theories. Thus, it is imperative to check CPT invariance wherever possible.

For the experimental verification of the validity of the CPT invariance we have to consider the following points:

- The most commonly used CPT tests, such us lifetime and mass equalities between particles and antiparticles, do not verify the full extent od CPT invariance (see for example [2]).

- Before the CPLEAR results, the set of measurements of all related variables was incomplete. Therefore, their claimed experimental accuracy very often depends on hidden theoretical assumptions [3], [4], [5], or experimental inter-correlations between measured quantities [6].

The neutral kaon system is a unique experimental environment to test discrete symmetries. Yet, it is the only system in nature where the violation of CP symmetry has been observed. In addition, experimental observation of interference effects allows direct measurement of T violation and CPT tests with a precision unachievable in other systems.

CPT invariance can be studied by studying the time evolution of a K^0 or $\bar{K}^0$ and the corresponding change in semileptonic decay rates. The lack of CPT invariance would appear as an asymmetry when comparing K^0 or $\bar{K}^0$ semileptonic rates.

Phenomenology

In the semileptonic decays of the neutral kaons there are four independent rates, depending on the strangeness of the neutral kaon (K^0 or $\bar{K}^0$) at the production time $t = 0$ and on the charge of the decay lepton (e^+ or e^-),

$$R_+(\tau) = R[K^0_{t=0} \to e^+\pi^-\nu_{t=\tau}], \quad \bar{R}_-(\tau) = R[\bar{K}^0_{t=0} \to e^-\pi^+\bar{\nu}_{t=\tau}],$$
$$R_-(\tau) = R[K^0_{t=0} \to e^-\pi^+\bar{\nu}_{t=\tau}], \quad \bar{R}_+(\tau) = R[\bar{K}^0_{t=0} \to e^+\pi^-\nu_{t=\tau}].$$

The above four rates can be parametrized as a function of the mixing parameters ϵ (T-violation parameter) and δ (CPT-violation parameter):

$$\epsilon = \frac{\Lambda_{\bar{K}^0,K^0} - \Lambda_{K^0,\bar{K}^0}}{2(\lambda_L - \lambda_S)} \quad \text{and} \quad \delta = \frac{\Lambda_{\bar{K}^0,\bar{K}^0} - \Lambda_{K^0,K^0}}{2(\lambda_L - \lambda_S)}.$$

Here, $\Lambda_{i,j}$ are the elements of the effective Hamiltonian Λ and $\lambda_{L,S} = m_{L,S} - \frac{i}{2}\Gamma_{L,S}$ its eigenvalues. $m_{L,S}$ and $\Gamma_{L,S}$ are the mass and decay width for the K_L and K_S states, $\Delta m = m_L - m_S$ and $\Delta\Gamma = \Gamma_S - \Gamma_L$. The K_L mixing parameter is defined as $\epsilon_L = \epsilon - \delta$.

The decay amplitudes corresponding to the four rates can be written as [7], [8]

$$< e^+\pi^-\nu| \Lambda |K^0 > = a + b, \quad < e^-\pi^+\bar{\nu}| \Lambda |\overline{K}^0 > = a^* - b^*,$$
$$< e^-\pi^+\bar{\nu}| \Lambda |K^0 > = c + d, \quad < e^+\pi^-\nu| \Lambda |\overline{K}^0 > = c^* - d^*.$$

The amplitudes b and d are CPT-violating, c and d describe possible violations of the $\Delta S = \Delta Q$ rule and the imaginary parts are all T-violating. The quantities

$$x = \frac{c^* - d^*}{a + b} \quad \text{and} \quad \bar{x} = \frac{c^* + d^*}{a - b}$$

describe the violation of the $\Delta S = \Delta Q$ rule in decays into positive and negative leptons, respectively, while $y = -b/a$ describes CPT violation in semileptonic decays in the case where $\Delta S = \Delta Q$ rule holds. The parameters $x_+ = (x + \bar{x})/2$ and $x_- = (x - \bar{x})/2$ describe therefore violation of the $\Delta S = \Delta Q$ rule in CPT conserving and CPT violating amplitudes, respectively. We assume $x, \bar{x}$ and $y \ll 1$.

The CPLEAR Experiment

Experimental tests of CPT invariance require pure K^0 and $\overline{K}^0$ beams. In addition, strangeness of the kaon at the decay time has to be known on the event by event basis. These facts influence the design of the experiment as follows.

Initially pure K^0 and $\overline{K}^0$ states are produced concurrently in the $p\bar{p}$ annihilation channels $\pi^+ K^- K^0$ and $\pi^- K^+ \overline{K}^0$ each with a branching ratio of $\approx 2 \times 10^{-3}$. The strangeness of the neutral kaon is tagged by the charge of the acompanying kaon and is therefore known for each event. The mass and momentum of the produced neutral kaons are obtained from measurement of the $\pi^\pm K^\mp$ pair kinematics. At subsequent time $t = \tau$ neutral kaons are detected and strangeness-tagged if they undergo a semileptonic decay ($\Delta S = \Delta Q$ rule), by the charge of the decay electron.

The experimental apparatus [9] is shown in Fig. 1. The detector had a typical near-4π geometry and was mounted inside a solenoid magnet with a field of 0.44 T. The 200 MeV/c $\bar{p}$ provided by LEAR were stopped in a pressurized hydrogen gas target. A series of cylindrical tracking devices provided information about the trajectories of charged particles. The spatial resolution $\sigma \approx 300\mu m$ was sufficient to locate the annihilation vertex, as well as the decay vertex of neutral kaons decaying to charged particles with a precision of a few mm in the transverse plane. Together with the momentum resolution of $\sigma_p/p \approx (5-10)\%$, this enabled a lifetime resolution of $\sigma \approx (5-10) \times 10^{-12} s$. The tracking detectors were followed by the particle identification detector (PID) which comprised a threshold Čerenkov detector, mainly effective for K/π separation above 350 MeV/c momentum and scintillators which measured the energy loss (dE/dx) and the time of flight

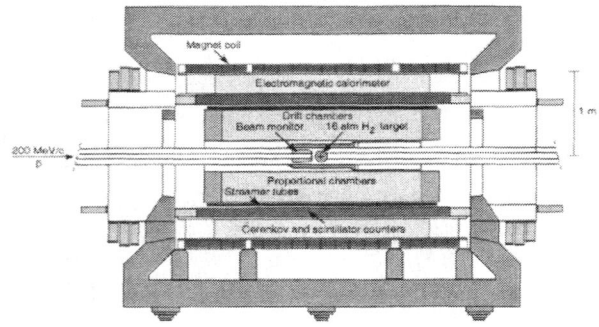

FIGURE 1. The general layout of the CPLEAR detector.

of charged particles. The PID was also used to separate electrons from pions below 350 MeV/c. The outermost detector was a lead/gas sampling calorimeter designed to detect photons of the neutral kaon decays to neutral pions. It also provided e/π separation at higher momenta (p > 300 MeV/c). An online event reconstructiom was performed by a sophisticated multilevel trigger system allowing the detector to operate at a $\bar{p}$ rate of 1 MHz.

The detector has run smoothly during five years and accumulated nearly 2×10^8 decays of strangeness-tagged neutral kaons entering the final data sample with a decay time $\tau > 1\tau_S$. The presented results correspond to 1.3×10^6 reconstructed $e\pi\nu$ decays.

Direct test of CPT invariance

To extract the CPT violation parameter $\mathcal{R}e(\delta)$ in an optimal way we build the time-dependent decay-rate asymmetry A_δ, defined as

$$A_\delta(\tau) = \frac{\bar{R}_+ - R_-(1 + 4\mathcal{R}e(\epsilon_L))}{\bar{R}_+ + R_-(1 + 4\mathcal{R}e(\epsilon_L))} + \frac{\bar{R}_- - R_+(1 + 4\mathcal{R}e(\epsilon_L))}{\bar{R}_- + R_+(1 + 4\mathcal{R}e(\epsilon_L))}. \tag{1}$$

By considering $\mathcal{R}e(a)$ to be of the order of unity and taking into account relation $\mathcal{R}e(\epsilon_L) = \mathcal{R}e(\epsilon) - \mathcal{R}e(\delta)$, Eq. (1) can be written, to first order in the small parameters, as

$$A_\delta(\tau) = 2\frac{\mathcal{I}m(x_+)e^{-\frac{1}{2}(\Gamma_S+\Gamma_L)\tau}\sin(\Delta m\tau) + \mathcal{R}e(x_-)E_-(\tau)}{E_+(\tau) - e^{-\frac{1}{2}(\Gamma_S+\Gamma_L)\tau}\cos(\Delta m\tau)}$$

$$+ \frac{-4\mathcal{R}e(\delta)E_-(\tau) - 2\mathcal{R}e(x_-)E_-(\tau) + [2\mathcal{I}m(x_+) + 4\mathcal{I}m(\delta)]e^{-\frac{1}{2}(\Gamma_S+\Gamma_L)\tau}\sin(\Delta m\tau)}{E_+(\tau) + e^{-\frac{1}{2}(\Gamma_S+\Gamma_L)\tau}\cos(\Delta m\tau)}$$

$$+ 4\mathcal{R}e(\delta) \quad (2)$$

where

$$E_\pm(\tau) = \frac{e^{-\Gamma_S\tau} \pm e^{-\Gamma_L\tau}}{2}$$

We note that the asymmetry $A_\delta(\tau)$ does not depend on the parameter y. For lifetimes comparable with $1/\Gamma_S$, A_δ is sensitive to $\mathcal{I}m(\delta)$, $\mathcal{I}m(x_+)$ and $\mathcal{R}e(x_-)$, while for long lifetimes it depends only on $\mathcal{R}e(\delta)$, becoming simply

$$A_\delta(\tau) = 8\mathcal{R}e(\delta), \quad (3)$$

thus allowing $\mathcal{R}e(\delta)$ to be measured without any assumption on the $\Delta S = \Delta Q$ rule.

The detection efficiencies of $K^+\pi^-$ and $K^-\pi^+$ pairs, used to tag $\bar{K}^0$ and K^0, respectively, are not identical due to the different strong interaction cross-sections of opposite-charge kaons and pions with matter. The difference of the two efficiencies is of the order of 12 %. To restore the initial symmetry at the production of K^0 and $\bar{K}^0$, we introduce a normalization factor $\xi = \epsilon(K^+\pi^-)/\epsilon(K^-\pi^+)$ which is the ratio of the detection efficiencies for the $K^+\pi^-$ and $K^-\pi^+$ pairs. Similarly, the detection efficiencies of the two final states $e^-\pi^+\bar{\nu}$ and $e^+\pi^-\nu$ are not identical. To take this into account, we introduce a second normalization factor $\eta = \epsilon(\pi^+e^-)/\epsilon(\pi^-e^+)$. Correcting the measured numbers of semileptonic decays with these factors, we construct the asymmetry

$$A_\delta^{exp}(\tau) = \frac{\eta\bar{N}_+(\tau) - \alpha_{2\pi}N_-(\tau)}{\eta\bar{N}_+(\tau) + \alpha_{2\pi}N_-(\tau)} + \frac{\bar{N}_-(\tau) - \eta\alpha_{2\pi}N_+(\tau)}{\bar{N}_-(\tau) + \eta\alpha_{2\pi}N_+(\tau)} \quad (4)$$

where $N_+(N_-)$ and $\bar{N}_+(\bar{N}_-)$ stand for the observed numbers of initial K^0 and $\bar{K}^0$ accompanied by a decay $e^+(e^-)$.

Due to its construction with $\alpha_{2\pi}$ instead of ξ, the additional factor $[1+4\mathcal{R}e(\epsilon-\delta)]$ causes a cancellation of terms with $\mathcal{R}e(\epsilon)$ and $\mathcal{R}e(y)$ in A_δ and the long-lifetime expansion reads $A_\delta \xrightarrow{\tau \gg \tau_S} 8\mathcal{R}e(\delta)$.

The experimental asymmetry is shown in Fig. 2. Fitting A_δ with the complete phenomenological expression [10], depending on $\mathcal{R}e(\delta)$ and $\mathcal{I}m(\delta)$, $\mathcal{I}m(x_+)$, $\mathcal{R}e(x_-)$, yields the result

$$\mathcal{R}e(\delta_{CPT}) = (3.0 \pm 3.3_{stat} \pm 0.6_{syst}) \times 10^{-4}$$
$$\mathcal{I}m(\delta_{CPT}) = (-1.5 \pm 2.3_{stat} \pm 0.3_{syst}) \times 10^{-2}$$
$$\mathcal{R}e(x_-) = (0.2 \pm 1.3_{stat} \pm 0.3_{syst}) \times 10^{-2}$$
$$\mathcal{I}m(x_+) = (1.2 \pm 2.2_{stat} \pm 0.3_{syst}) \times 10^{-2} \quad (5)$$

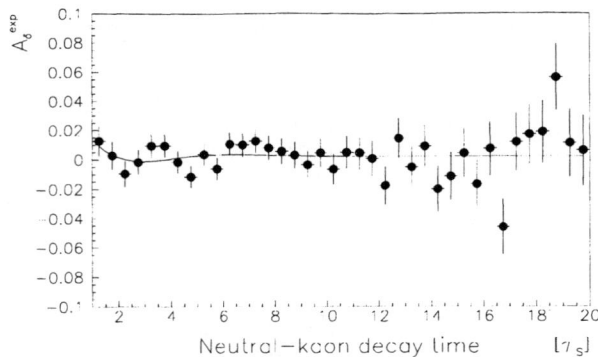

FIGURE 2. The asymmetry A_δ^{exp} versus the neutral-kaon decay time. The solid line represents the result of the fit.

with a high degree of correlation between $\mathcal{I}m(\delta), \mathcal{I}m(x_+)$ and $\mathcal{R}e(x_-)$.

When $\Delta Q = \Delta S$ ($x = 0$) is assumed in the fit, the error on $\mathcal{R}e(\delta)$ decreases by 25 % and by an order of magnitude on $\mathcal{I}m(\delta)$. The precision on $\mathcal{R}e(\delta)$ is thus improved by a factor of 75 and that on $\mathcal{I}m(\delta)$ by a factor of 12 compared to previous measurements performed under the same conditions [3].

Indirect test of CPT invariance

CPT invariance can be tested also indirectly by studying the constraints on our results derived from the Bell-Steinberger relation

$$\mathcal{R}e(\epsilon) + i\mathcal{I}m(\delta) = \frac{\Gamma + i\Delta m}{2\Gamma^2 + \Delta m^2} \sum A_{L,f}^* A_{S,f} \qquad (6)$$

which relates all decay channels of neutral kaons to the parameters describing T and CPT non-invariance. With the present precision of the two-pion decay parameters the dominant uncertainties arise from the three-pion and semileptonic decays. Moreover, the semileptonic decays enter the relation with the parameter $\mathcal{R}e(y)$, describing CPT violation in semileptonic decays and yet not measured. Having improved the precision of three-pion decays [11], [12] and measured precisely the semileptonic decay rates [13], [10] we obtain

$$\mathcal{R}e(y) = (0.3 \pm 3.1) \times 10^{-3}$$
$$\mathcal{I}m(\delta) = (2.4 \pm 5.0) \times 10^{-5} \qquad (7)$$

Details of the analysis can be found elsewhere [14]. Our result on $\mathcal{I}m(\delta)$ is almost one order of magnitude more accurate than that of a previous similar analysis [4] owing to improvements in the accuracy of various measurements where CPLEAR has made significant contributions. In addition, $\mathcal{R}e(y)$ was determined for the first time: a result which could not be achieved from semileptonic data alone.

K^0 and $\bar{K}^0$ mass and decay width differences

The CPT violation parameter δ can be written as

$$\delta = |\delta| \times e^{i(\Phi_{SW} - \Phi_{CPT} - \frac{\pi}{2})} \qquad (8)$$

with $\Phi_{CPT} = arctan[\frac{1}{2}(\Gamma_{\bar{K}^0\bar{K}^0} - \Gamma_{K^0K^0})/(M_{\bar{K}^0\bar{K}^0} - M_{K^0K^0})]$. It is conveniently represented in the complex plane [7] by the projections along Φ_{SW} axis ($\delta_{\parallel}$) and its normal ($\delta_{\perp}$):

$$\delta_{\parallel} = \frac{1}{4} \frac{\Gamma_{K^0K^0} - \Gamma_{\bar{K}^0\bar{K}^0}}{\sqrt{\Delta m^2 + (\frac{\Delta \Gamma}{2})^2}} \quad \text{and} \quad \delta_{\perp} = \frac{1}{2} \frac{M_{K^0K^0} - M_{\bar{K}^0\bar{K}^0}}{\sqrt{\Delta m^2 + (\frac{\Delta \Gamma}{2})^2}} \qquad (9)$$

The parameters $\delta_{\parallel}$ and $\delta_{\perp}$ can be expressed as functions of the measured quantities $\mathcal{R}e(\delta)$, $\mathcal{I}m(\delta)$ and Φ_{SW} as

$$\delta_{\parallel} = \mathcal{R}e(\delta)cos(\Phi_{SW}) + \mathcal{I}m(\delta)sin(\Phi_{SW}),$$
$$\delta_{\perp} = -\mathcal{R}e(\delta)sin(\Phi_{SW}) + \mathcal{I}m(\delta)cos(\Phi_{SW}), \qquad (10)$$

and allow in turn the K^0 - $\bar{K}^0$ decay-width and mass differences to be determined as

$$\Gamma_{K^0K^0} - \Gamma_{\bar{K}^0\bar{K}^0} = \delta_{\parallel} \times \frac{2\Delta\Gamma}{cos(\Phi_{SW})}$$
$$M_{K^0K^0} - M_{\bar{K}^0\bar{K}^0} = \delta_{\perp} \times \frac{\Delta\Gamma}{cos(\Phi_{SW})} \qquad (11)$$

Thus the evaluation of the K^0 - $\bar{K}^0$ mass and decay-width differences is straightforward once the CPT violation parameters $\mathcal{R}e(\delta)$ and $\mathcal{I}m(\delta)$ are known.

With the values of $\mathcal{R}e(\delta)$ and $\mathcal{I}m(\delta)$ given in (5) and (7), together with $\Phi_{SW} = (43.5 \pm 0.1)^{\circ}$ [3] we obtain from Eqs. (10)

$$\delta_{\parallel} = (1.9 \pm 2.0) \times 10^{-4}$$
$$\delta_{\perp} = (-1.5 \pm 2.0) \times 10^{-4} \qquad (12)$$

and subsequently from Eqs. (11)

$$\Gamma_{K^0K^0} - \Gamma_{\bar{K}^0\bar{K}^0} = (3.9 \pm 4.2) \times 10^{-18} GeV$$
$$M_{K^0K^0} - M_{\bar{K}^0\bar{K}^0} = (-1.5 \pm 2.0) \times 10^{-18} GeV \qquad (13)$$

with a correlation coefficient of -0.95. Fig. 3 shows the error elipses corresponding to 1σ, 2σ and 3σ. Our result on the mass difference is a factor of two better than the one obtained with a similar calculation in Ref. [15]. The improvement is mainly due to $\mathcal{R}e(\delta)$ being now known with a smaller error.

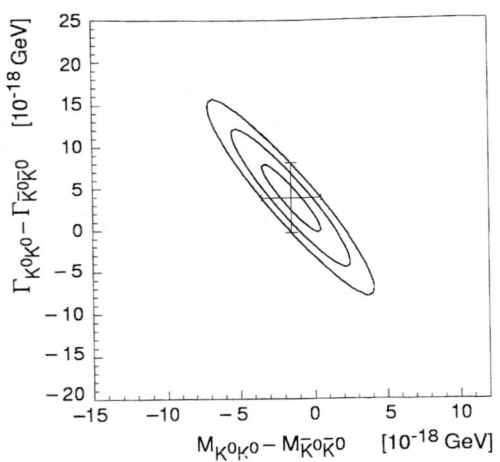

FIGURE 3. The K^0 - $\bar{K}^0$ decay-width versus mass difference. The 1σ, 2σ and 3σ are also shown.

Summary

We have directly tested the CPT invariance through the measurement of $\mathcal{R}e(\delta)$ and $\mathcal{I}m(\delta)$ with a precision of the order of 10^{-4} and 10^{-2}, respectively. Moreover, we have evaluated $\mathcal{I}m(\delta)$ from various measurements, many of which come from CPLEAR, using the unitarity relation. The error is one order of magnitude smaller than for $\mathcal{R}e(\delta)$. We have shown that K^0 - $\bar{K}^0$ mass and decay-width differences are consistent with CPT invariance within a few 10^{-18} GeV. The present precision is limited by the error of $\mathcal{R}e(\delta)$ and the next limiting parameter comes from the error of η_{000}.

REFERENCES

1. J.S. Bell, Proc. Royal Soc. **A231** (1955) 479;
   G.L. Lüders, Ann. Phys. **2** (1957) 1;
   R. Jost, Helv. Phys. Acta **30** (1957) 409.
2. T.D. Lee, Particle physics and introduction to field theory, Harwood Academic Publishers, New York (1981).
3. C. Caso et al., Particle Data Group, Eur. Phys. J. **C3** (1998) 1.
4. K.R. Schubert et al., Phys. Lett. **B31** (1970) 662.
5. G.P. Thomson and Y. Zou, Phys. Rev. **D51** (1995) 1412.
6. B. Schwingenheuer et al., Phys. Rev. Lett. **74** (1995) 4376.
7. C.D. Buchanan et al., Phys. Rev. **D45** (1992) 4088.

8. L. Maiani, CP and CPT Violation in Neutral Kaon System, in: L. Maiani et al. (Eds.), The Second DAΦNE Physics Handbook, INFN, Frascati, 1995, p. 3. The parametrizations presented there are equivalent to ours but formulated ia a slightly different notation.
9. R. Adler et al., CPLEAR Collaboration, Nucl. Instr. Methods **A379** (1996) 76.
10. CPLEAR Collab.: A. Angelopoulos et al., Phys. Lett. **B444** (1998) 52.
11. CPLEAR Collab.: A. Agelopoulos et al., Eur. Phys. J. **C5** (1998) 389.
12. CPLEAR Collab.: A. Angelopoulos et al., Phys. Lett. **B425** (1998) 391.
13. CPLEAR Collab.: A. Angelopoulos et al., Phys. Lett. **B444** (1998) 43.
14. CPLEAR Collab.: A. Apostolakis et al., Phys. Lett. **B456** (1999) 297.
15. E. Shabalin, Phys. Lett. **B369** (1996) 335.

B Physics Prospect at Tevatron Run II

Kin Yip[1]

Fermilab, P.O. Box 500 - MS 352, Batavia, IL 60510-0500, U.S.A.

Abstract. We give an account of various possible measurements in B physics for the collider experiments, CDF and DØ, in the coming Tevatron Run II at Fermilab. This includes the measurements of $\sin(2\beta)$ and other angles in the unitarity triangle related to the CKM matrix, which may lead to observation of CP violation outside the kaon system. Measurements of mass and lifetime differences, especially for B_s, will be particularly interesting at Tevatron as this probably would not be studied in the present B factories.

I B PHYSICS AT HADRON COLLIDER

At hadron colliders such as Tevatron at Fermilab, b (bottom) quarks are produced copiously by the strong interaction which fragment into all kinds of b-hadron states such as $B^{\pm}$, B^0, B_s, B_c, Λ_b and their excited states. These B mesons and baryons then decay to lowest energy states via weak interactions which provide the opportunities for us to study weak decays. These weak decays also provide information on the CKM (Cabibbo-Kobayashi-Maskawa) matrix elements such as V_{tb}, V_{ts} and V_{td} to test the electroweak theory.

Moreover, since particles like B_s, B_c, Λ_b are not produced at B factories running at the $\Upsilon(4S)$ resonance, hadron colliders provide unique capabilities for studying these particles and their interactions.

The hadron collider also allows us to test the QCD (Quantum Chromodynamics) theory. Since the mass of the b quark is much larger than the QCD dimensional parameter, Λ_{QCD}, various perturbation theories can be applied to calculate the properties of the inclusive b production and this would allow us a chance to test the perturbative QCD theory. The production of $b\bar{b}$ at the energy scale of Tevatron is dominated by gluon processes and so the gluon structure functions can also be probed by the experimenters.

---

[1] for the CDF and DØ Collaborations.

II EXPECTED DETECTOR PERFORMANCE FOR CDF AND DØ

The $b\bar{b}$ production cross-section at Tevatron is several orders of magnitudes higher than the e^+e^- machines, but the inelastic scattering cross-section is even larger. Typically, the signal to background ratio is only about 10^{-3}. In order to study b decays which belong to the lower end of the energy and momentum spectra at Tevatron, specialized lepton and di-lepton triggers such as $J/\psi \rightarrow \mu^+\mu^-$ are required. In Run II, both the CDF and DØ experiments will move the track triggers to Level 1 and will also have displaced track triggers at Level 2 using the silicon vertex detectors (SVX). This will allow both experiments to trigger and study purely hadronic B decays such as $B^0 \rightarrow \pi^+\pi^-$ and $B_s^0 \rightarrow D_s^-\pi^+$. Precise secondary vertex reconstruction is necessary for any of these B decay studies to succeed.

The Run II detector upgrade information for CDF and DØ experiments can be found in [1]. Only some highlights that are most relevant to the B physics studies are mentioned here.

CDF will have a new silicon vertex detector in Run II which makes use of the new SVX-III chip for electronics readout. The essentially deadtime-less feature of the SVX-III chip would increase the trigger bandwidth, which is particularly important for collecting B physics events because they have the highest trigger rates. CDF will also install a low-mass radiation-hard single-sided axial-strip silicon detector, called "Layer 00", at the very small radius of about 1.4 cm, just outside the beampipe. This silicon detector is expected to achieve an impact parameter resolution of ($5.5 + 17/P_T$) μm, where P_T is the transverse momentum.

Another new detector of CDF called "Time of Flight" (TOF) is also added between the Central Outer Tracker and the superconducting solenoid magnet. This detector would employ precision timing electronics and achieve a timing resolution for an individual particle of about 100 ps. This translates into a $K - \pi$ separation at about 1.2 σ (standard deviation) level for all momentum and about 2 σ level for momentum p < 1.6 GeV/c.

As an important part of the Run II upgrade, DØ now has a 2.8 m long superconducting solenoid magnet with a magnetic field strength of 2 Tesla, inside the calorimeter. With the scintillating fiber tracker and the silicon vertex detector inside the solenoid, the momentum resolution dP_T/P_T^2 will be about 0.002. The tracking efficiency is expected to be about 95% and the tracking system has a wide coverage in both barrel and forward regions, with pseudo-rapidity $|\eta|$ down to 3.5. The primary vertex resolutions in $r - \phi$ and $r - z$ are about 15 - 30 μm and 50 μm respectively; whereas the secondary vertex resolutions are about 40 μm in $r - \phi$ and 100 μm in $r - z$.

The stronger side of DØ is in lepton including excellent coverage, trigger efficiency and identification capability. The addition of central and forward preshower in Run II upgrade adds to the strength of electron triggering and identification. There is also a major upgrade in the muon system, especially in its trigger front-end electronics. The momentum threshold of the muon trigger can be as low as 1.5 GeV/c.

Furthermore, DØ also plans to have impact parameter trigger at Level 2, using the new silicon tracker. This will further enhance DØ's reach in B physics studies.

III MEASUREMENT OF $\sin(2\beta)$ FROM $B \to J/\psi K_S^0$

The possible manifestations of CP violation may be generally classified as follows:

1. CP violation in the decay or also often called direct CP violation, which occurs in both charged and neutral decays;

2. CP violation in the mixing or also often referred to as indirect CP violation, when the neutral mass eigenstates are not CP eigenstates;

3. CP violation in the interference between decays with and without mixing, which occurs in decays into final states that are common to B^0 and $\overline{B}^0$.

What we are interested in here is the 3rd type of CP violation, in which the final states are J/ψ and K_s. The flavor eigenstates oscillate and mix and the mass eigenstates can be expressed:

$$|B_L\rangle = p\,|B^0\rangle + q\,|\overline{B}^0\rangle,$$
$$|B_H\rangle = p\,|B^0\rangle - q\,|\overline{B}^0\rangle, \qquad (1)$$
$$\Delta m_B = M_{B_H} - M_{B_L}$$

where B_L and B_H stand for the light and heavy mass eigenstates respectively, and Δm_{B_H} is the mass difference between these two mass eigenstates and is in fact also the frequency of oscillation between the two flavor eigenstates, B^0 and $\overline{B}^0$. p and q are the complex coefficients which obey the normalization condition $|p|^2 + |q|^2 = 1$.

As depicted in Fig. 1, both B^0 and $\overline{B}^0$ may decay to the same final states, J/ψ and K_s, but their decay rates may be different. The CP violation manifests itself in the time dependent asymmetry a_F:

$$a_F = \frac{\Gamma(\overline{B}^0 \to F) - \Gamma(B^0 \to F)}{\Gamma(\overline{B}^0 \to F) + \Gamma(B^0 \to F)} \qquad (2)$$

where F stands for the final state, $J/\psi\, K_s$.

By convention, the coupling between the mass eigenstates (d,s,b) and flavor eigenstates (d', s', b') is often expressed in terms of the 3×3 unitary CKM matrix:

$$\begin{pmatrix} d' \\ s' \\ b' \end{pmatrix} = \begin{pmatrix} V_{ud} & V_{us} & V_{ub} \\ V_{cd} & V_{cs} & V_{cb} \\ V_{td} & V_{ts} & V_{tb} \end{pmatrix} \begin{pmatrix} d \\ s \\ b \end{pmatrix} \qquad (3)$$

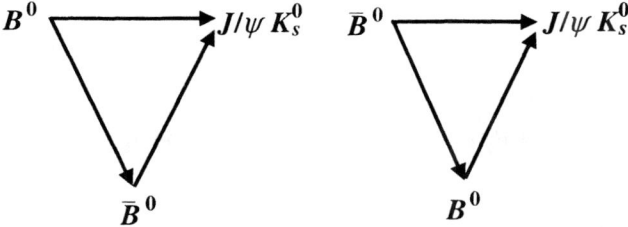

$$\Gamma(B^0 \to J/\psi K_s^0) \neq \Gamma(\bar{B}^0 \to J/\psi K_s^0)$$

FIGURE 1. Both B^0 and $\bar{B}^0$ may decay to the common final states, J/ψ and K_s, but their decay rates (Γ) may be different.

In the Standard Model, CP violation arises from a single complex phase in the CKM matrix. It is customary to express the CKM matrix $\mathcal{W}$ in terms of four Wolfenstein parameters (λ, A, ρ, η):

$$\mathcal{W} \simeq \begin{pmatrix} 1 - \lambda^2/2 & \lambda & A\lambda^3(\rho - i\eta) \\ -\lambda & 1 - \lambda^2/2 & A\lambda^2 \\ A\lambda^3(1 - \rho - i\eta) & -A\lambda^2 & 1 \end{pmatrix} + \mathcal{O}(\lambda^4) \quad (4)$$

The unitarity of the CKM matrix implies various relations among its elements. One of the most useful unitary conditions is:

$$V_{tb}^* V_{td} + V_{cb}^* V_{cb} + V_{ub}^* V_{ud} = 0 \quad (5)$$

Typically, the constraint of this condition is visualized in the unitary triangle as in Fig. 2.

From the unitarity triangle, we have $\beta \equiv arg\left[-\frac{V_{cd} V_{cb}^*}{V_{td} V_{tb}^*}\right]$. For the decay mode $B_d \to J/\psi + K_s$, one finds:

$$\sin(2\beta) = Im\left[-\left(\frac{V_{tb}^* V_{td}}{V_{td} V_{tb}^*}\right)\left(-\frac{V_{cs}^* V_{cb}}{V_{cs} V_{cb}^*}\right)\left(-\frac{V_{cd}^* V_{cs}}{V_{cd} V_{cs}^*}\right)\right] = \frac{2\eta(1-\rho)}{\eta^2(1-\rho)^2} \quad (6)$$

The first part in the product term is due to $B^0 - \bar{B}^0$ mixing, the second due to the ratio of the transition amplitudes $\bar{A}/A$ (as appeared in Equation (4)) and the third due to $K^0 - \bar{K}^0$ mixing. In the Standard Model, the time dependent asymmetry is directly related to $\sin(2\beta)$ by the following relationship in leading order:

$$\frac{\Gamma(\bar{B}^0 \to J/\psi K_s) - \Gamma(B^0 \to J/\psi K_s)}{\Gamma(\bar{B}^0 \to J/\psi K_s) + \Gamma(B^0 \to J/\psi K_s)} \cong \sin(2\beta)\sin(\Delta m_d t) \quad (7)$$

$B_d \to J/\psi + K_s$ is said to be a gold-plated channel because

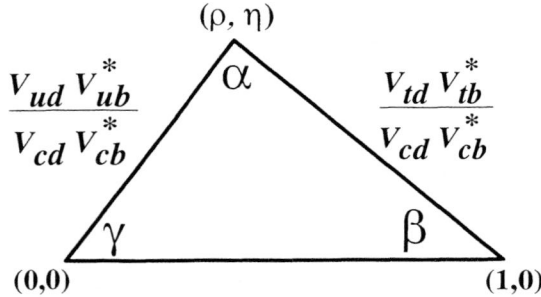

FIGURE 2. The unitarity triangle indicating the relationship between the CKM matrix elements.

- it has readily accessible final states ($K_s \to \pi^+\pi^-$ and $J/\psi \to \mu^+\mu^-$) with very small background;
- it has relatively large branching ratio;
- it has negligible theoretical uncertainty. Because the penguin contribution is expected to be very small since the $c\bar{c}$ must be popped from vacuum — mass suppression. And the dominant penguin diagram contribution to the asymmetry has the same phase as the tree level diagram.

CDF in Run I has published a measurement of

$$\sin(2\beta) = 0.79^{+0.41}_{-0.44} \tag{8}$$

in which the errors include both the statistical and systematic contributions [2]. This is the most precise experimental measurement of $\sin(2\beta)$ so far and it seems to favor a non-zero value of $\sin(2\beta)$.

In order to measure $\sin(2\beta)$, one needs to reconstruct the $B^0/\overline{B}^0$ at production from its decay products, ie., J/ψ to di-leptons and K_s to $\pi^+\pi^-$. This is briefly summarized in Fig. 3. One has to reconstruct the two vertices of J/ψ and K_s in each event. It is also necessary to tag the flavor of the B meson at production and measure its decay length precisely.

One of the most difficult parts of this analysis is to reconstruct K_s from its decay products, charged pions. Because it has relatively longer lifetime and may decay far away from the beamspot. If K_s decays far into the detector, it may miss some or all parts of the tracking system. DØ has done studies using full GEANT simulation [4] and offline reconstruction (which is being actively developed) to investigate the feasibility of the K_s reconstruction using the silicon and fiber trackers. The track reconstruction efficiency using the current offline reconstruction codes can be found

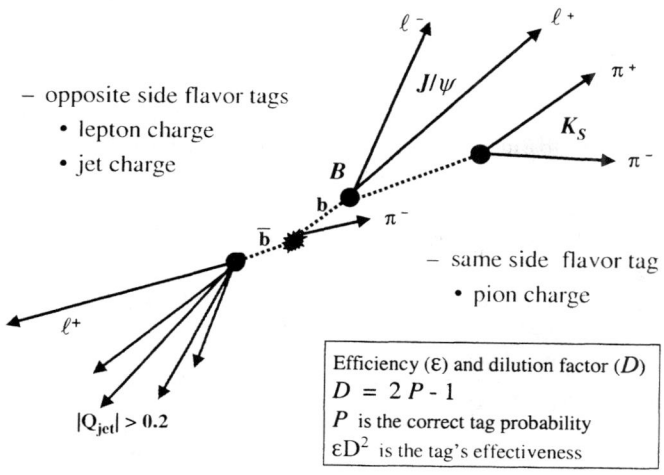

FIGURE 3. Illustration of the topology involved in the process $B^0 \to J/\psi + K_s$ and various B tagging methods.

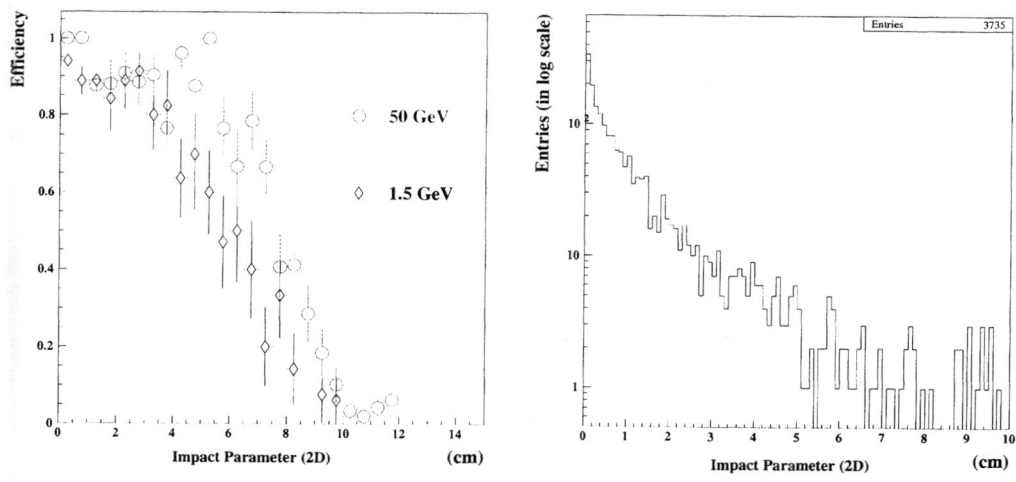

FIGURE 4. The plot on the left shows the efficiency of track reconstruction vs impact parameter of the charged particle tracks. The fluctuation is due to limited Monte Carlo statistics. The plot on the right shows the distribution of the impact parameters of charged pions from K_s.

in Fig. 4. Compared with the impact parameter distribution of the charged pion tracks from K_s, it seems that DØ should be able to reconstruct K_s rather efficiently.

One same-side flavor tag (SST) and two opposite-side flavor tags (OST) have been used by CDF in Run I in the measurement of $\sin(2\beta)$. The SST method makes use of the correlation between the charge of the nearby pion and the b quark charge due to fragmentation or B^{**} production [3]. This requires reconstruction of the soft pions. The OST methods rely on the lepton charge (if available) or the overall jet charge (if there is no lepton reconstructed) to determine the flavor of the b quark at production.

The tagging effectiveness of a tagging method in a measurement depends on ϵD^2, where ϵ is the tagging efficiency and D is the dilution factor. The dilution factor D is equal to $2P-1$, where P is probability that the method tags the flavor correctly.

The tagging effectiveness, ϵD^2, for CDF and DØ experiments are expected to be about the same, at $\sim 10\%$. DØ has larger lepton and forward tracking coverage than CDF; whereas CDF may be able to employ an additional opposite side kaon tag that DØ would not be able to do.

Using the time dependent analysis, the accuracy of the $\sin(2\beta)$ is given by

$$\sigma(\sin 2\beta) \approx e^{x_d^2 \Gamma^2 \sigma_t^2} \sqrt{\frac{1+4x_d^2}{2x_d}} \frac{1}{\epsilon D^2 N} \sqrt{1 + \frac{B}{S}} \qquad (9)$$

where x_d is the mixing parameter (which will be discussed a bit more in the next section), Γ is the decay width of B^0, σ_t is the time resolution of the decay time measurement (which is about 100 fs in DØ), ϵD^2 is the tagging effectiveness, N is the number of events, S/B is the signal to background ratio (expected to be about 0.75 in DØ).

Assuming a luminosity of 2 fb^{-1}, for the process $B^0 \to J/\psi + K_s$, DØ is expected to collect about 40,000 events in which J/ψ decays to $\mu^+\mu^-$ and about 30,000 events in which J/ψ decays to e^+e^-. Combining all these events, the overall $\sigma(\sin 2\beta)$ is expected to be as low as 0.03. CDF also expects to have similar precision in this measurement of $\sin(2\beta)$. If these predictions can be realized, they will be very competitive with even the future B factories.

IV NEUTRAL B FLAVOR OSCILLATIONS

There are two neutral B mesons in which two CP-conjugate states exist, B_d and B_s. In both of these systems, the mass eigenstates are not CP (flavor) eigenstates, but mixtures of the two CP-conjugate quark states. These neutral B mesons oscillate via the second order weak interaction involving charged current transition ($\Delta B = 2$ and $\Delta Q = 0$) as shown in the box diagrams in Fig. 5. There is a CP-violating phase (η in Equation (4)) that enters in the amplitude for these diagrams.

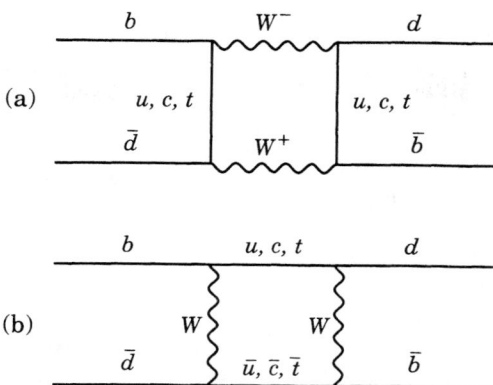

FIGURE 5. Box diagrams for $B^0 - \overline{B}^0$ mixing through the second order weak charged current transition ($\Delta B = 2$ and $\Delta Q = 0$).

The decay width differences $\Delta \Gamma_B = \Gamma_L - \Gamma_H$, where L and H have the same meanings as in Equation (1), of the neutral B mesons are expected to be much smaller than Δm_B. And, $\Delta \Gamma_B / \Gamma_B$ is $\approx \mathcal{O}(10^{-2})$.

The mixing of the B_d and B_s is governed by Δm_B and $\Delta \Gamma_B$. Different from the case of kaon, in the neutral B meson systems, it is the mass differences that dominate the physics and the two states have nearly equal predicted widths (and thus lifetimes). A useful parameter called "mixing parameter", x_B, is defined as

$$x_B = \frac{\Delta m_B}{\Gamma_B} \qquad (10)$$

The recent world averages of the experimental measurements of Δm_d (ie., the mass difference for B_d) and its lifetime τ_d (ie. $1/\Gamma_d$) are shown in Fig. 6. This roughly corresponds to a mixing parameter of $x_d = 0.754 \pm 0.029$ for B_d.

The oscillation frequency Δm_q, where q is either the d quark or the s quark, is proportional to $|V_{tb}^* V_{tq}|^2$. Since $V_{ts} \gg V_{td}$, the oscillation frequency (also the mass difference) Δm_s is expected to be much larger than Δm_d. The larger oscillation frequency of B_s makes it more difficult to measure compared to that of B_d. The relationship between the CKM matrix elements, the B meson masses and mass differences is given by [5]:

$$\frac{|V_{td}|}{|V_{ts}|} = (1.16 \pm 0.06) \cdot \sqrt{\frac{\Delta m_d \cdot m(B_s)}{\Delta m_s \cdot m(B_d)}} \qquad (11)$$

All the quantities on the right-handed side of the above equation have been quite

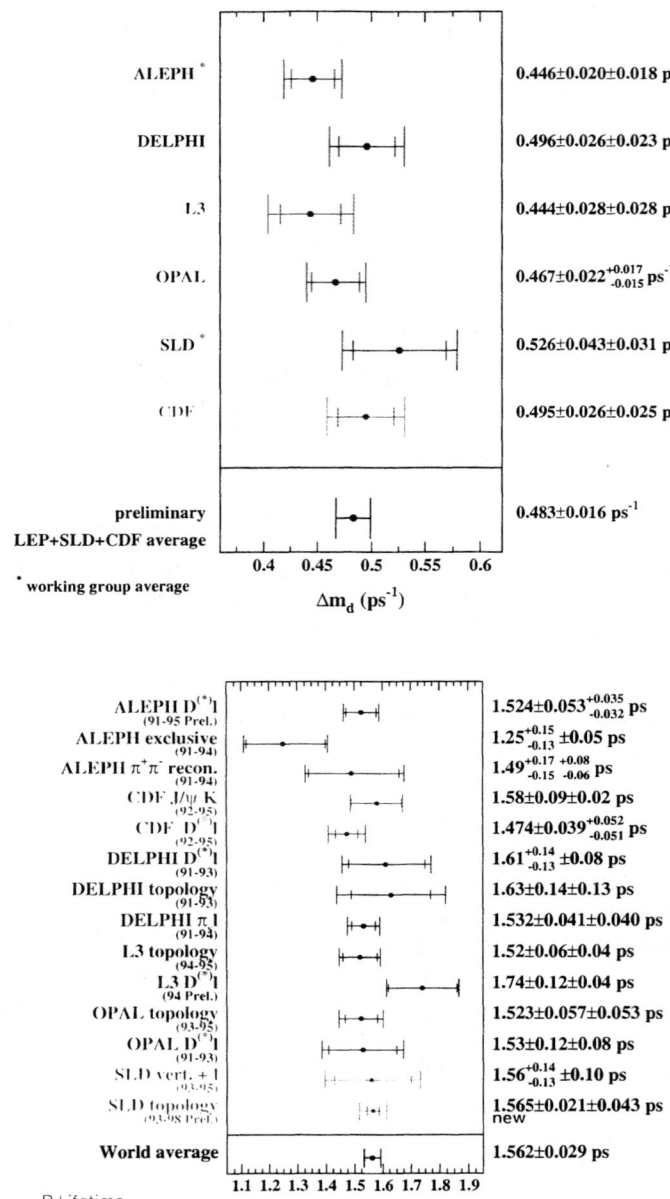

FIGURE 6. The world average of Δm_d and τ_d (ie. $1/\Gamma_d$) measurements for B_d from the LEP B Lifetimes and Oscillations Working Groups.

accurately measured except Δm_s. Improved measurement of Δm_s will give us further insight towards the CKM matrix elements.

Currently, there are only upper limits for the mass difference Δm_s, and the mixing parameter x_s for B_s [6]:

$$\Delta m_s > 9.8 \text{ ps}^{-1}, \ x_s > 14.6, \text{with confidence level} > 95\% \qquad (12)$$

A recent Standard Model global fit gives $m_s < 20$ ps^{-1} at 95% confidence level. [7]

Not very long time ago, some time integrated measurements were done to extract the mass difference Δm_d from the quantity $\chi_d \simeq \frac{1}{2}x_d^2/(1+x_d^2)$, which is the total probability that a created B_d decays as a $\overline{B}_d$. However, because of the large value of x_s, the quantity χ_s will be close to its upper limit of 0.5. This means that we cannot determine x_s accurately by measuring χ_s. Therefore, time dependent measurements are needed to determine Δm_d. This requires excellent time resolution to measure the proper decay time of the B_s meson in order to resolve the time dependent mixing of B_s system.

For studying B_s mixing, CDF would look at the process

$$B_s \to D_s^- \pi^+, \ D_s^- \pi^+ \pi^+ \pi^- \quad \text{with} \quad D_s^- \to \phi \pi^-, \ K^* K^- \qquad (13)$$

CDF expects to collect about 20,000 signal events in these channels. With the help of the TOF detector, the flavor tagging effectiveness, ϵD^2, is expected to be about 11%. The proper time resolution (σ_t) is expected to be

$$\sigma_t = 0.045 \text{ ps} \oplus t \cdot \sigma_{P_T}/P_T \qquad (14)$$

assuming the presence of "Layer 00" silicon vertex detector. The overall sensitivity of the CDF experiment to the measurement of x_s is given in Fig. 7. For a conclusive 5 σ measurement, CDF believes to be able to measure x_s up to the value of 63 (56) if the signal to background ratio S/B is 2/1 (1/2), assuming a luminosity of 2 fb^{-1}.

Though without the presence of specific hadron identification capability, DØ may also be able to measure x_s up a value of 20 to 30. DØ expects to collect about 1000 signal events in similar channels as CDF (as stated above). Moreover, DØ also looks to collect about 1000 signal events from the channel $B_s \to J/\Psi + K^*$ with a decent tagging effectiveness about 7%.

V OTHER INTERESTING B PHYSICS MEASUREMENTS

A B hadron lifetimes

B lifetime measurement will continue to play a very important role in the B physics program at Tevatron in Run II. B lifetime measurement is a good meeting point between theories and experiments as theoretical calculation of the widths of

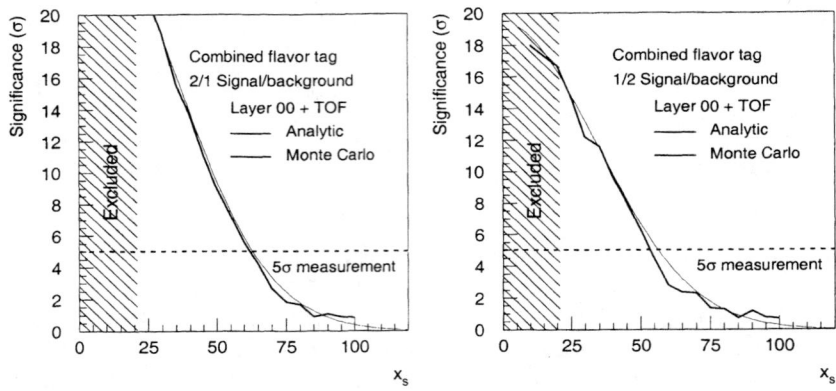

FIGURE 7. Expected reach of x_s that CDF may be able to attain in Run II. The significance on the y-axis corresponds to how many σ of a measurement that CDF may be able to do given a value of x_s.

inclusive decays remains a challenge. Hopefully, Run II upgrades and datasets will enhance lifetime measurements of sufficient precision to promote fruitful comparison with theory.

It is most convenient to study lifetime ratios because in this way, one cancels the dependence on quantities which are poorly known, such as the b-quark mass. A few example of these lifetime ratio experimental measurements are given below:

$$\begin{aligned}
\tau(B_s)/\tau(B) &= 0.95 \pm 0.04 \\
\tau(B^+)/\tau(B^0) &= 1.07 \pm 0.03 \\
\tau(\Lambda_b)/\tau(B^0) &= 0.78 \pm 0.04
\end{aligned} \quad (15)$$

It is said that none of these ratios have been measured with sufficient precision to test the theories. Moreover, not all lifetime comparisons which can be made have been made.

At the moment, the lifetime ratio $\tau(\Lambda_b)/\tau(B^0)$ is particularly interesting and it perhaps draws the most attention. Because this lifetime ratio is experimentally measured to be significantly different from 1. This is very hard (if possible) for the theorists to explain. So far this has been measured through the semi-leptonic decays. It will be very interesting for CDF and DØ to measure $\tau(\Lambda_b)$ at Run II using exclusive and full hadronic decay modes. In this way, we may get rid of the dependence on the K factor (which was needed to allow for the missing energy due to the presence of neutrinos in semi-leptonic decays). Full hadronic decay modes would also allow a full reconstruction of the vertex of the B hadron and this would help reject the backgrounds significantly and reduce the experimental uncertainties.

It is perhaps also possible to measure the lifetime differences in B mesons. The dominant Feynman diagram for $\Delta\Gamma_q$ (where q is d or s) involves $(V_{cb}V_{cq}^*)$. Looking at the Equation (4), one can see that $\Delta\Gamma_d$ is smaller than $\Delta\Gamma_s$ by a factor of λ^2. Though $\Delta\Gamma_d$ is too small to measure, it may be possible for us to measure $\Delta\Gamma_s$. Observing a non-zero $\Delta\Gamma_s$ is not an observation of CP violation, but it does imply a natural separation of CP-even and CP-odd eigenstates by their respective lifetimes, rather like the case of K_s and K_L in the kaon system. Consequently, this may help measure CP violation in the B_s system. Since B_s mesons are not produced in the e^+e^- machines of the B factories running on the $\Upsilon(4S)$ resonance, the studies of B_s will be unique to the hadron colliders such as Tevatron.

A theoretical calculation [8] has shown that

$$\frac{\Delta\Gamma_s}{\Gamma_s} = 0.18(\frac{f_{B_s}}{200 MeV})^2 \sim \mathcal{O}(0.1) \qquad (16)$$

CDF predicts that they can measure this ratio with a precision of $\sigma(\Delta\Gamma_s/\Gamma_s) \sim 0.065$ in Run II.

B α, γ and CP violation

Owing to the large trigger bandwidth available ($\sim$ 50 kHz) at Level 1, CDF also may trigger on the events of $B^0 \to \pi^+\pi^-$. This is a channel that may give us a measurement of $\sin(2\alpha)$, where α is one of the three angles of the unitarity triangle defined in Fig. 2, though penguin contributions are expected to affect this determination severely. CDF has estimated that they may be able to get as many as 15200 events in this channel if the branching ratio is 1×10^{-5}. With a sample of only 5000 events, the tagging effectiveness, ϵD^2, is estimated to be about 9.1% which corresponds to $\delta A(\pi^+\pi^-) \approx 0.1 - 0.15$.

With more restricted Level 1 trigger bandwidth, DØ probably would not collect as many events of $B_d \to \pi^+\pi^-$ and $B_s \to K^+K^-$ as CDF. But DØ should be able to complement CDF's measurement in these channels. The overall detection efficiency in these channels is about 0.25 to 0.5%. DØ may collect about a thousand events in each of these channels.

CDF also considers to use the channel $B_s/\overline{B}_s \to D_s^\pm K^\mp$ to measure γ, the remaining angle in the unitarity triangle as in Fig. 2. CDF has estimated that they may collect about 700 events in this channel.

There have been also some suggestions that one may be able to extract $\sin(\gamma)$ from the samples of $B_d \to \pi^+\pi^-$ and $B_s \to K^+K^-$, provided that $\sin(2\beta)$ is determined separately and accurately. [9] The proposed strategy makes use of the U-spin flavor symmetry of strong interactions and the strategy is affected neither by penguin contributions nor by any final-state-interaction effects. Preliminary attempt to apply this strategy in the experiment has shown to be quite promising but more work are required to make firm conclusions. [10].

Finally, CP violation in B_s, which can be measured by studying the asymmetry of the B_s decays as in the process $B_s \to J/\psi + \phi$, is expected to be small according to the Standard Model. Any significant asymmetry measured will provide an unambiguous sign of new physics. DØ expects to collect as many as 35,000 signal events. For lesser extent of lepton and forward tracking coverage, CDF expects to collect about 18,000 events but probably with better time resolution in the decay time (length).

VI SUMMARY

There is certainly a long way to go before we may have many fruitful results from the B physics program at Tevatron Run II. Nevertheless, the prospect for successful B physics studies at Run II is excellent. CDF and DØ should have precise measurements of $\sin(2\beta)$ from the process $B_d \rightarrow J/\psi + K_s$ in Run II, which may provide a conclusive observation of CP violation outside the kaon system. Dedicated experimental studies of the masses and lifetimes in the B systems will enable us to have stringent tests on the Standard Model and various theoretical calculations. As B factories operating at the resonance of $\Upsilon(4S)$ will not produce B_s, experiments at Tevatron have the fortunate privilege to perform accurate studies of the neutral B_s mesons such as their mass and lifetime differences. Successful measurement of the lifetime difference in the neutral B_s system will give us a natural separation between the CP-even and CP-odd eigenstates. This will help study the CP violation in the B_s system, from which any large CP violation will show a definite sign of new physics beyond the Standard Model.

ACKNOWLEDGMENTS

The author acknowledges many useful discussions with colleagues from the CDF and DØ collaborations. Thanks especially go to Rick Jesik for his help in providing me with some of the material presented in the Workshop and in this report.

REFERENCES

1. CDF detector upgrade information may be found at http://www-cdf.fnal.gov/upgrades/upgrades.html ; and DØ detector upgrade information may be found at http://www-d0.fnal.gov/hardware/upgrade/upgrade.html .
2. "Measurement of $\sin2\beta$ from $B \rightarrow J/\psi K_s^0$ with the CDF detector", T. Affolder *et al.*, (CDF Collaboration), Phys. Rev. D **61**, 072005 (2000).
3. "Method for flavor tagging in neutral B meson decays", M. Gronau, A. Nippe, and J. L. Rosner, Phys. Rev. D **47**, 1988 (1993); "Identification of neutral B mesons using correlated hadrons", M. Gronau and J. L. Rosner, *ibid.* **49**, 254 (1994).
4. The DØ Run II GEANT simulation can be found at http://www-d0.fnal.gov/newd0/d0atwork/computing/MonteCarlo/d0gstar_history.html .

5. Presentation given by Joseph Kroll, from the CDF Collaboration, at "2000 Aspen Winter Conference on Particle Physics", Aspen, Jan. 20, 2000. A copy of the transparencies of his presentation may be obtained from http://hep.physics.wisc.edu/aspen2000/kroll.ps.gz .
6. The information is extracted from the web page at http://lepbosc.web.cern.ch/LEPBOSC/combined_results/moriond_2000 .
7. "Constraints on the parameters of the CKM matrix by End 1998", F. Parodi, P. Roudeau and A. Stocchi, hep-ex/9903063, Nuovo Cim. **A112** 833-854 (1999).
8. BABAR Collaboration, "Report of the BABAR Physics Workshop", Equation (11.91), Chap. 11, p. 770 (1999). The electronic copy may be obtained from http://www.slac.stanford.edu/pubs/slacreports/slac-r-504.html .
9. "New strategies to extract β and γ from $B_d \to \pi^+\pi^-$ and $B_s \to K^+K^-$", R. Fleischer, Phys. Lett. B **459** 306 (1999).
10. Presentation given by F. Würthwein, from the CDF collaboration, in the second (and final) meeting of the "Workshop on **B physics at the Tevatron**, Run II and beyond", Fermilab, Feb. 24th, 2000. A copy of the transparencies of his presentation may be obtained from http://www-theory.fnal.gov/people/ligeti/Brun2/febr/wurthwein.ps.gz .

SEARCHES FOR NEW PHYSICS IN LEP AND TEVATRON

Searches for New Physics at NuTeV

Janet Conrad
for the NuTeV Collaboration

P.O. Box 137, Irvington NY 10533

Abstract. The NuTeV Experiment at Fermilab has searched for indications of physics beyond the standard model using indirect and direct search methods. The indirect method compares the NuTeV measurement of $\sin^2 \theta_W$ to the world measurement for this parameter. The direct measurement searches for decays of exotic neutral particles in a light mass, instrumented region upstream of the NuTeV calorimeter. No evidence for new physics has been observed. Therefore, NuTeV has placed constraints on extensions to the Standard Model which include certain Z'''s, neutral heavy leptons and the particle responsible for the Karmen Timing Anomaly.

The NuTeV experiment can search for indications of physics beyond the standard model using both indirect and direct searches. In this article, we first report on our indirect searches using the precision measurement of $\sin^2 \theta_W$. Deviations from the value obtained from neutrino scattering, when compared to measurements from other processes, could give evidence for Z'''s or neutrino oscillations. In the second half of these proceedings, we report on the NuTeV direct searches for Neutral Heavy Leptons and and other "heavy-neutrino-like" particles. Both types of searches have given a null result. Based on this, NuTeV has placed stringent limits on several important extensions to the standard model.

I   THE NUTEV DETECTOR

NuTeV is a neutrino scattering experiment which took data during the 1996-97 Fixed Target Run at Fermi National Accelerator Laboratory. This experiment is the last in a 15-year series which made use of the Lab E detector. Its immediate predecessor was the CCFR experiment which took data from 1987-90. The NuTeV results are substantially more precise than CCFR due to better control of systematics, as explained below.

Table I describes the sources of neutrinos in the NuTeV neutrino beam. An important upgrade for the beam from CCFR to NuTeV was the introduction of the sign-selection [1]. The 800 GeV proton beam was targeted upward on BeO at an 7.8 mrad angle. The resulting pion and kaon beam was then bent toward Lab E by two

Source	ν Mode	$\bar{\nu}$ Mode
$\pi^{\pm}_{\mu 2}$	0.786	0.856
$K^{\pm}_{\mu 2,\mu 3}$	0.201	0.133
$K^{\pm}_{e3}$	0.0124 ± 0.00015	0.0081 ± 0.00012
K_{Le3}	0.00047 ± 0.00009	0.00153 ± 0.00030
$\mu \to \nu_e$	0.00020 ± 0.00002	0.00096 ± 0.00010
Charm Meson $\to \nu_e$	0.00021 ± 0.00004	0.00069 ± 0.00012
$\Lambda_c \to \nu_e$	0.00007 ± 0.00004	0.00022 ± 0.0001

TABLE 1. Sources of neutrinos in the NuTeV beam

dipole magnets. This had two important consequences: 1) depending on the setting of the dipole magnet, the beam was almost purely neutrinos or antineutrinos (with contamination of the opposite particle $\sim 0.1\%$); and 2) the neutral hadrons were directed away from the detector, significantly decreasing the neutrinos from neutral kaon decay compared to CCFR. The sign-selection allowed a new technique for the measurement of $sin^2\theta_W$ which is unique to NuTeV and substantially improves on the result from CCFR. During the run, NuTeV observed ~ 4.5 million interactions, which is comparable to CCFR.

NuTeV used a refurbished Lab E detector, which consisted of a target calorimeter followed by a muon spectrometer [2]. The target calorimeter consisted of 168 3m×3m×5.1cm iron plates, with 84 scintillation counters and 42 drift chambers interspersed between the plates. The scintillation counters provided triggering and energy measurement, and the drift chambers permitted muon tracking. The toroid spectrometer had a 15kG field. The toroid was divided longitudinally into three regions with sets of drift chambers interspersed for muon tracking. The target and toroid regions were calibrated continuously throughout the run using a test-beam which provided muons, pions and electrons over a wide range of energies [2]. This was an important upgrade from the CCFR experiment, which calibrated the detector during short running periods at the midpoint and end of the run.

Both neutral current (NC) and charged current (CC) interactions are observed in the Lab E detector. Typically CC events are characterized by the long distribution of hits in the scintillation counters from the outgoing muon. The muon can also be tracked using the drift chambers. On the other hand, NC events will typically have deposits of energy over a short longitudinal range of scintillation counters.

The NuTeV Decay Channel was a new light-mass detector built for the 1996 running period. [3] It was located in front of the Lab E detector. It consisted of a veto wall, followed by a helium-filled decay region, with total volume of 3m×3m×40m. Two stations of 3m×3m drift chambers, identical to those in the Lab E toroid, were interspersed within the decay region. A third station of drift chambers was located just downstream of the decay region and immediately in front of the calorimeter. The purpose of this detector was to look for decays of new particles which accompany the neutrino beam. The drift chamber stations provided tracking and the

	Short (NC) Events	Long (CC) Events	$R_{20} \equiv$ Short/Long
ν	386K	919K	0.4198 ± 0.0008(stat)
$\bar{\nu}$	88.7K	210K	0.4215 ± 0.0017(stat)

TABLE 2. Number of events obtained by NuTeV as a function of event length. The cut between short and long events is 20 counters.

Lab E detector provided particle identification.

II PRECISION MEASUREMENT OF $\sin^2 \theta_W$

In this section, I briefly report on the NuTeV precision electroweak measurement and its consequences for beyond-the-standard-model searches.

The premise of this analysis [4] is that cross-comparisons of the measurement of $\sin^2 \theta_w$ from different processes give sensitivity to new physics. Neutrino experiments measure $\sin^2 \theta_{W\,on\,shell}$, which is related in the standard model to the collider measurements of M_Z and M_W by the formula: $\sin^2 \theta_{W\,on\,shell} = 1 - (M_Z/M_W)^2$. This comparison is sensitive to the top mass through the radiative corrections to the propagator in neutrino scattering. A measurement of $\sin^2 \theta_{W\,on\,shell}$ from neutrino scattering can also be compared to the corrected value of $\sin^2 \theta_{W\,eff}$, which is what is extracted from e^+e^- collider asymmetry analyses. If there is poor agreement from the comparison, then this would indicated new physics. If there is good agreement, then this technique can be used to refine our limits on the mass of the Higgs. Thus, this is a very powerful technique for understanding the standard model and beyond.

NuTeV is the first experiment to measure $\sin^2 \theta_W$ using the Paschos-Wolfenstein method:

$$R^- = \frac{\sigma^\nu_{NC} - \sigma^{\bar{\nu}}_{NC}}{\sigma^\nu_{CC} - \sigma^{\bar{\nu}}_{CC}}$$

$$= \left(\frac{1}{2} - \sin^2 \theta_W\right)$$

This is clearly only possible using a sign-selected beam. It has the advantage of being manifestly insensitive to the sea quark distributions and charm production at leading order. These were the source of the largest systematic error in the CCFR measurement. The above equation can be rewritten in the form:

$$R^- = \frac{\sigma^\nu_{NC} - \sigma^{\bar{\nu}}_{NC}}{\sigma^\nu_{CC} - \sigma^{\bar{\nu}}_{CC}} = \frac{R^\nu - rR^{\bar{\nu}}}{1 - r}$$

where $R^{\nu/\bar{\nu}} = \sigma^{\nu/\bar{\nu}}_{NC}/\sigma^{\nu/\bar{\nu}}_{CC}$, and $r = \sigma^{\bar{\nu}}_{CC}/\sigma^\nu_{CC}$.

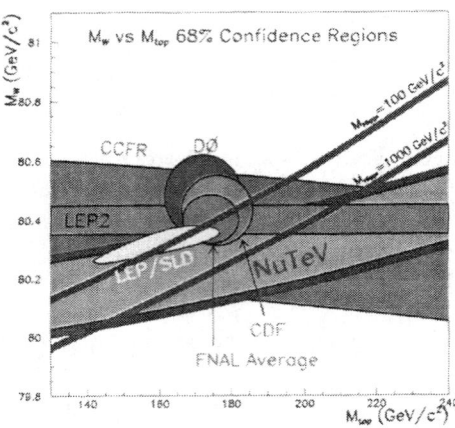

FIGURE 1. NuTeV measurement of $\sin^2 \theta_W$ recast in terms of the W and top mass dependence. Results from the previous neutrino experiment, CCFR, and direct measurements from collider experiments are also shown. Within errors, the experimental results agree.

In practice, NuTeV uses a variant of the Paschos-Wolfenstein concept designed to give the most accurate measurement of $\sin^2 \theta_W$ possible with our detector. First, an event length cut of 20 counters is used to separated events into "short" and "long" categories. The statistics for each category is given in table II. The connection of $\sin^2 \theta_W$ to R_{20}^ν and $R_{20}^{\bar{\nu}}$ is made using the Monte Carlo, which corrects for CC events which appear short, from muons which exit the detector to the side or range-outs, and NC events which appear long, due to charm production or meson decays to muons.

NuTeV can measure any linear combination of R^ν and $R^{\bar{\nu}}$. Therefore, we choose that combination which is least sensitive to our largest systematic: charm. Hence what is actually measured is $R^- = R_{20}^\nu - x R_{20}^{\bar{\nu}}$. The optimum choice indicated by the Monte Carlo, $x = 0.514$, is actually very close to what one expects from the Paschos-Wolfenstein relation.

Using this method, NuTeV has made the preliminary measurement: $\sin^2 \theta_{W\,on\,shell} = 0.2253 \pm 0.0019(\text{stat}) \pm 0.0010(\text{syst})$, assuming a top mass of 175 GeV and Higgs mass of 150 GeV. This is currently the most precise measurement from νN scattering and the first based on the Paschos-Wolfenstein ratio. This technique yields a precision that is comparable to collider measurements. Having chosen the "on-shell" convention, given the very precise measurement of the mass of the Z from LEP, this result implies a mass of the W of $M_W = 86.26 \pm 0.11$ GeV.

An interesting comparison to other measurements can be obtained when the

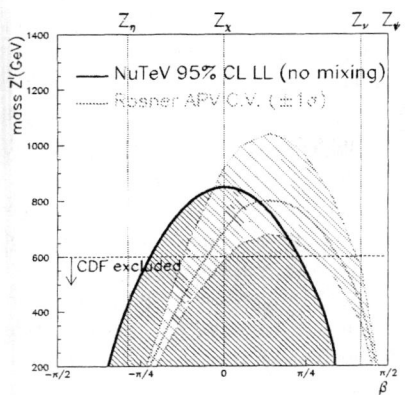 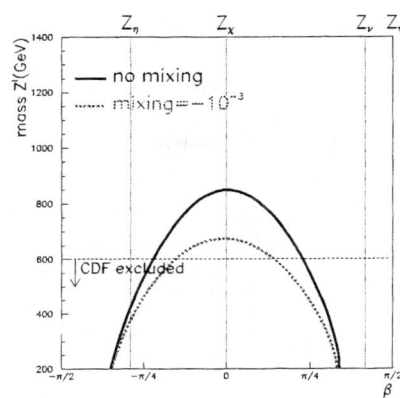

FIGURE 2. Left: The APV allowed region (narrow gray band) for Z' production as a function of mass and mixing angle β (see text). NuTeV excludes the hatched region below the thick line, assuming no $Z - Z'$ mixing, in accordance with the APV result. The CDF direct search excludes masses below 600 GeV, indicated by the dashed line. Right: In the general case where $Z - Z'$ mixing is allowed, the NuTeV limit is somewhat weaker.

result is presented as a function of m_t and m_W as shown in Fig. 1. The width of the grey region on either side of the NuTeV band indicates the weak sensitivity of this measurement to the choice of M_H (a range of m_H from 80 GeV to 1 TeV was considered). This figure also shows the result from CCFR as well as direct measurements of m_t vs m_W from the TeVatron and extracted values from the LEP and SLD measurements. Two bands indicate the theoretical standard model dependence of m_H on m_t and m_W given the precise M_Z measurements from LEP. One can see that, taken as a whole, the data are in good agreement and favor a light Higgs mass. A global analysis of all of the electroweak data, including the NuTeV result, indicates $m_H = 98^{+57}_{-38}$ GeV [5].

Using the good relative agreement of the data, these results can be used to set limits on extensions to the standard model. One example is the NuTeV search for a difference between $\nu_\mu \to \nu_\tau$ and $\bar{\nu}_\mu \to \bar{\nu}_\tau$, which would indicate CP violation in the lepton sector [6]. In these proceedings, NuTeV reports on a second example: the limits which can be set on Z''s.

There are both good theoretical and experimental reasons for considering the existence of extra neutral heavy gauge bosons. Extra Z bosons appear in various GUT and string-motivated extensions to the standard model. For example, the $E(6)$ breakdown to $SO(10) \times U(1)_\psi$ results in the Z_ψ and the $SO(10)$ breakdown to $SU(5) \times U(1)_\chi$ yields the Z_χ. Thus the new exchange boson could be: $Z' = Z_\chi \cos\beta + Z_\psi \sin\beta$, where the mixing angle β is an arbitrary parameter. The experimental motivation comes from a recent Atomic Parity Violation (APV) measurement [7] which indicates a 2.5σ deviation from the Standard Model prediction based on all other electroweak data. [8,9] This deviation could be improved by the inclusion of a Z' into the theory. The mass of the Z' allowed by the APV data as a function of the mixing angle β is shown in Fig.2 (left) by the light colored band [8]. Note that direct searches have already excluded masses below $\sim$600 GeV [10].

A new Z' will appear in the NC NuTeV sample through the direct exchange diagram as well as through mixing with the Standard Model Z. The NuTeV $\sin^2\theta_W$ result is 1.3σ above the Standard Model, as defined by the collider measurements. In general, the NuTeV data disfavor introducing a new Z' as an explanation for the APV data. The region which is excluded by the NuTeV is the shaded area below the bold line indicated on Fig. 2 (left). One can see that NuTeV excludes a substantial portion of the APV allowed space at 95% CL.

The APV result assumes no $Z - Z'$ mixing, but in the general case you can have small mixing. As discussed at this conference and in theoretical papers [9], if one allows a small (-10^{-3}) level of $Z - Z'$ mixing, this weakens the constraint on the mass limits. This is shown on Fig. 2 (right).

In summary, NuTeV has performed a precision measurement of $\sin^2\theta_W$ using a technique inspired by the Paschos-Wolfenstein relation. The result has comparable uncertainty to meaurements from the hadron and e^+e^- colliders. Cross-comparisons have indicated good agreement among the various experimental results. Therefore, we have used the NuTeV data to set limits on possible new physics. In particular, NuteV excludes a large region of the parameter space for the APV-inspired Z'.

III DIRECT SEARCHES FOR NEW PHYSICS USING THE NUTEV DECAY CHANNEL

The light mass NuTeV Decay Channel allows the direct search for decays of beyond-the-standard-model particles which may accompany the neutrinos in the beam. These particles must be massive, neutral, relatively long-lived, and weakly interacting. Beyond these requirements, this search is essentially model-free. Examples of what might be observed include "neutral heavy leptons" (NHL's), predicted by various GUT-inspired extensions to the Standard Model. [11] These would be produced in the decay of the secondary mesons (pions, kaons and charm mesons) in the beamline. There is also an unexplored window for light-mass, long-lived neutralinos which are produced in pairs from the proton interactions with the BeO target. [12] The neutral particle will be referred to as the N^0 below.

NuTeV is sensitive to decays of the N^0 to two charged particles and possibly also a neutrino. Search modes include $\mu\mu\nu$, $\mu e\nu$, $\mu\pi$, $ee\nu$ and $e\pi$. At this point, searches from 0.02 to 2 GeV have been completed. These are the possible mass ranges for N^0's produced by decays of secondary mesons. The initial search from 0.2 to 2 GeV resulted in no candidate events and allowed NuTeV to set limits on

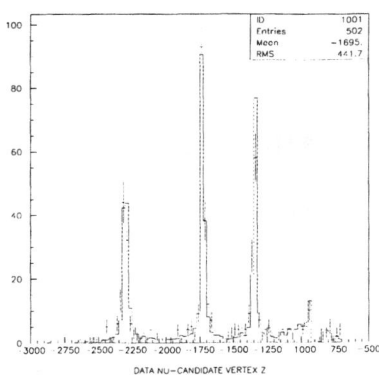 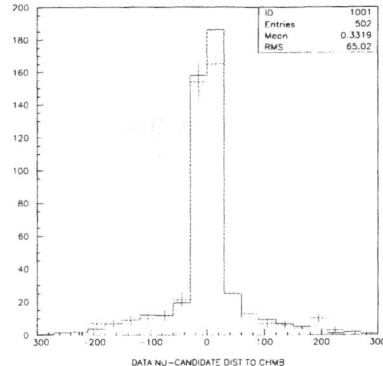

FIGURE 3. DIS event reconstruction in the decay channel, data are points, Monte Carlo is the histogram. Left: Shown as a function of longitudinal position in inches. Peaks indicate material in the decay channel (see text). Right: Shown as a function of distance from nearest chambers.

NHL's which extend almost an order of magnitude beyond the past experiments [3]. The second serach extended into the range of the Karmen Timing Anomaly (33.9 MeV) [13].

The reconstruction code for these searches was tuned using Monte Carlo and then tested using Deep Inelastic Events. Fig. 3 (left) shows reconstructed Deep Inelastic Events as a function of the longitudinal vertex position in inches. The data indicated by the points and the Monte Carlo by the histogram are in good agreement. The peaks are, from left to right, neutrino interactions in a testbeam chamber, the most upstream channel chamber, the middle channel chamber and the most downstream set of channel chambers. The region between is populated by events in the helium as well as misreconstructed events in the chambers. In general, misreconstructed events which appear in the helium are due to hit-confusion in events with many tracks. Fig. 3 (right) shows the position of the vertices with respect to the nearest chamber.

The initial search concentrated on NHL's from 0.2 to 2 GeV in mass decaying to $\mu\mu\nu$, $\mu e\nu$, and $\mu\pi$ via coupling with a light neutrino. The sensitivity of the experiment depends upon the mass of the NHL and the coupling, $|U|^2$. The lower mass limit of this search was set at the threshold for 2 muon production. The upper mass sensitivity is limited by the production mechanism, which is via decays of mesons in the beamline. The long distance from production to detector (1 km) makes NuTeV most sensitive to long-lived NHL's. Hence, within the mass range of 0.2 to 2 GeV, this implies very small couplings.

This is a blind analysis in the sense that the cuts were selected based strictly on the Monte Carlo. Events were required to have two well-reconstructed tracks and a high-quality vertex within the decay channel fiducial region. The vertex was required to be 1 m or 3σ, whichever was larger, from the position of the chambers

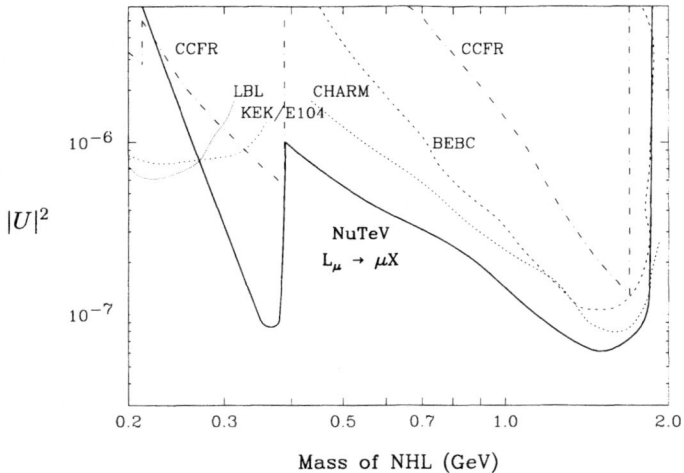

FIGURE 4. NuTeV limits (above solid line) on neutral heavy lepton production as a function of mass and coupling, $|U|^2$, compared to limits from previous experiments.

within the decay channel. In order to assure good particle identification, μ's were required to have 2.2 GeV, and e and π showers to have 10 GeV. In order to reduce the background from Deep Inelastic Scatters (the main source), we made kinematic cuts to isolate a region where the DIS rate will be low. Typically, for large invariant mass (W) DIS events, the fractional momentum carried by the struck quark (x_{bj}) will be large. Therefore, to remove DIS events while maintaining the NHL signal, we required $x_{bj} < 0.1$ and $W > 2$ GeV. After all cuts, the total background was estimated to be 0.57 ± 0.15 events.

No events were observed in this analysis. Therefore NuTeV has set a limit on neutral heavy leptons in the .2 to 2 geV mass region, as shown in Fig. 4. We have excluded a substantial new portion of parameter space, for the case of decays to at least 1 muon.

A second search extended to lower mass regions. This was motivated largely by the Karmen Timing Anomaly [14] which could be interpreted as a 33.9 MeV neutral particle produced in π decay which sunsequently decays to $ee\nu$. This signal is observed in Karmen beam dump experiment, in which the particles would be produced nearly at rest. The allowed region is a band which covers a wide range of lifetimes $vs.$ branching ratio.

If this particle were produced in the NuTeV beam, it would be highly boosted. As a result, NuTeV has access to only the short-lifetime solutions for the Karmen Timing anomaly. The high boost, combined with small mass, implies that decays will have a very small opening angle. As a result, the vertex resolution is poor and

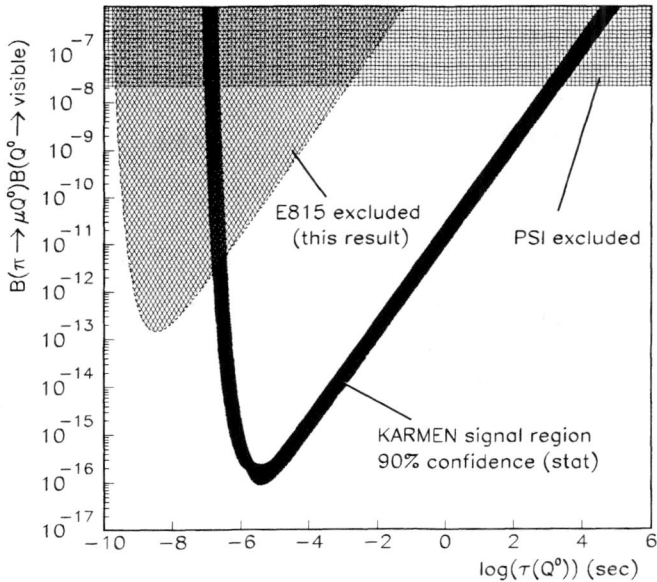

FIGURE 5. The allowed region for the Karmen Timing Anomaly as a function of branching ratio and lifetime (narrow band). NuTeV excludes four orders of magnitude of the short lifetime solution (shaded region at left). The PSI direct search excludes branching ratios larger than 2×10^{-8} (shaded region at top).

it is not feasible to make a cut around the chamber region for this analysis.

Again, this was a blind analysis with cuts based on the Monte Carlo. Events were required to have good track and vertex χ^2. The vertex was constrained to be within the upper and lower limits of the decay region. Due to the small opening angle, the showers associated with the two tracks were allowed to be merged. A likelihood function was used to determine if the shower was electromagnetic.

Because the π decay which produces this hypothetical particle has such small Q-value, the direction of the particle has a deviation from the pion which is on the order of μradians. This represents an advantage for NuTeV because the SSQT clusters the pions so that they point at the center of the detector. On the other hand, photon conversion background to the $ee\nu$ decay signal is populated nearly isotropically across the face of the detector.

No events were observed in a search from 0.02 to 0.2 GeV. Therefore, NuTeV has set a limit which excludes over four orders of magnitude of the short lifetime solution to the Karmen Timing Anomaly. This is presented in Fig. 5.

REFERENCES

1. Bernstein, *et al.*, FNAL TM-1884
2. Harris, Yu, *et al.*, NIM A447, 377.
3. Vaitaitis, *et al.*, Phys. Rev. Lett. 83 (1999) 4943.
4. Zeller, to be submitted to the Proceedings of the 7th Conf. on the Intersections of Particle and Nuclear Physics, Quebec City, Canada, 2000.
5. Groom, *et al.*, Eur. Phy J. C15 (2000)
6. Harris, submitted to Proceedings of the Int. Europhysics Conf. on HEP, Tampere, Finland, 1999.
7. Bennett, *et al.* Phys. Rev. Let., (1999) 2482.
8. Rosner, hep-ph/9907524.
9. Langacker and Erler, PRL $\underline{84}$, 212-215 (2000); Casalbuoni, *et al.*, HEP-PH/0001215.
10. Abe, *et al.*, Phys. rev. Lett., 79, (1997) 2192.
11. Gronau, *et al.*, Phys. Rev. D29 (1984) 2539.
12. Borissov, *et al.*, to be submitted to Phys. Rev. Letters.
13. J. A. Formaggio, *et al.*, Phys. Rev. Lett. 84, 4043 (2000).
14. Armbruster, *et al.*, Phys. Lett. B348 (1995) 19.

Supersymmetry searches at LEP

Laurent Duflot

Laboratoire de l'accélérateur Linéaire
BP 34 91898 ORSAY CEDEX FRANCE
duflot@lal.in2p3.fr

Abstract. The status of the searches for supersymmetry using the LEP data taken in 1999 and earlier and their consequence will be given.

I INTRODUCTION

In a little more than four years of running at centre-of-mass energies above the Z mass, the four LEP experiments have collected nearly $2\,\text{fb}^{-1}$ at centre-of-mass energies ranging from 130 to 202 GeV. Almost half of that luminosity was collected in 1999, as shown in table 1.

The results and interpretations of supersymmetric particle searches will be presented, most of them being based on the latest datasets. The limits shown will be 95% C.L. limits and these results are preliminary. The results from the four LEP experiments were not combined yet.

4 experiments - 1999 dataset				
$\sqrt{s}$ (GeV)	192	196	200	202
$\mathcal{L}$ (pb^{-1})	113	313	328	155

TABLE 1. Luminosity collected by the four LEP experiments in 1999 at the various centre-of-mass energies.

Most results will be presented in the minimal supersymmetric extension of the standard model (SM), the MSSM. Two Higgs doublets are introduced to give mass to the up- and down-type quarks without introducing anomalies. After the breaking of the electroweak symmetry, four physical states remain: the neutral scalars h and H, the neutral pseudoscalar A and the charged Higgs $H^{\pm}$. Supersymmetry associates to every degree of freedom of the SM fields a supersymmetric degree of freedom with a spin differing by 1/2. Leptons and quarks have spin 0 partners (sleptons and squarks) for each chirality state, gauge bosons and Higgs fields have spin 1/2 partners (gauginos and higgsinos) and a spin 3/2 gravitino is introduced. The neutral gauginos and higgsinos mix to form four eigenstates called neutralinos (χ_i^0) and similarly the charged gauginos and higgsinos forms the two charginos states ($\chi_i^\pm$). For simplicity, the lightest states will be noted χ and $\chi^\pm$. Mixing can occur in the "sfermion" sector. It is proportional to the fermion mass, hence relevant only for the third generation sfermions.

Since e.g. no selectron with the electron mass has been detected, SUSY must be broken at our energy scale. Unfortunately, there are no compelling models of SUSY breaking now and the soft SUSY breaking is parametrised by a number of parameters: gaugino mass terms M_1, M_2 and M_3, scalar masses $m_{\tilde{f}}$ and trilinear couplings A_f. Two additional parameters are of relevance: the SUSY Higgs mass term μ and the ratio of the two Higgs v.e.v $\tan\beta$.

The interpretation of LEP searches are usually performed in a semi-constrained model (SCMSSM) inspired from minimal SUGRA (CMSSM): the gaugino masses are supposed to be the same at the GUT scale as well as the sfermion masses, but the trilinear couplings and the Higgs masses are left non-universal, such that μ is considered to be a free parameter. The low energy parameters are then extracted from the Renormalisation Group Equations, which leads to heavy gluinos and squarks, and left-sleptons heavier that right-sleptons.

In the following, the case of SUSY breaking driven by the gravitational interaction at high energy will be discussed first, assuming that R-parity (a multiplicative quantum number with value $+1$ for standard particles and -1 for SUSY particles) is conserved. The case of non conservation of the R-parity will then be discussed. The gauge mediated SUSY breaking models will be discussed last.

II STANDARD SUSY SEARCHES

In that case, R-parity is conserved and the Lightest Supersymmetric Particle is supposed to be the lightest neutralino χ which is weakly interacting and escape detection, leading to the well-known signature of SUSY: missing energy. Four main topologies arise from SUSY production at LEP: acoplanar leptons (from e.g. slepton pair production), acoplanar jets (from e.g. squark production), lepton + jets + missing energy and multijet + missing energy (from e.g. chargino production). The signal topology depend crucially on $\Delta M = m_{NLSP} - m_\chi$ (NLSP: Next to Lightest Supersymmetric Particle) which control the among of visible energy. The main SM background at low ΔM is from $\gamma\gamma$ production (with a large cross-section, leading to reduced sensitivities) while four fermion processes dominate the background for large ΔM.

The excluded regions from the sleptons searches are shown in figure 1. For low ΔM values, the lepton from $\tilde{l}_R$ decay is very soft and almost undetectable. The associated $\tilde{e}_R\tilde{e}_L$ production leads to a "single electron" signature that is used to cover that region. A somewhat significant excess of events over the estimated SM background is observed by the LEP experiments in the stau channel. However, can-

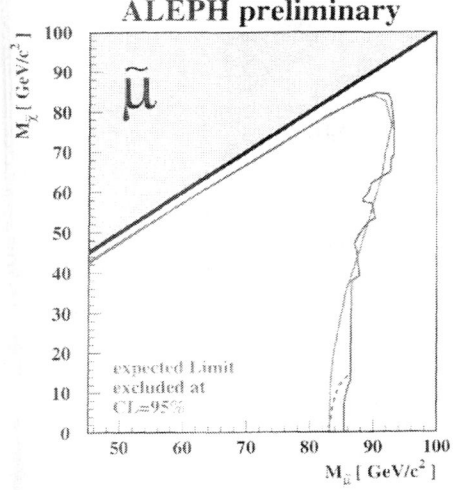 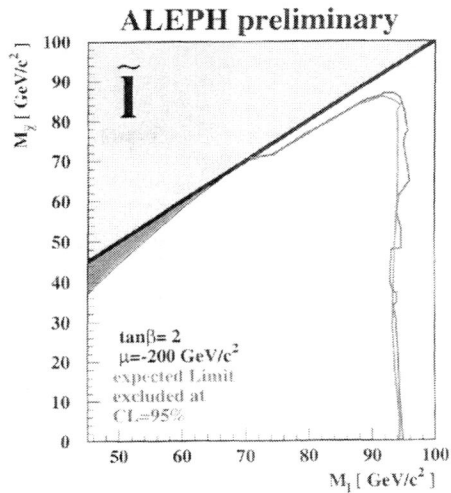

FIGURE 1. Excluded region in the smuon mass - neutralino mass plane (left) and the slepton mass - neutralino mass plane (right) assuming universal slepton masses at the GUT scale.

didates contribute to only part of the $m_{\tilde{t}}$-m_χ plane. Moreover, part of systematics on the background estimates are correlated between experiments since the same generators are used. Detailed studies are on the way.

Due to the large top mass, large mixing is quite natural in the stop sector and the stop might well be within the reach of LEP. Various analyses have been performed to cover the expected decay channels: $\tilde{t} \to c\chi$ and $\tilde{t} \to bl\tilde{\nu}$ (assuming that $\tilde{t} \to \chi^+ b$ is forbidden, given the limits on the chargino mass, see later). The excluded region region is shown in figure 2 together with the results from sbottom searches. Although the mass reach of LEP is significantly lower than that of the TeVatron, LEP is able to cover the region of medium to small mass differences, inaccessible at the TeVatron. Again, standard analyses become inefficient for very low mass differences, but the stop hadron lifetime become then significant and a combination of standard analyses, searches for large impact parameter tracks and searches for heavy quasi-stable charged particle allow ALEPH to set a lower limit on the stop mass of $63\,\text{GeV}/c^2$, independent of the mass difference.

In the case of large sfermion masses, the chargino searches allow to exclude chargino masses almost up to the kinematic limit but for low mass differences. Since in the SCMSSM, neutralino and chargino masses depend on the same SUSY parameters, the searches for neutralinos complement the searches for charginos and improve over then in some regions of the parameter space. When the mass difference

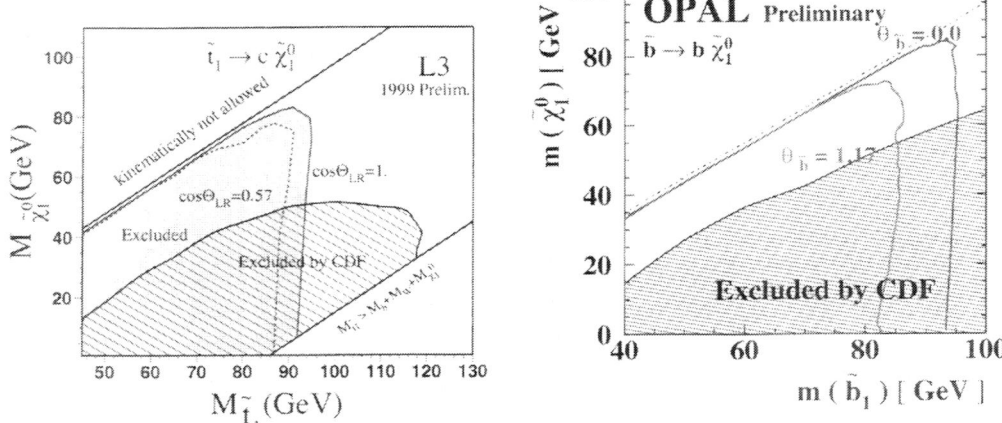

FIGURE 2. Excluded region in the stop mass - neutralino mass plane (left) and the sbottom mass - neutralino mass plane (right).

between the chargino and the LSP is less than $\sim 150\,\text{MeV}/c^2$, the chargino becomes long-lived and search for quasi-stable charged particle allow to exclude a large region of the corresponding parameter space. For mass differences up to a few GeV/c^2, the standard searches are inefficient and an ISR photon is searched for in addition to the soft decay products of the charginos in order to fight the background. Combining these searches, a lower limit of $69.4\,\text{GeV}/c^2$ is set on the chargino mass by L3, as can be seen in figure 3.

The combination of the chargino and neutralino searches allow to set a lower limit on the LSP mass in the case of heavy sleptons. If sleptons are light, the chargino cross-section decreases and the leptonic branching ratios increases, while the leptonic modes suffer from the large WW background. Moreover, a hole in the exclusion region, called the "corridor", opens up for sneutrino slightly lighter than the chargino: the two body decay $\chi^+ \to l\tilde{\nu}$ dominates, the lepton is very soft if the mass difference is small, and the signal cannot be disentangled from the background. However, within the SCMSSM, the slepton searches can be used to constrain the parameter space, which leads to an absolute limit on the neutralino mass of $37.5\,\text{GeV}/c^2$ from L3 as shown in figure 4.

It is well known that the MSSM predicts a low mass Higgs boson, hence the strong limits derived from LEP Higgs searches (in the SM, $m_H > 107.7\,\text{GeV}/c^2$) severely constrain the SUSY parameter space (especially for low $\tan\beta$ values) for the "benchmark" sets of parameters used at LEP. However, when larger sets of parameters were considered, loopholes were found. They were essentially of two

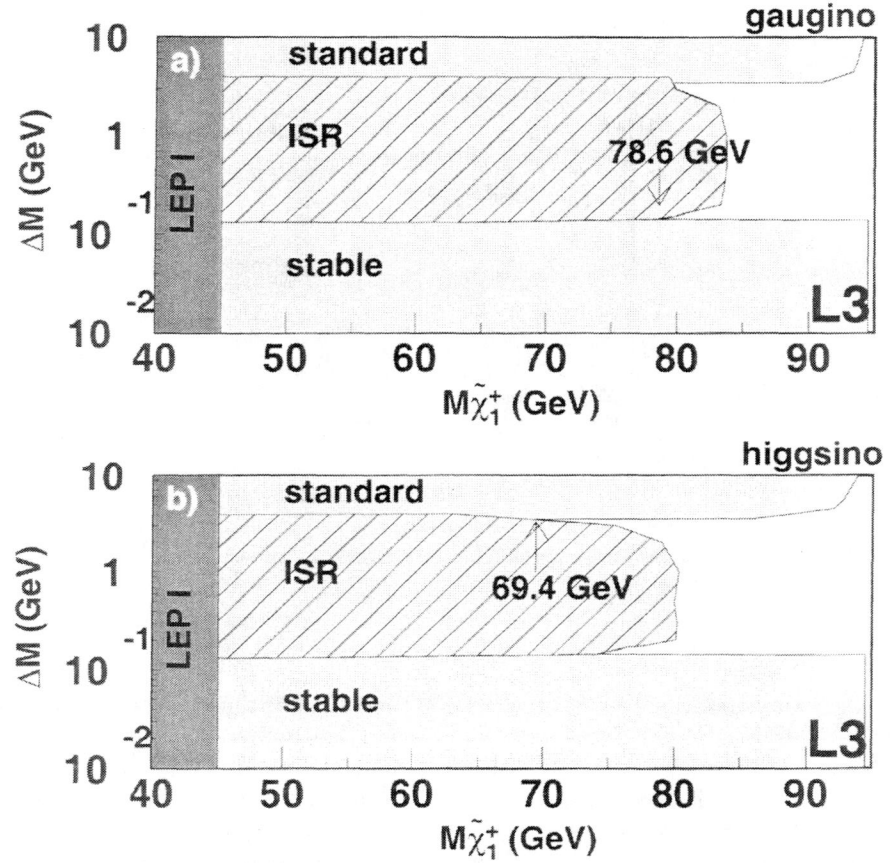

FIGURE 3. Excluded region in the ΔM-$m_{\chi^\pm}$ plane.

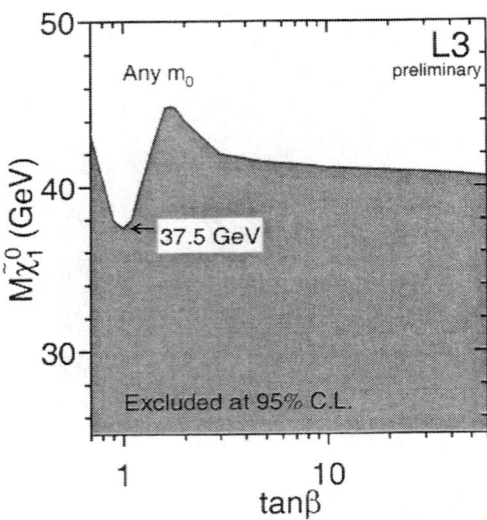

FIGURE 4. Limit on the mass of the lightest neutralino as a function of $\tan\beta$ for any m_0.

types: points where $\sin^2(\beta - \alpha)$ (α is the mixing angle in the neutral Higgs sector) is very small, hence the hZ cross-section is small, and m_A and m_H are too large for the HZ and hA processes to be accessible; points where $\sin\alpha$ is very small, hence the branching ratio of h to $b\bar{b}$ is suppressed while standard LEP searches are based on b-tagging. ALEPH has performed a large scan of the parameter set, introducing reinterpretations of the standard analyses and searches for charged Higgs. New analyses were developed, in particular a search for Higgses with no b-tagging (the result from a similar search in OPAL is shown in figure 5). Two analyses were developed to control the region of low m_A: a searches for HZ $\rightarrow$ H$\nu\bar{\nu} \rightarrow \tau\tau\tau\tau\nu\bar{\nu}$ and a search for hA $\rightarrow$ AAA with $m_A < 2m_\tau$ and at least one A to $\mu^+\mu^-$. These new analyses allowed to confirm that the mass limits derived in the "benchmark cases" are robust. These can then be used to constrain the SUSY parameter set and allow the lower limit on the neutralino mass to be improved: $m_\chi > 40\,\text{GeV}/c^2$.

The limits on the neutralino mass are strictly valid in the case of negligible mixing in the $\tilde{\tau}$ sector otherwise new loophole regions appear, in particular when the chargino decays to $\tilde{\tau}\nu$ and the mass difference between the $\tilde{\tau}$ and the LSP is small such that the $\tilde{\tau}$ is almost invisible. These configurations are under study but it seems that they can be excluded from other searches.

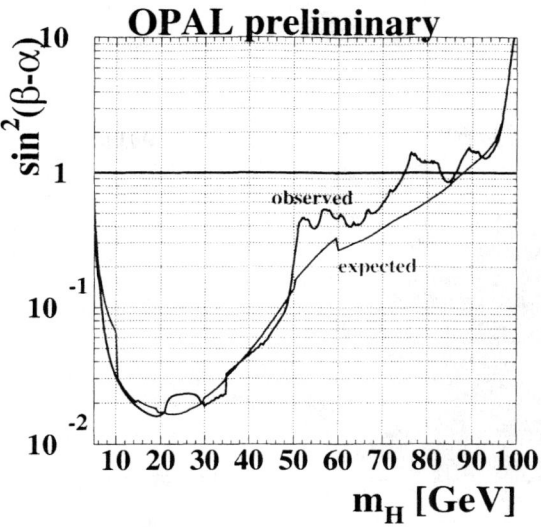

FIGURE 5. Observed and expected limit on $\sin^2(\beta - \alpha)$ from Higgs searches with no b-tagging in the 1998 OPAL dataset.

III SEARCH FOR SGOLDSTINOS

The goldstino is the Goldstone fermion appearing from the super-Higgs mechanism that is absorbed by the gravitino when it becomes massive. The golstino has scalar SUSY partners, a scalar S and a pseudoscalar P, that can be massive if the SUSY breaking scale is low, of the order of the EW scale. DELPHI has searched for the associated production of a S particle and a photon, with a decay of the S to gluons (the branching ratio ranges from 90 to 95%) or photons (with branching ratios of order 5 to 10%). The excluded regions are shown in figure 6.

IV R-PARITY VIOLATING SUSY SEARCHES

In its most general form, the MSSM allows for lepton and baryon number violating terms in the superpotential:

$$W_{\not R_p} = \lambda LLE + \lambda' LQD + \lambda'' UDD + \xi LH$$

where L (resp. E) is a left (resp. right) lepton superfield and U (resp D) is a left (resp right) quark superfield, with generation indices implied. This potential contains 45 new couplings that lead to direct decays of SUSY scalars to SM particles, breaking the R-parity symmetry. Some of these coupling are constrained by low

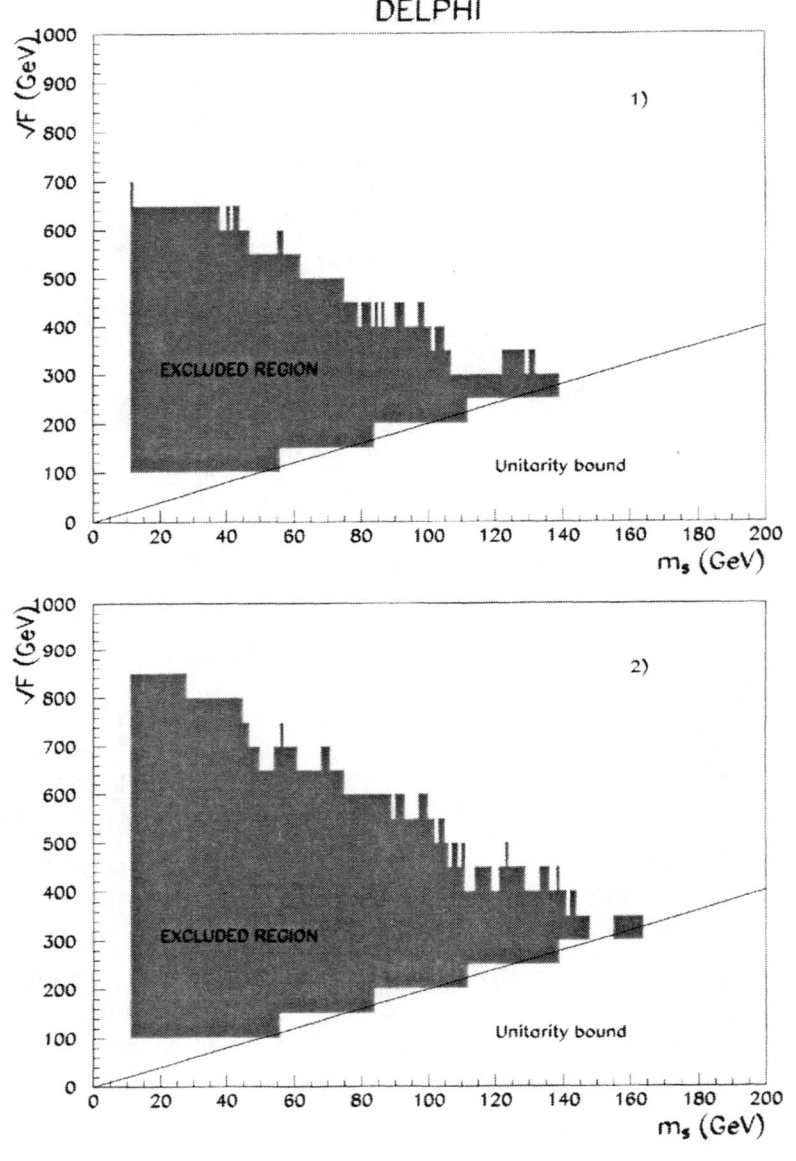

FIGURE 6. Excluded regions in the m_S-SUSY breaking scale plane from sgoldstino searches.

energy measurements (lepton number violation, proton lifetime, flavour physics, etc...). It is often said that the last term can be absorbed in a redefinition of the fields but this is only true at lowest order. However, in the following, this term will be ignored.

Practically, it is impossible to consider all these terms simultaneously in interpreting the results of the searches. It will be assumed that, similarly to the Yukawa couplings, one of these couplings will be completely dominant over the others.

Another consequence of R-parity violation is that the LSP is no longer stable: it decays to SM particles, either at tree level or via higher order processes. The cosmological constraints on relic LSP no longer apply and in principle any SUSY particle can be the LSP. This leads to complicated topologies to be searched for (e.g. 23 different analyses were developed by the ALEPH collaboration), ranging from acoplanar leptons (from e.g. slepton production and an LLE term) to many jets or many jets with leptons (up to ten quarks in chargino production and decay via an UDD term !).

It is beyond the scope of this paper to discuss in details the analyses and their results. As an illustration, the result from slepton searches in the case of an LLE model from the OPAL collaboration is shown in figure 7 in the slepton mass - neutralino mass plane. Both the case of slepton LSP and neutralino LSP are considered. In general, limits obtained in these scenarios are as strict or stricter than the limits obtained in the case of R-parity conservation.

V "GAUGE MEDIATED SUSY" SEARCHES

In these models (GMSB models), the breaking of SUSY is communicated to the visible sector via (new) gauge interactions. This leads to very specific pattern of SUSY breaking terms. There is no clear mechanism to generate the μ term, which is a severe drawback of the model, and this parameter is introduced "by hand". The simplest models considered here have only four additional parameters:

- $\sqrt{F}$, the SUSY breaking scale
- Λ, an universal mass term
- Q, the messenger scale
- n, the number of complete representations of messenger fields

In the GMSB models, the gravitino is naturally very light (sub-eV) and hence is the LSP. Since its coupling to other SUSY particles is small, the SUSY particles will decay to the NLSP. The NLSP is naturally either the lightest neutralino (which decays to gravitino-photon) or the sleptons (decaying to lepton-gravitino), with the possibility that the stau be lighter that the other sleptons, leading to a dominance of τ production. In both cases, the signatures are rather striking: presence of photons or leptons with missing energy. The NLSP lifetime is given by the SUSY breaking scale $\sqrt{F}$ or equivalently the gravitino mass:

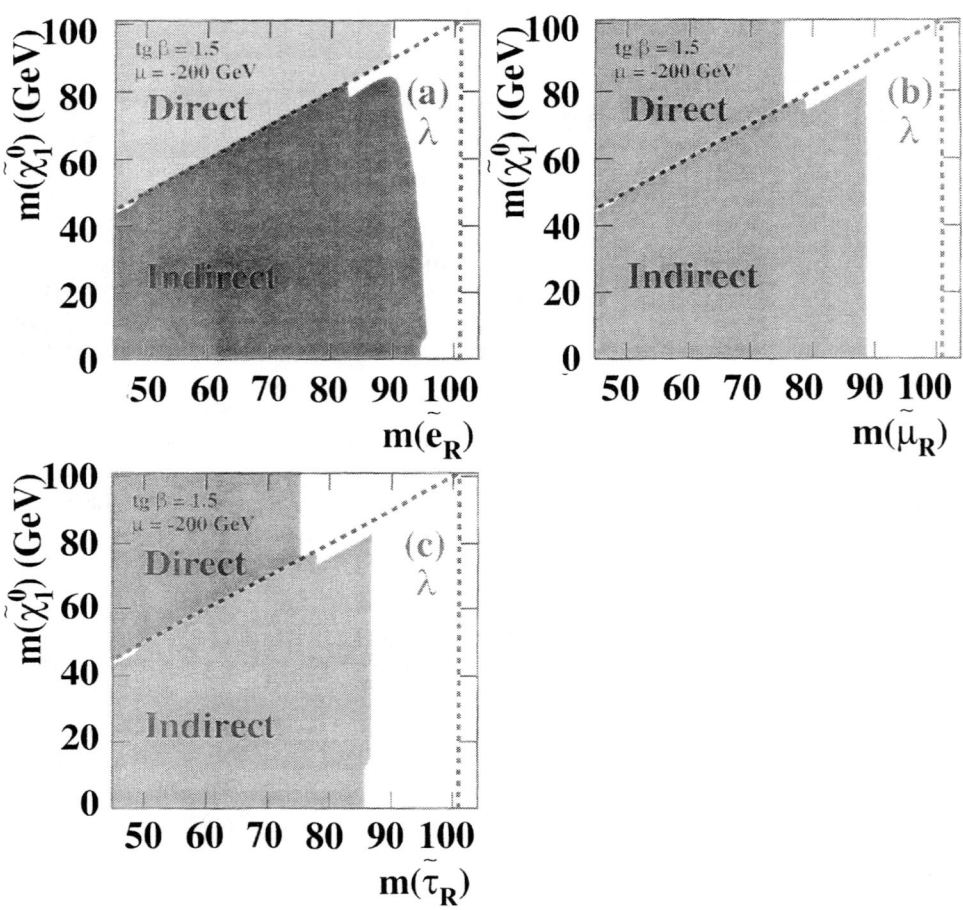

FIGURE 7. Excluded regions in the slepton mass - neutralino mass plane from slepton production in case of a dominant LLE R-parity violating coupling.

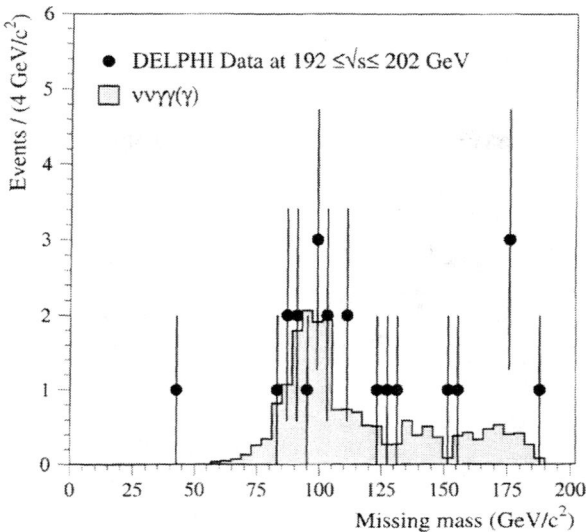

FIGURE 8. Missing mass distribution for the selected acoplanar photon events with the SM prediction.

$$\tau \propto m_{\tilde{G}}^2 \propto \left(\frac{F}{M_{pl}}\right)^2$$

In the case of neutralino LSP, the neutralino pair production leads to acoplanar photons, possibly not pointing to the interaction point if the LSP lifetime is non negligible. If the LSP lifetime is very large, the signatures are those of the gravity mediated SUSY breaking model and the corresponding limits apply. Figure 8 shows the results from the search for acoplanar photons in the DELPHI experiment.

In the case of slepton NLSP, short slepton lifetimes lead to the same acoplanar lepton topologies as those of the gravity mediated SUSY breaking models. The search for cascade decays of neutralino via slepton helps to further constrain the model. For intermediate lifetime, the slepton decays within the tracking volume. These configurations are explored by the search for tracks with large impact parameters and tracks with kinks. All these analyses are combined in order to cover the whole spectrum of slepton lifetimes as shown in figure 9.

As for the case of gravity mediated SUSY breaking models, a comprehensive study of the model is performed by combining the results of all the analyses in a scan of the parameter space performed by the ALEPH collaboration:

- $10^4\, GeV < Q < 10^{12}\, GeV$

- $0.1\, eV < m_{\tilde{G}} < 0.1\, MeV$

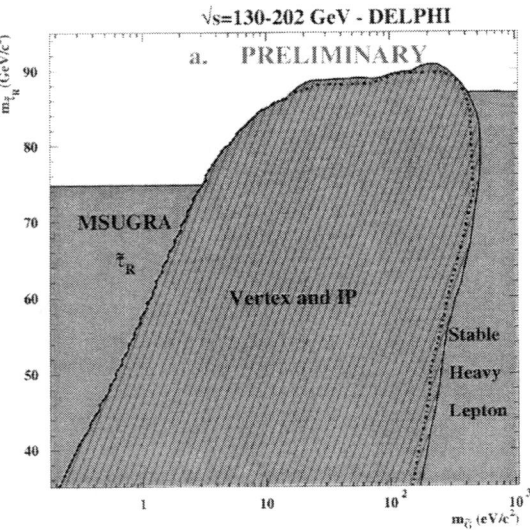

FIGURE 9. Excluded regions in the $\tilde{\tau}$ mass - gravitino mass plane.

- $1\,TeV < \Lambda < \min(\sqrt{F}, Q)$
- $n = 1, .., 5$
- $1.3 < \tan\beta < 38$
- $sign(\mu) = \pm$

Figure 10 shows that all analyses are indeed needed to constrain the parameter space. The results from the scan are illustrated in figure 11.

VI CONCLUSIONS

The results of the search for supersymmetric particles at LEP have been presented. Unfortunately, no signal of SUSY has been found. The results of searches in individual channels are combined to further constrain the parameter space. In the standard SUSY scenario, this lead to a limit on the neutralino mass of approximately $40\,\text{GeV}/c^2$. In the GMSB models, limits on the mass scale are set.

I would like to thanks the organisers for the friendly atmosphere of the conference and for the time devoted to discussions which I enjoyed.

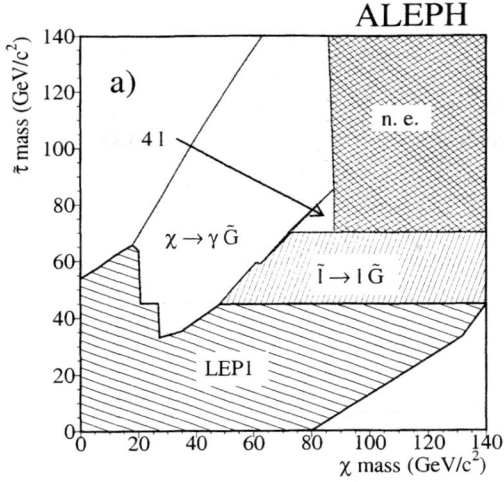

FIGURE 10. Regions of the $\tilde{\tau}$ mass - gravitino mass excluded by various analyses.

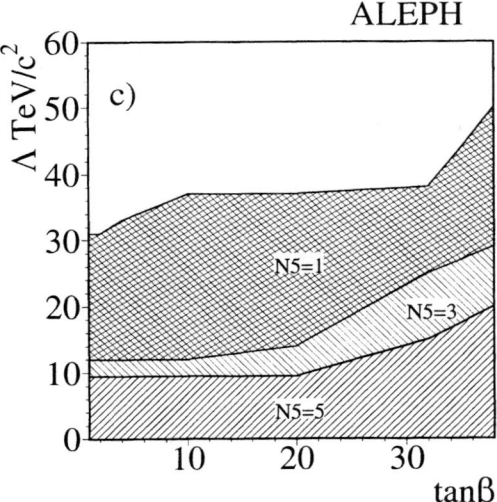

FIGURE 11. Excluded regions in the Λ - $\tan\beta$ plane depending on the number of complete messenger representations.

Limits on the size of extra dimensions from LEP experiments

V. Zhuravlov

Joint Institute for Nuclear Research, Dubna

Abstract. We give the summary of the bounds on extra dimensions obtained from LEP running at energy above Z-pole. We present limits on "effective" Planck mass M_D and radius of extra dimensions from direct graviton production, limits on cut-off parameter M_s from virtual graviton exchange contribution and limits on compactification scale M_c from electroweak observables and Bhabha scattering.

I INTRODUCTION

The idea that we live in more then 3 spatial dimensions has a long history. It was born in the 20's when Kaluza and Klein introduced one additional compact dimension. After that the possibility of extra dimensions was being considered by physicist for a long time, and in the 80's the string theory appeared. The theory required ten or eleven dimensions compactified with the radius $\sim 1/M_{Planck}$. At the end of the 90's it was recognized that the radius of compactification could be much higher then Planck scale, which gives an opportunity to experimentalists to test the consequences of extra dimensions at present colliders.

In this note we review the limits on extra dimensions for the following two theoretical models. The first proposed by Arkani-Hamed, Dimopoulos and Dvali assumes that only gravitons live in extra dimensions. In the second model gauge bosons are also allowed to propagate in extra dimensions. In the first model collider signatures can be divided into direct graviton production (γ and missing energy or Z and missing energy) and virtual graviton exchange in the fermion and boson pairs production. For the second model the constrains can be obtained from electroweak precision measurements and from Bhabha scattering mediated by Kaluza-Klein modes of gauge bosons.

II THEORETICAL MODELS

In the model proposed by Arkani-Hamed, Dimopoulos and Dvali [1] all Standard Model particles live in 3-brane of (3+N)-space, while gravitons can propagate in

N extra dimensions. Extra dimensions should be hidden for us – and the good possibility to provide this is to compactify them into a torus. To obey the Newton law the radius of compactification should be less then 1 mm and the number of extra dimension should be greater then one. One of the main advantage of this model is a natural solution of "hierarchy problem". One can introduce "effective" (3+N)-dimensional Planck mass which obeys the relation

$$M_{Pl}^2 = (M_D)^{2+N} \cdot r^N$$

where M_{Pl} is the ordinary Planck mass and r is the radius of compact dimensions. The "effective" Planck mass can be chosen as small as $\sim 1\,\mathrm{TeV}$ if number of extra dimensions greater then one.

Due to compact dimensions a tower of massive Kaluza-Klein (KK) gravitons arise with the mass spectrum

$$m_k = 2\pi \frac{k}{r}, k = 1, 2, ...$$

and the usual small coupling to SM particles $\sim \frac{1}{M_{Pl}}$. Summation over all KK states gives the coupling $\sim \frac{1}{M_D}$ which is on the order of electroweak scale. Using this fact one can make an optimistic conclusion: present colliders are sensitive to extra dimensions and quantum gravity.

Massive KK states of gravitons can be directly produced by colliders. As gravitons escape detectors without an interaction the experimental signature is the missing energy. Another possibility to look for extra dimensions is measurements of cross section of fermion and boson pairs production, which is subject to the influence of virtual graviton exchange.

In the second theoretical model SM fermions live in 3-brane and SM bosons reside on a p-brane, where $3 \leq p \leq D$, D - full number of dimensions. As in the first model, gravitons are allowed to to propagate in all dimensions. In this case the main influence of extra dimension arise from KK modes of gauge bosons with a common mass spectrum

$$M_k = kM_c, k = 1, 2, ...$$

where $M_c = 1/r$ is compactification scale. The effect of gravity is negligible in comparison with the effect of gauge bosons. It was shown in [2] that gauge couplings could be unified at the TeV scale.

Experimental manifestation of this model at the present colliders can be detected as (i) deflection of electroweak precision measurements from their SM values due to mixing of "ordinary" gauge bosons with their KK excitations and (ii) modification of the cross section of fermion pair production due to exchange of a KK mode of γ and Z^0.

Both models can be considered as realizations of the string theory and, in contrast to "conservative" string theory, the prediction of both models can by tested at present colliders and in the nearest future.

III EXPERIMENTAL SIGNATURES FOR ADD MODEL

A $e^+e^- \to \gamma G$

Experimental signature for this channel is one (or more) γ and missing energy. The same signature has Standard Model process $e^+e^- \to \nu\bar{\nu}\gamma(\gamma)$. The main background is produced by processes $e^+e^- \to e^+e^-\gamma$, where both electrons escape detection and $e^+e^- \to \gamma\gamma$ with one detected photon.

The differential cross section is given by [3]

$$\frac{d^2\sigma}{dx d\cos\theta} = \frac{\alpha}{32s} \frac{\pi^{N/2}}{\Gamma(N/2)} \left(\frac{\sqrt{s}}{M_D}\right)^{N+2} f(x, \cos\theta) \qquad (1)$$

where

$$f(x, \cos\theta) = \frac{2(1-x)^{\frac{N}{2}-1}}{x(1-\cos^2\theta)} \times$$

$$[(2-x)^2(1-x+x^2) - 3x^2(1-x)\cos^2\theta - x^4 \cos^4\theta]$$

$$x = \frac{E_\gamma}{E_{beam}}$$

The production of graviton and photon has been searched for at LEP II. Experiments used slightly different techniques to derive $e^+e^- \to \gamma G$ cross section. ALEPH collaboration made binned maximum likelihood fit of Standard Model prediction added to graviton production cross section to the 2-dimensional (E_{miss}, $\cos\theta$) distribution for all γ candidates. Value of $(\frac{1}{M_D})^{N+2}$ was used as a fit parameter. DELPHI found the kinematic region which is more sensitive to the signal: $6 < E_\gamma < 50$ GeV for the barrel calorimeter and $18 < E_\gamma < 50$ GeV for the forward calorimeter. Then a comparison of the full number of observed events in the described kinematic region with the integrated predicted cross section has been done. L3 collaboration performed the fit of the SM prediction plus graviton production to the E_γ spectrum of γ candidates. OPAL used "cut based" method: cut $E_\gamma < 34$ GeV was applied to minimize expected upper limit on signal cross section. The results of LEP experiments in the form of 95% C.L. limits on the "effective" Planck mass M_D are presented in table 1

An illustration of ALEPH analysis is presented in fig. III A, where on the plot (a) the invariant mass distribution of the system recoiling against the γ candidate is shown and on the plot (b) the polar angle distribution of photon candidate is shown.

Fig. III A presents the 95% C.L. cross section limit for $e^+e^- \to \gamma G$ production and expected cross sections with 2 and 4 extra dimensions.

Value of $(\frac{1}{M_D})^{N+2}$ was used to find the average of the LEP experiments. Results of averaging and results of the LEP experiments are shown in fig. 3.

TABLE 1. 95% C.L. lower limits on mass scale M_D, TeV

		$N=2$	$N=4$	$N=6$
ALEPH	189 - 202 GeV	1.10	0.70	0.52
DELPHI	192 - 202 GeV	1.250	0.792	0.57
L3	189 GeV	1.018	0.674	0.506
OPAL	189 GeV	1.086	0.710	0.528

95% C.L. lower limits on the "effective" Planck mass and upper limits on radius of compact dimensions derived from averaging of ALEPH, DELPHI and OPAL data are presented in table 2

TABLE 2. Average 95% C.L. limits from ALEPH 183 - 202 GeV, DELPHI 189 - 202 GeV and OPAL 189 GeV

	$N=2$	$N=4$	$N=6$
M_D, TeV	1.29	0.80	0.58
r, cm	$2.9 \cdot 10^{-2}$	$1.4 \cdot 10^{-9}$	$4.5 \cdot 10^{-12}$

B $e^+e^- \to Z^0 G$

Gravitons in ADD model can also be produced in association with Z bosons, but phase space available for graviton is less then in the case of $e^+e^- \to \gamma G$ because of the Z boson mass. Therefore experimental sensitivity to extra dimensions in this channel is suppressed. Experimental signature of this processes is a single Z^0 and missing energy. A search for this processes done only by L3 collaboration with hadronic decayed Z's. Obtained upper limits on the cross section and corresponding common radius of extra dimensions and lower limit on the "effective" Planck mass are shown in the table III B. Visible mass distribution of the candidates together with the Standard Model expectation and its modification by graviton production is shown in fig. 4.

C $e^+e^- \to Z^0 \to Z^* G_0$

Projection of the spin of a graviton propagating in extra dimensions onto world brane have tree possible values: 1, 2 and 3. From the 3-dimensional point of view it means that we have three types of gravitons. Coupling of spin-0 gravitons to the Standard Model particles is proportional to their masses, spin-1 gravitons

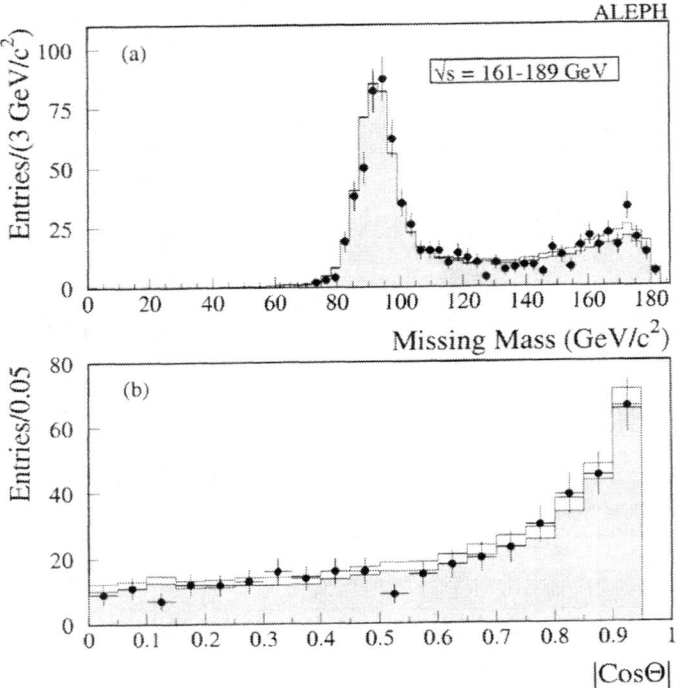

FIGURE 1. a - the invariant mass distribution of the system recoiling against the γ candidate, b - the polar angle distribution of photon candidate. Points stand for real data, dashed histograms are Standard Model MC and open histograms corresponds to graviton production with two extra dimensions and "effective" Planck mass $M_D = 0.99$ TeV.

are decoupled from matter and coupling of spin-2 gravitons is proportional to the energy-momentum tensor. The most prominent channel for a search for spin-0 gravitons at LEP is the graviton emission by Z boson: $e^+e^- \to Z \to Z^*G_0$ and $Z^* \to f\bar{f}$. Z branching ratio is proportional to $(M_Z/M_D)^{N+2}$, where N is the number of extra dimension. Topology of this channel is identical to invisible Higgs decay. Search for spin-0 gravitons has been done by ALEPH collaboration with LEP1 data. Hadronic decayed Z's were used, no candidate events were found and 1.4 events were expected from background. The branching ratios and limits on M_D from this analysis are shown in table 4

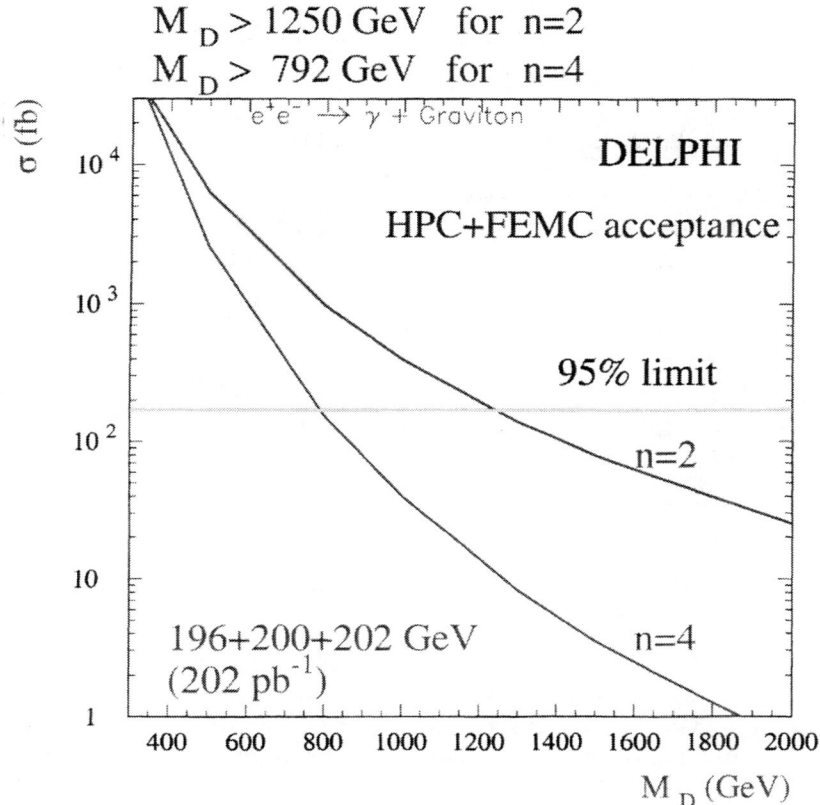

FIGURE 2. 95% C.L. cross section limit for $e^+e^- \to \gamma G$ production and expected cross sections for 2 and 4 extra dimensions

D $\quad e^+e^- \to \gamma\gamma(\gamma)$

Beside the direct production of gravitons extra dimensions can manifest themselves by modification of the cross sections of Standard Model processes by the graviton exchange. We start with two gamma process.

Modification of differential cross section by s-channel exchange of a spectrum of virtual gravitons is given by [4]

$$\frac{d\sigma(e^+e^- \to \gamma\gamma)}{d\cos\theta} = \frac{\pi}{s}[\alpha F_1(-\sin^2\frac{\theta}{2}) - \lambda\frac{4s^2}{\pi M_s^4}F_2(-\sin^2\frac{\theta}{2})]^2$$

$$F_1(x) = \sqrt{\frac{1+x+2x^2}{-x(1+x)}} \quad . \tag{2}$$

$$F_2(x) = \sqrt{\frac{-x(1+x)(1+2x+2x^2)}{16}}$$

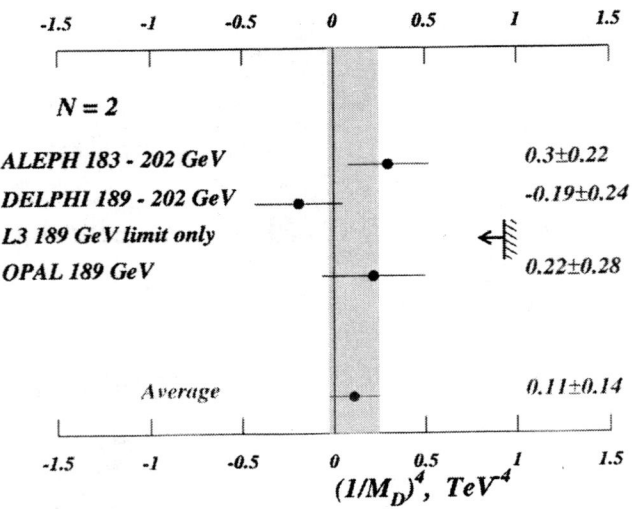

FIGURE 3. Value of $(\frac{1}{M_D})^{N+2}$ obtained from LEP experiments data and average over ALEPH, DELPHI and OPAL.

Cut off parameter M_s was introduced to regularise ultraviolet divergence due to infinite KK tower of massive gravitons, parameter λ describes the topology of extra dimensions and can not be calculated without full knowledge of the model. As

$$\lambda = \mathcal{O}(\infty)$$

we will consider two cases: $\lambda = 1$ with positive interference and $\lambda = -1$ with negative interference.

In table 5 we present the lower limits on cutoff parameter M_s obtained by LEP collaborations. Angular distribution of two gamma candidates obtained from data collected by L3 experiment is shown on fig. 5. Also shown are the Standard Model expectation and its distortion by the graviton exchange with $M_s = 0.65$ TeV.

$$\mathbf{E} \quad e^+e^- \to \mu\mu, \tau\tau$$

Differential cross section of fermion pair production is given by [5]

TABLE 3. Upper limit on the cross section σ_{ZG} and r and lower limit on M_D, 95% C.L., 189 GeV L3 collaboration, hadronic Z^0 decays, $N = 2$

σ_{ZG}, pb	M_D, TeV	r, cm
0.29	0.42	0.27

TABLE 4. Branching ratios and 95% C.L. lower limits on M_D ALEPH collaboration (LEP1 data): Hadronic Z^0 decays

	$N = 2$	$N = 4$	$N = 6$
$BR(Z \to Z^*G_0)$, M_D in TeV	$\frac{4.0 \cdot 10^{-8}}{M_D^4}$	$\frac{4.5 \cdot 10^{-11}}{M_D^6}$	$\frac{7.5 \cdot 10^{-14}}{M_D^8}$
M_D (TeV)	0.35	0.17	0.12

$$\frac{d\sigma}{dz} = \frac{\pi\alpha^2}{2s}P_{ij}[A_{ij}^2(1+z^2)+2B_{ij}^2 z] - \frac{\lambda}{M_s^4}\frac{s\alpha P_i}{4}[2z^3 v_i^2 - (1-3z^2)a_i^2] + \frac{\lambda^2}{M_s^8}\frac{s^3}{32\pi}[1 - 3z^2 + 4z^4] \quad (3)$$

where i, j are summed over γ and Z exchange
$z = cos\theta$
$A_{ij} = v_i v_j + a_i a_j$, $B_{ij} = v_i a_j + v_j a_i$
P_{ij}, P_i – propagator factors

The change of the total cross section given by the graviton exchange is negligible since integration of the interference term in eq. 3 over full range of the polar angle is vanishing. Consequently one should consider angular distribution of the fermion pairs. Therefore in this analysis one can use fermions with well reconstructed charge, i.e. muon and tau pairs. Bhabha scattering is a special case and will be considered in the following sections.

In contrast to $e^+e^- \to \gamma G$ the dependence on the number of extra dimensions is weak and included into λ parameter. "Non-radiative" events are studied.

The lower limits on the cutoff parameter M_s obtained by LEP experiments are shown in table 6

Differential cross section of muon pairs production and data – Monte Carlo ratio for 183 GeV and 189 GeV data collected by OPAL experiment are shown in fig. III E.

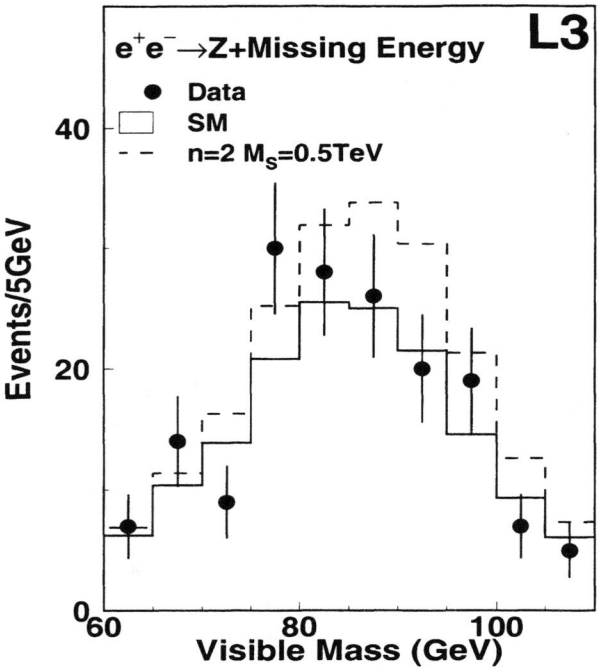

FIGURE 4. Visible mass distribution of Zg candidates. Points are real data, solid line is Standard Model expectation and dashed line is modification due to graviton production. L3 collaboration, 189 GeV data, hadronic Z decays.

F Bhabha scattering

Differential cross section of $e^+e^- \to e^+e^-$ can be presented in the a form

$$\frac{d\sigma}{d\cos\theta} = SM(s,t) + \frac{\lambda}{M_s^4} \cdot A_{int}(s,t) + (\frac{\lambda}{M_s^4})^2 \cdot B_{graviton}(s,t)$$

In contrast to muon and tau pairs production integrated interference is not vanishing because of t-channel contribution. This fact and much higher statistics of Bhabha scattering lead to the best sensitivity of this process to gravitons in extra dimension.

Theoretical error in the Standard Model cross section of Bhabha forward scattering is a limiting factor for the constrains on extra dimensions, while in the case of $e^+e^- \to \gamma\gamma$, $\mu\mu$ and $\tau\tau$ statistical error is dominated.

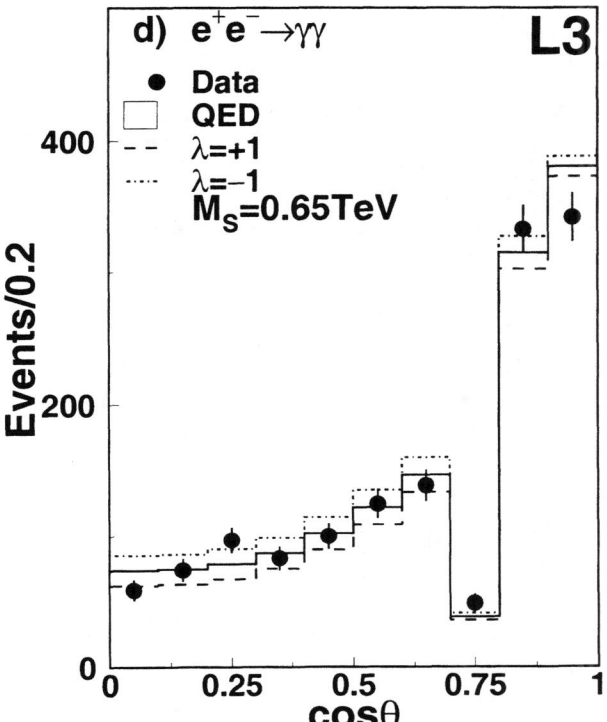

FIGURE 5. Angular distribution of $e^+e^- \to \gamma\gamma(\gamma)$ candidates. Points stands for 189 GeV data, solid line - Standard Model expectation, dashed (dotted) line - deflection by graviton exchange with positive (negative) interference, $M_s = 0.65$ TeV, L3 collaboration.

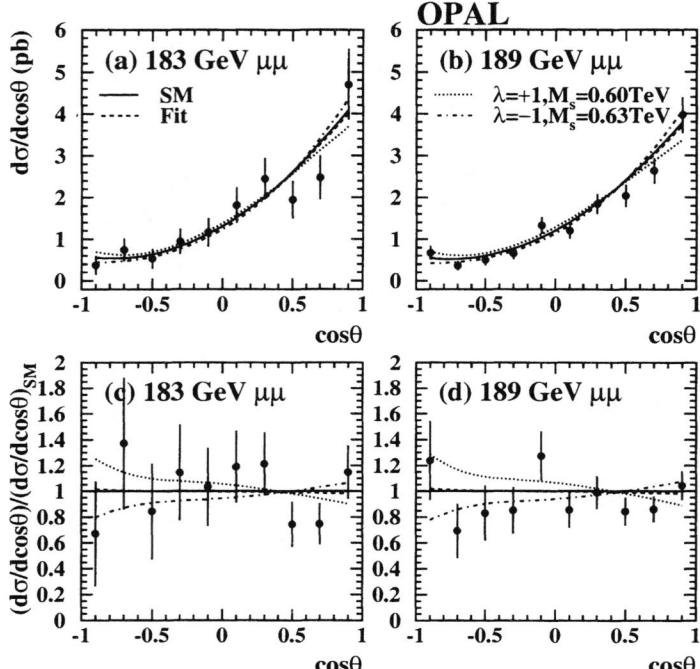

FIGURE 6. Differential cross section of muon pair production for 183 GeV (a) and 189 GeV (b) and ratio data – Monte Carlo ratio for 183 GeV (c) and 189 GeV (d). Points are real data, solid line – Standard Model prediction, dashed line - best fit of the cutoff parameter M_s to data, dotted (dashed-dotted) line – graviton exchange with positive (negative) interference

TABLE 5. Lower limits on M_s for $e^+e^- \to \gamma\gamma(\gamma)$ process

		M_s, GeV $\lambda = 1$	M_s, GeV $\lambda = -1$
ALEPH	189-202 GeV	810	813
DELPHI	130-202 GeV	713	691
L3	183-189 GeV	770	760
OPAL	183-189 GeV	660	634

TABLE 6. Lower limits on M_s for $e^+e^- \to \mu\mu$ and $e^+e^- \to \tau\tau$ processes

		$\mu\mu$		$\tau\tau$	
		$\lambda = 1$	$\lambda = -1$	$\lambda = 1$	$\lambda = -1$
ALEPH	130-189 GeV	607	563	527	509
DELPHI	183-202 GeV	725	598	645	557
L3	183-189 GeV	690	560	540	580
OPAL	183-189 GeV	600	630	630	500

In order to set the limit on M_s Global fit to 4 LEP experiments data was done in [6]. In this paper the following data set was taken: differential cross sections measured by ALEPH and L3 experiments for 183 and 189 GeV data and total cross sections and forward-backward assimetries measured by DELPHI and OPAL experiments for 183 and 189 GeV. Obtained 95 % C.L. lower limits on the cutoff parameter M_s are: $M_S > 1.261$ TeV for $\lambda = 1$ and $M_S > 0.962$ TeV for $\lambda = -1$.

Ratio of observed and predicted cross section measured by ALEPH experiment together with an expectation for graviton exchange is shown on fig. 7.

Figure 8 presents values $\frac{\lambda}{M_s^4}$ obtained by LEP experiments (95% C.L. limits only available for L3 collaboration) from γ, muon and tau pairs. Averaged value for γ, muon and tau pairs and averaged value for electron pairs production are also shown.

Average 95% C.L. limits from $e^+e^- \to \mu\mu$, $e^+e^- \to \tau\tau$, $e^+e^- \to \mu\mu$, $e^+e^- \to \gamma\gamma$ and $e^+e^- \to e^+e^-$ are $M_s > 1.27$ TeV for $\lambda = 1$ and $M_s > 0.98$ TeV for $\lambda = -1$.

G $e^+e^- \to ZZ$ and $e^+e^- \to WW$

Massive gauge boson pairs production is also sensitive channel to extra dimensions. In contrast to fermion pair production, in the case of boson pair production additional graviton exchange diagram changes not only differential cross section but also full cross section. Searches for extra dimensions in W pairs and Z pairs

FIGURE 7. Observed / Predicted cross section, ALEPH 189 GeV Data, inner bars – data errors, outer bars – data+theory errors. Dashed (dotted) line stands for positive (negative) interference with graviton exchange. Parameter Λ is connected to M_s as $\Lambda = (2\pi)^{1/4} M_s$.

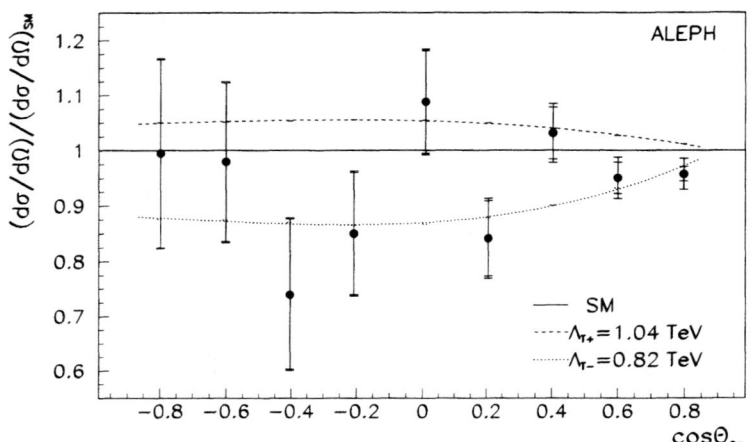

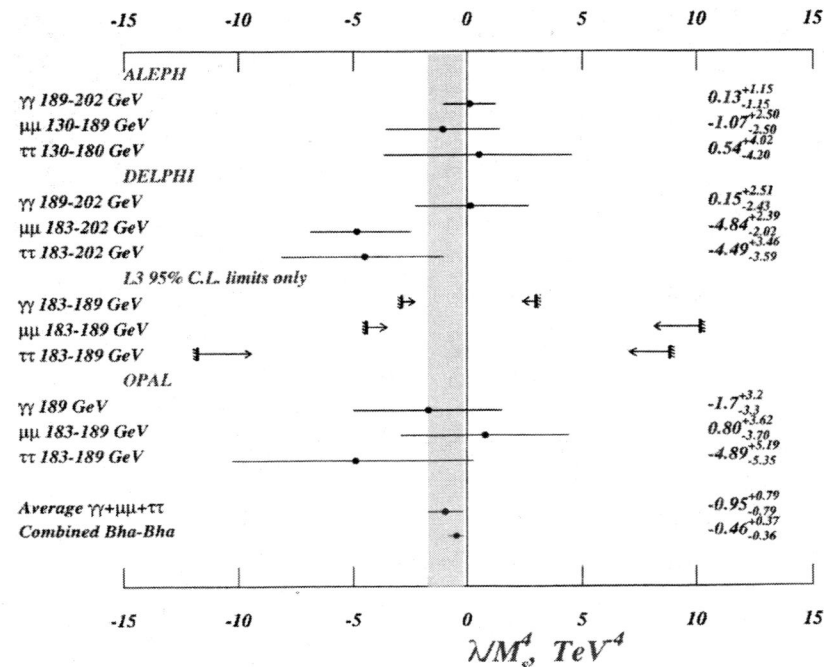

FIGURE 8. $\frac{\lambda}{M_s^4}$ obtained from γ, muon and tau production, average of $\frac{\lambda}{M_s^4}$ over ALEPH, DELPHI and OPAL data for γ, muon and tau production, average of $\frac{\lambda}{M_s^4}$ over ALEPH, DELPHI, L3 and OPAL data for Bhabha scattering

was done by L3 collaboration. Fig. 9 shows the reconstructed Z boson mass for $e^+e^- \to ZZ \to q\bar{q}l^+l^-$ channel and fig. 10 presents the polar angle distribution for $e^+e^- \to WW$ candidates with hadronic and semileptonic decays of W's.

Inspite of small number of events observed for these channels, they give quite strong 95% limits on M_s: $M_s > 0.77$ TeV ($\lambda = 1$) and $M_s > 0.76$ TeV ($\lambda = -1$) for Z boson pairs production and $M_s > 0.79$ TeV ($\lambda = 1$) $M_s > 0.68$ TeV ($\lambda = -1$) for W boson pairs production.

IV GAUGE BOSONS IN EXTRA DIMENSIONS

The second theoretical model with gauge bosons propagating in extra dimensions is hardly constrained by electroweak observables measurements. It was shown in [7] that the global fit of the electroweak observables with the compactification scale as a free parameter and one extra dimension give the limits $M_c > 3.8$ TeV with Higgs propagating in extra dimension and $M_c > 3.3$ TeV with Higgs confined in "3-brane". Transforming these numbers to the limits on radius of extra dimension one can obtain $r < 5.2 \cdot 10^{-5}$ fm and $r < 6.0 \cdot 10^{-5}$ fm.

Constrains on compactification scale also can be found from Bhabha scattering. Cross section modification is given by

$$\frac{d\sigma}{d\cos\theta} = (\frac{d\sigma}{d\cos\theta})_{SM} \cdot |S(s,t)|^2$$

$$S(s,t) = \frac{\Gamma(1 - \frac{s}{M_c^2})\Gamma(1 - \frac{t}{M_c^2})}{\Gamma(1 - \frac{s}{M_c^2} - \frac{t}{M_c^2})} \approx (1 - \frac{\pi^2}{6}\frac{st}{M_c^4})$$

$s, t \ll M_c^2$

In [8] the limit of $M_c = 0.631$ TeV , $r < 3. \cdot 10^{-4}$ fm was found from ALEPH, L3 and OPAL 183 and 189 GeV e^+e^- differential cross section.

V CONCLUSION

In this talk we present the following limits on extra dimensions.
Limits on M_D and r in from direct graviton production

	$N = 2$	$N = 4$	$N = 6$
M_D, TeV	1.28	0.80	0.58
r, cm	$2.9 \cdot 10^{-2}$	$1.4 \cdot 10^{-9}$	$4.5 \cdot 10^{-12}$

Limits on cut-off parameter M_s from virtual graviton exchange contribution

$M_s > 1.27$ TeV $\lambda = 1$
$M_s > 0.98$ TeV $\lambda = -1$

Limits on compactification scale M_c from electroweak observables

FIGURE 9. Reconstructed Z mass for $e^+e^- \to ZZ \to q\bar{q}l^+l^-$, L3 data at 189 GeV. Points stand for real data, solid line – Standard Model expectation, dashed (dotted) line – graviton exchange with $\lambda = 1$ ($\lambda = -1$).

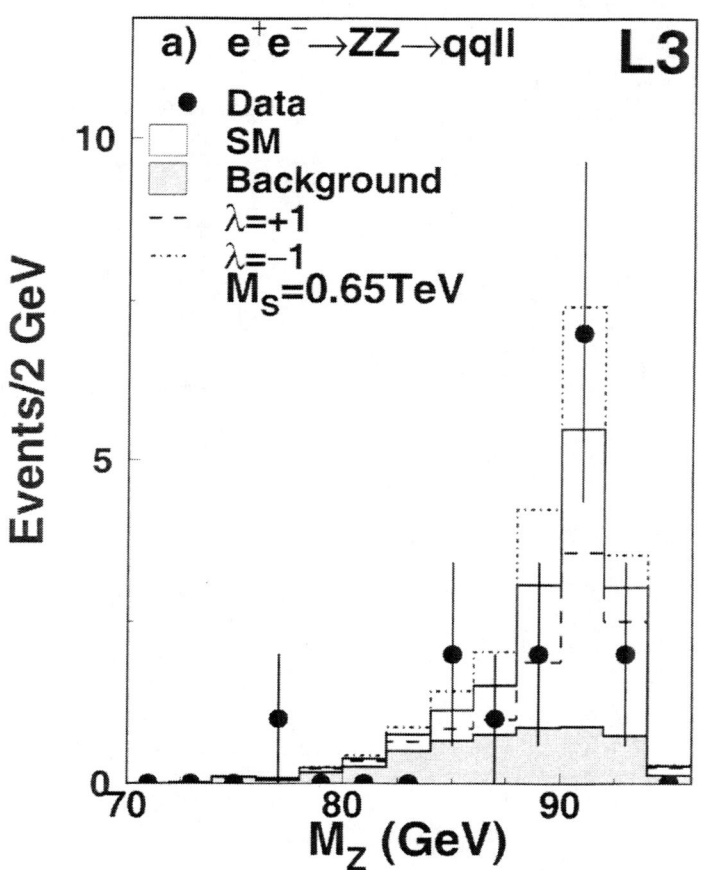

FIGURE 10. Polar angle distribution for $e^+e^- \to WW$, L3 data at 189 GeV. Points stand for real data, solid line – Standard Model expectation, dashed (dotted) line – graviton exchange with $\lambda = 1$ ($\lambda = -1$)

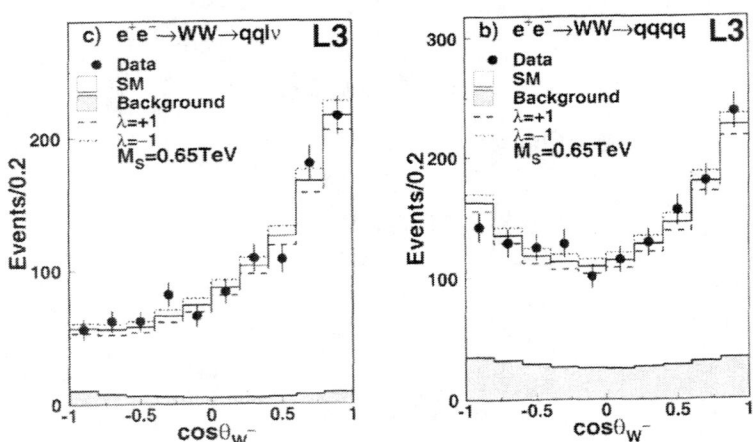

$M_c > 3.8$ TeV   Higgs in extra dimensions
$M_c > 3.3$ TeV   Higgs confined in "3-brane"

Limit on compactification scale M_c from Bhabha scattering

$$M_c = 0.631 \text{ TeV}$$

REFERENCES

1. N.Arkani-Hamed, S.Dimopoulos, G. Dvali, Phys. Lett. **B429**, 263 (1998); I.Antoniadis, N.Arkani-Hamed, S.Dimopoulos, G. Dvali, Phys. Lett. **B436**, 257 (1998); N. Arkani-Hamed, S.Dimopoulos, G. Dvali, Phys. Rev. D**59**, 086004 (1999);
2. K.Dienes, E.Dudas, T.Gherphetta, Nucl. Phys **B537**, 47 (1999)
3. G.Giudice, R. Rattazzi and D.Wells hep-ph/9811291, E.Mirabelli, M.Perelstein and M.Peskin hep-ph/9811337, K.Cheung and W.-Y.Keung hep-ph/9903294
4. G.Giudice, R.Rattazzi, J.Wells hep-ph/9811291
   K.Agashe and N.Deshpande OITS-669
5. J.Hewett hep-ph/9811356, T.Rizzo hep-ph/9901209, K.Cheung and W.-Y.Keung hep-ph 9903294, K.Agashe and N.Deshpande OITS-669
6. D.Bourilkov hep-ph/9907380
7. T.Rizzo and J.Wells hep-ph/9906234
8. D.Bourilkov hep-ph/0002172

Neutrino Dispersion in Intense Magnetic Field[1]

E. Elizalde

Institute for Space Studies of Catalonia, CSIC, Barcelona 08034, Spain
University of Barcelona, Dept. of Structure and Const. of Matter, Spain

E. J. Ferrer

Institute for Space Studies of Catalonia, CSIC, 08034, Barcelona, Spain
Phys. Dept., SUNY at Fredonia, NY 14063, USA

V. de la Incera

University of Barcelona, Dept. of Structure and Const. of Matter, Spain
Phys. Dept., SUNY at Fredonia, NY 14063, USA

The effect of external magnetic fields on neutrino propagation has recently received increasing attention. Moreover, taking into account that in the early Universe the existence of sufficiently strong primordial magnetic fields seems to be a very plausible idea, it is interesting to investigate the neutrino self-energy and dispersion relation at sufficiently strong magnetic fields ($m_e^2 \ll eB \ll M_W^2$). At such strong fields, since the gap between the electron Landau-levels is larger than the electron mass square, we are allowed to use the lower Landau level (LLL) approximation for the electron.

Calculating the self-energy in the LLL approximation we obtain in the momentum representation

$$\Sigma(p) = \left[a\slashed{p}_\| + b\slashed{u} + c\widehat{\slashed{B}} \right] L \qquad (1)$$

where the coefficients a, b and c are Lorentz-invariant functions which depend on the momentum and magnetic field. Their leading contributions in powers of $1/M_W^2$ are given by

---

[1] This work has been suported in part by NSF grants PHY-9722059 (VI & EF), POWRE-PHY-9973708 (VI) and PB96-0925 DGICYT and 1999SGR-00257 CIRIT (EE).

$$a = \frac{g^2}{(4\pi)^2}\lambda, \qquad b = a(p \cdot \hat{B}), \qquad c = a(p \cdot u), \qquad \lambda = \frac{|eB|}{M_W^2} \qquad (2)$$

In (2) u_μ is the four-velocity ($u_\mu u^\mu = 1$) and $\hat{\mathbf{B}} = \vec{B}/|\vec{B}|$.

Using the result (1) for $\Sigma(p)$ in the dispersion equation for neutrinos propagating in the external magnetic field one arrives to the following energy-momentum relation

$$E \simeq |\vec{p}|(\pm 1 + a\sin^2\alpha) \qquad (3)$$

where the positive and negative signs correspond to neutrino and antineutrino energies. In Eq. (3), α is the angle between the direction of the neutrino momentum and that of the applied magnetic field.

To obtain the neutrino index of refraction n, we substitute (3) into the formula $n \equiv |\vec{p}|/E$ to find

$$n \simeq 1 - a\sin^2\alpha \qquad (4)$$

From Eqs. (3) and (4) it is clear that neutrinos moving with different direction in the magnetized space will have different dispersion relations and consequently, different indices of refraction.

To conclude, we have that the dispersion relation (3), although independent of the mass of the charged lepton, may contribute to the neutrino oscillation through the following new mechanism: For flavor oscillations it is well known that any variation in the dispersion relations corresponding to neutrinos with different flavors is significative. For magnetic fields $m_e^2 \ll eB \ll m_\mu^2, M_W^2$ (m_μ is the muon mass), the muon-neutrino self-energy, which corresponds to the weak-field approximation [1], [2] ($eB \ll m_\mu^2$), is analytically different from the one we are reporting for the electron-neutrino, which corresponds to the strong-field approximation ($m_e^2 \ll eB$). Hence, the corresponding dispersion relations associated to these two different flavors (electron and muon) will be significantly different, thus contributing to the oscillation process.

REFERENCES

1. J. C. D'Olivo, J. F. Nieves and P. B. Pal, Phys. Rev. D**40** (1989) 3679.
2. A. Erdas and G. Feldman, Nucl. Phys. B**343** (1990) 597.

SCHEDULE

April 30 - Arrival

Monday 1 May

8:30 – 8:55   Registration

9:00 – 9:15   Welcome

Moderator: Leung
9:15 – 10:05   M. Vagins - Notes from the Underground: Super-Kamiokande's Past, Present, and Future (I)
10:15 – 10:45   Coffee Break
10:50 – 11:30   P. Q. Hung - Teeny, Tiny Dirac Neutrino Mass: An Unorthodox Point of View
11:40 – 12:20   R. Volkas - Relic Neutrino Asymmetries
12:30 – 4:00   Lunch (and informal discussions)

Moderator: Shafi
4:15 – 5:05   P. Ramond - Neutrinos as Probes of New Theories
5:15 – 5:35   Coffee Break
5:40 – 6:20   K. S. Babu - A Mass Relation for Neutrinos

Tuesday 2 May

8:45 – 9:00   Welcome, Dean Brad Weiner

Moderator: Halprin
9:00 – 9:50   P. Rosen - What's New in Washington?
10:00 – 10:30   Coffee Break
10:45   Departure for Visit to Arecibo
12:30   Arrival in Arecibo
12:30 – 1:30   Lunch

Moderator: Pantoja
1:40 – 1:55   Daniel Altschuler - Welcome and Brief History of the Arecibo Observatory
2:00 – 2:25   D. Lorimer - What's New in the Pulsar World
2:30 – 2:55   M. Nolan - Radar Obsevations of Asteroids
3:00 – 3:25   Break

	Moderator: Shafi
3:30 – 4:30	A. Filippenko - Evidence from Type Ia Supernovae for an Accelerating Universe
4:40 – 5:30	P. Nugent - Measurement of the Cosmological Parameters with Supernovae
5:40 – 7:00	Scientific and Recreational Activities (Tour and Cocktail)
7:15	Return from Arecibo
8:30	Arrival in San Juan

Wednesday 3 May

	Moderator: Leung
9:00 – 9:50	M. Vagins - Notes from the Underground: Super-Kamiokande's Past, Present, and Future (II)
10:00 – 10:25	Coffee Break
10:30 – 11:00	K. Yip - B Physics at Tevatron Run II
11:10 – 12:00	J. Rosner - CP Violation
12:10 – 4:00	Lunch (and informal discussions)

	Moderator: Halprin
4:15 – 5:05	M. Shaevitz - The Femilab Neutrino Oscillation Program
5:15 – 5:35	Coffee Break
5:40 – 6:10	D. Caldwell - Neutrinos in Supernovae and Extra Dimensions
6:20 – 6:50	A. Filipcic - First Direct Observation of T Violation

Thursday 4 May

	Moderator: Shafi
9:00 – 9:50	R. Van de Water - The SNO Detector Status
10:00 – 10:25	Coffee Break
10:30 – 11:00	P. Frampton - Minimal Model of Neutrino Masses and Mixings
11:10 – 11:40	D. Zavrtanik - Experimental Tests of CPT: CPLEAR Contribution
11:50 – 12:20	L. Duflot - SUSY Searches at LEP2
12:30	Lunch (and informal discussions)

Afternoon free

Workshop Dinner

Friday 5 May

9:40 – 9:55   Coffee Served

Moderator: Halprin
10:00 – 10:50 H. Nielsen - Fitting Finally Neutrino Masses and Mixings in Extended AGUT Model
11:00 – 11:25 Coffee Break
11:30 – 12:00 V. Zhuravlov - Limits on the Size of Extra Dimensions from LEP Experiments
12:10 – 4:00 Lunch (and informal discussions)

Moderator: Leung
4:15 – 5:05 J. Conrad - The Search for New Physics at NuTeV
5:15 – 5:35 Coffee Break
5:40 – 6:10 W. Louis - LSND Oscillation Results

6:30   Farewell Cocktail

LIST OF PARTICIPANTS

ALTSCHULER, Daniel
daniel@naic.edu
Arecibo Observatory

BABU, K. S.
babu@osuunx.ucc.okstate.edu
Oklahoma State University, USA

CALDWELL, David O.
caldwell@slac.stanford.edu
University of California (Santa Barbara), USA

CONRAD, Janet
conrad@fnal.gov
Columbia University, USA

CONSTANTIN, Lucian
lucian@ltp1.upr.clu.edu
University of Puerto Rico, Rio Piedras

DUFLOT, Laurent
duflot@lal.in2p3.fr
Laboratoire de l'Accelerateur Lineaire, France

FERRER, Efrain J.
ferrer@fredonia.edu
SUNY-Fredonia/IEEC-Barcelona, Spain

FILIPCIC, Andrej
andrej.filipcic@ijs.si
Jozef Stefan Institute, Slovenia

FILIPPENKO, Alex
alex@astro.berkeley.edu
University of California Berkeley, USA

FRAMPTON, Paul
frampton@physics.unc.edu
University of North Carolina, USA

GUPTA, Virendra
virendra@aruna.cieamer.conacyt.mx
CINVESTAV, Merida, Mexico

HALPRIN, Arthur
halprin@udel.edu
University of Delaware, USA

HUNG, P. Q.
pqh@virginia.edu
University of Virginia, USA

INCERA, Vivian
incera@fredonia.edu
SUNY-Fredonia/Univ. de Barcelona, Spain

LEUNG, Chung Ngoc
leung@physics.udel.edu
University of Delaware, USA

LORIMER, Duncan
lorimer@naic.edu
Arecibo Observatory

LOUIS, William
louis@lanl.gov
LANL, USA

NIELSEN, H. B.
holger.nielsen@cern.ch
Niels Bohr Institute, Copenhagen

NIEVES, Jose F.
nieves@ltp.upr.clu.edu
University of Puerto Rico, Rio Piedras

NOLAN, Michael
nolan@naic.edu
Arecibo Observatory

NUGENT, Peter
penugent@LBL.gov
Lawrence Berkeley National Laboratory, USA

PANTOJA, Carmen
cpantoja@astro2.cnnet.clu.edu
University of Puerto Rico, Rio Piedras

PONCE DE LEON, Jaime
jpdel@ppg.cnnet.clu.edu
University of Puerto Rico, Rio Piedras

RAMOND, Pierre
ramond@phys.ufl.edu
University of Florida, USA

ROSEN, Peter
peter.rosen@science.doe.gov
U.S. Department of Energy

ROSNER, Jonathan L.
rosner@hep.uchicago.edu
University of Chicago, USA

SAHU, Sarira
sarira@nuclecu.unam.mx
Instituto de Ciencias Nucleares, UNAM, Mexico

SELSBY, Ronald
selsby@ppg.cnnet.clu.edu
University of Puerto Rico, Rio Piedras

SHAEVITZ, Mike
shaevitz@fnal.gov
Fermilab/Columbia University, USA

SHAFI, Qaisar
shafi@bartol.udel.edu
Bartol Research Institute, USA

THARRATS, Jesus
tharrats@ppg.cnnet.clu.edu
University of Puerto Rico, Rio Piedras

TORRUELLA, Alfredo J.
torruell@ppg.cnnet.clu.edu
University of Puerto Rico, Rio Piedras

UBRIACO, Marcelo R.
ubriaco@ltp.upr.clu.edu
University of Puerto Rico, Rio Piedras

VAGINS, Mark
mvagins@uci.edu
University of California, Irvine, USA

VAN DE WATER, Richard
vdwater@surf.sno.laurentian.ca
University of Pennsylvania, USA

VOLKAS, Ray
r.volkas@physics.unimelb.edu.au
The University of Melbourne, Australia

YIP, Kin
kinyip@fnal.gov
Fermilab, USA

ZAVRTANIK, Danilo
danilo.zavrtanik@ses-ng.si
Nova Gorica Polytechnic, Slovenia

ZHURAVLOV, Vadym
juravlev@nusun.jinr.ru
Joint Institute for Nuclear Research, Dubna

Author Index

B

Babu, K. S., 3

C

Caldwell, D. O., 203
Conrad, J., 331

D

de la Incera, V., 372
Duflot, L., 341

E

Elizalde, E., 372

F

Ferrer, E. J., 372
Filippenko, A. V., 227
Frampton, P. H., 18
Froggatt, C. D., 35

H

Hung, P. Q., 24

L

Lorimer, D. R., 247
Louis, W. C., 93

N

Nielsen, H. B., 35
Nugent, P., 263

R

Ramond, P., 75
Riess, A. G., 227
Rosner, J. L., 283

S

Shaevitz, M. H., 105

T

Takanishi, Y., 35

V

Vagins, M. R., 122
Van de Water, R. G., 193
Volkas, R. R., 213

Y

Yip, K., 314

Z

Zavrtanik, D., 305
Zhuravlov, V., 354